METHODS IN MOLECULAR BIOLOGY

For further volumes:
http://www.springer.com/series/7651

Recombinant Glycoprotein Production

Methods and Protocols

Edited by

Virgínia Picanço-Castro

Regional Blood Center of Ribeirão Preto, Ribeirão Preto Medical School, University of São Paulo, Brazil

Kamilla Swiech

School of Pharmaceutical Sciences of Ribeirão Preto, University of São Paulo, Brazil

Humana Press

Editors
Virgínia Picanço-Castro
Regional Blood Center
of Ribeirão Preto,
Ribeirão Preto Medical School
University of São Paulo
Brazil

Kamilla Swiech
School of Pharmaceutical Sciences
of Ribeirão Preto
University of São Paulo
Brazil

ISSN 1064-3745 ISSN 1940-6029 (electronic)
Methods in Molecular Biology
ISBN 978-1-4939-7311-8 ISBN 978-1-4939-7312-5 (eBook)
DOI 10.1007/978-1-4939-7312-5

Library of Congress Control Number: 2017951847

Printed on acid-free paper

This Humana Press imprint is published by Springer Nature
The registered company is Springer Science+Business Media LLC
The registered company address is: 233 Spring Street, New York, NY 10013, U.S.A.

Preface

The majority of recombinant protein approved by regulatory agencies or in clinical trials have some form of post-translational modification (PTM), which can profoundly affect protein properties. Among the PTMs, glycosylation is the most common modification and one of the most important quality attributes as it can affect pharmacokinetics, immunogenicity, biological and clinical activity. For this reason, the expression system that will be used for protein production needs to be carefully defined, taking into consideration a plethora of aspects. This is the first step of the bioprocess development, being part of the upstream processing, which includes the steps associated with the culture and maintenance of cells. Once produced, the protein is subjected to a series of purification and polishing steps to meet purity and quality requirements, configuring the downstream processing.

This book, organized into 22 chapters, covers a wide spectrum of techniques and approaches that have been used in the upstream and downstream processing for recombinant glycoproteinsssss production. Chapter 1 gives the state of the art of therapeutic recombinant glycoproteins and the platforms that are being employed for their production. Chapter 2 instructs readers to explore the patent literature, which is a useful source of technical information. Chapters 3–14 describe the different expression systems used for glycoproteins production. Chapters 15–18 describe some methods employed in the downstream processing of different glycoproteins. Finally, Chapters 19–22 are useful to find information about analytical tools and formulation strategies.

We believe that this *Recombinant Glycoprotein Production* volume will be useful and inspiring for those who are interested in the production of this important class of recombinant protein, especially for therapeutic purposes.

University of São Paulo, Brazil

Virgínia Picanço-Castro
Kamilla Swiech

Contents

Contributors

MÁRIO SOARES ABREU-NETO • *Regional Blood Center of Ribeirão Preto, University of São Paulo, Ribeirão Preto, Brazil*

MIKAEL RØRDAM ANDERSEN • *Department of Biotechnology and Biomedicine, Technical University of Denmark, Kongens Lyngby, Denmark; Novo Nordisk Foundation Center for Biosustainability, Technical University of Denmark, Lyngby, Denmark*

BRUNA SAMHAM ARCHANGELO • *School of Pharmaceutical Sciences of Ribeirão Preto, University of São Paulo, Ribeirão Preto, Brazil*

RENATO M. ASTRAY • *Laboratório de Imunologia, Viral Instituto Butantan, São Paulo, Brazil*

RAFAEL TAGÉ BIAGGIO • *School of Pharmaceutical Sciences of Ribeirão Preto, University of São Paulo, Ribeirão Preto, Brazil*

ALINE DE SOUSA BOMFIM • *Center for Cell-based Therapy CTC and School of Pharmaceutical Sciences of Ribeirão Preto, University of São Paulo, Ribeirão Preto, Brazil*

CARLOS ALEXANDRE BREYER • *Biosciences Institute, São Paulo State University, Coastal Campus, São Vicente, São Paulo, Brazil*

JOHANNES FELIX BUYEL • *Fraunhofer Institute for Molecular Biology and Applied Ecology IME, Aachen, Germany; Institute for Molecular Biotechnology, RWTH Aachen University, Aachen, Germany*

ANGELO LUIS CARON • *Center for Cell-based Therapy CTC, Ribeirão Preto Medical School, University of São Paulo, Ribeirão Preto, SP, Brazil; School of Pharmaceutical Sciences of Ribeirão Preto, University of São Paulo, São Paulo, Brazil*

RIMENYS J. CARVALHO • *COPPE, Chemical Engineering Program, Cell Culture Engineering Laboratory (LECC), Federal University of Rio de Janeiro, Rio de Janeiro, RJ, Brazil*

NATALIA CEAGLIO • *CONICET, Facultad de Bioquímica y Ciencias Biológicas, Universidad Nacional del Litoral, Santa Fe, Argentina*

LAURA CERVERA • *Bioengineering Department, McGill University, Montreal, Canada*

MING-TANG CHEN • *Biologics Discovery, MRL, Boston, USA*

DIMAS TADEU COVAS • *Center for Cell-based Therapy CTC, Regional Blood Center of Ribeirão Preto, Ribeirão Preto Medical School, University of São Paulo, Ribeirão Preto, Brazil*

ANDREIA CRUZ • *International Iberian Nanotechnology Laboratory (INL), Braga, Portugal*

THAYANA A. CRUZ • *Chemical Engineering Program/COPPE, Federal University of Rio de Janeiro, Rio de Janeiro, Brazil; Federal University of Rio de Janeiro (UFRJ), IQ, Biochemistry Program, Rio de Janeiro, RJ, Brazil*

MARCOS ANTONIO DE OLIVEIRA • *Instituto de Biociências, Universidade Estadual Paulista, São Vicente, Brazil*

JOHN DRESIOS • *Leidos, Inc., San Diego, CA, USA*

JOSIMAR O. ELOY • *School of Pharmaceutical Sciences, UNESP, Araraquara, SP, Brazil; UNL, CONICET, FBCB, Cell Culture Laboratory, de Santa Fe, Argentina*

MARINA ETCHEVERRIGARAY • *UNL, CONICET, FBCB, Cell Culture Laboratory, de Santa Fe, Argentina*

VITOR MARCEL FAÇA • *Department of Biochemistry and Immunology, Ribeirão Preto Medical School, University of São Paulo, Ribeirão Preto, Brazil*

YUZHOU FAN • *Department of Biotechnology and Biomedicine, Technical University of Denmark, Kongens Lyngby, Denmark*
DAIANNE MACIELY CARVALHO FANTACINI • *Center for Cell-based Therapy CTC, Ribeirão Preto Medical School, University of São Paulo, Ribeirão Preto, Brazil*
EDSON GALDINO DO NASCIMENTO FILHO • *Department of Biochemistry and Immunology, Ribeirão Preto Medical School, University of São Paulo, Ribeirão Preto, Brazil*
PEDRO FONTE • *Instituto de Investigação e Inovação em Saúde, Universidade do Porto, Porto, Portugal*
MARCELA CRISTINA CORRÊA FREITAS • *Center for Cell-based Therapy CTC, Ribeirão Preto Medical School, University of São Paulo, Ribeirão Preto, Brazil*
VLADIMIR GRANOVSKI • *Center for Cell-based Therapy CTC, Ribeirão Preto Medical School Medical School, University of São Paulo, Ribeirão Preto, Brazil*
AGUSTINA GUGLIOTTA • *UNL, CONICET, FBCB, Cell Culture Laboratory, Santa Fe, Argentina*
TANJA HOLLAND • *Fraunhofer Institute for Molecular Biology and Applied Ecology IME, Aachen, Germany*
AMINE A. KAMEN • *Bioengineering Department, McGill University, Montreal, Canada*
PATRICK J. KENNEDY • *i3S—Instituto de Investigação e Inovação em Saúde, Universidade do Porto, Porto, Portugal; INEB—Instituto Nacional de Engenharia Biomédica, Universidade do Porto, Porto, Portugal; ICBAS—Instituto Ciências Biomédicas Abel Salazar, Universidade do Porto, Porto, Portugal; IPATIMUP—Instituto de Patologia e Imunologia Molecular da Universidade do Porto, Porto, Portugal*
RICARDO KRATJE • *UNL, CONICET, FBCB, Cell Culture Laboratory, Santa Fe, Pcia. de Santa Fe, Argentina*
GUILHERME PAUPERIO LANFREDI • *Department of Biochemistry and Immunology, Ribeirão Preto Medical School, University of São Paulo, Ribeirão Preto, Brazil*
ROBERT J. LEE • *College of Pharmacy, The Ohio State University, Columbus, USA*
DANIEL LEY • *Department of Biotechnology and Biomedicine, Technical University of Denmark, Kongens Lyngby, Denmark*
WAN-CANG LIU • *State Key Laboratory of Bioactive Substance and Function of Natural Medicines, Institute of Materia Medica, Chinese Academy of Medical Sciences, Peking Union Medical College, Beijing, People's Republic of China; Institute of Medicinal Biotechnology, Chinese Academy of Medical Sciences, Peking Union Medical College, Beijing, People's Republic of China; Biotechnology Core Laboratory, National Institute of Diabetes and Digestive and Kidney Diseases, National Institutes of Health, Bethesda, MD, USA*
RENATA F.V. LOPEZ • *School of Pharmaceutical Sciences of Ribeirão Preto, University of Ribeirão Preto, Ribeirão Preto, SP, Brazil*
AMANDA MIZUKAMI • *Center for Cell-based Therapy CTC, Ribeirão Preto Medical School, University of São Paulo, Ribeirão Preto, SP, Brazil*
ADAM NYLEN • *Biologics Discovery, MRL, Palo Alto, USA*
MARCOS OGGERO • *UNL, CONICET, FBCB, Cell Culture Laboratory, Santa Fe, Pcia. de Santa Fe, Argentina*
ERIK D. PENDLETON • *Kyowa Kirin Pharmaceutical Research, Inc/Leidos, Inc, La Jolla, USA*
CARLOS A. PEREIRA • *Laboratório de Imunologia Viral, Instituto Butantan, São Paulo, Brazil*

CRISTIANO GONÇALVES PEREIRA • *School of Economics, Business Administration and Accounting, University of São Paulo, Ribeirão Preto, Brazil*
ADALBERTO PESSOA • *School of Pharmaceutical Sciences, University of São Paulo, São Paulo, Brazil*
RAQUEL PETRILLI • *School of Pharmaceutical Sciences of Ribeirão Preto, University of São Paulo, Ribeirão Preto, Brazil*
VIRGÍNIA PICANÇO-CASTRO • *Center for Cell-based Therapy CTC, Regional Blood Center of Ribeirão Preto, Ribeirão Preto Medical School, University of São Paulo, Ribeirão Preto, Brazil*
INÊS MENDES PINTO • *International Iberian Nanotechnology Laboratory (INL), Braga, Portugal*
GECIANE SILVEIRA PORTO • *School of Economics, Business Administration and Accounting, University of São Paulo, Ribeirão Preto, Brazil*
ELISA MARIA DE SOUSA RUSSO • *School of Pharmaceutical Sciences of Ribeirão Preto, University of São Paulo, Ribeirão Preto, Brazil*
BRUNO SARMENTO • *i3S—Instituto de Investigação e Inovação em Saúde, Universidade do Porto, Porto, Portugal; INEB—Instituto Nacional de Engenharia Biomédica, Universidade do Porto, Porto, Portugal; ICBAS—Instituto Ciências Biomédicas Abel Salazar, Universidade do Porto, Porto, Portugal*
FLÁVIA SOUSA • *Instituto de Investigação e Inovação em Saúde, Universidade do Porto, Porto, Portugal*
CHALLISE J. SULLIVAN • *Leidos, Inc., San Diego, CA, USA*
KAMILLA SWIECH • *School of Pharmaceutical Sciences of Ribeirão Preto, University of São Paulo, Ribeirão Preto, Brazil*
DANIELLA C. VENTINI-MONTEIRO • *Laboratório de Imunologia, Viral Instituto Butantan, São Paulo, Brazil*
PING ZHU • *State Key Laboratory of Bioactive Substance and Function of Natural Medicines, Institute of Materia Medica, Chinese Academy of Medical Sciences, Peking Union Medical College, Beijing, People's Republic of China*

Chapter 1

Platforms for Recombinant Therapeutic Glycoprotein Production

Amanda Mizukami, Angelo Luis Caron, Virgínia Picanço-Castro, and Kamilla Swiech

Abstract

The majority of FDA-approved biology-derived products are recombinant glycoproteins. These proteins have been used for the treatment of several diseases, with numerous products currently approved for clinical use. The choice of the expression system is a key step toward a successful functional protein production, since glycosylation influences yield, pharmacokinetics, biological activity, and immunogenicity. This chapter covers the general aspects of therapeutic recombinant glycoproteins and the platforms that are being employed for their production.

Key words Recombinant glycoprotein, Glycosylation, Expression systems, Mammalian cells, Human cells, Insect cells, Plant cells, Bacteria, Yeasts

1 Introduction

Recombinant therapeutic proteins include several protein classes and have been widely explored over the years for the treatment of a broad range of diseases, including metabolic disorders, autoimmune diseases, and cancer [1]. Leader and coworkers suggested a classification for the therapeutic recombinant proteins according to their application and function. Four groups were categorized: Group I—comprises protein with enzymatic activity; Group II—protein with targeting activity; Group III—protein vaccines; and Group IV—protein diagnostics [2]. This diversity of protein functions reflects the multiple possibilities of pharmacological actions and highlights the potential of these proteins in the biomedical arena. Currently, more than 246 biopharmaceutical products have received licenses in the US and European Union markets, being the vast majority recombinant protein products. The growing number of these products emphasizes their importance in the pharmaceutical industry pipeline [3]. Compared with synthetic drugs,

Virgínia Picanço-Castro and Kamilla Swiech (eds.), *Recombinant Glycoprotein Production: Methods and Protocols*, Methods in Molecular Biology, vol. 1674, DOI 10.1007/978-1-4939-7312-5_1, © Springer Science+Business Media LLC 2018

therapeutic recombinant proteins present some advantages: high specificity and complexity; are well-tolerated and less-immunogenic; and importantly, clinical development and regulatory approval could be faster than small-molecule drugs [2].

A few years ago, proteins with recognized therapeutic potential were derived from human sources, such as insulin from pancreas and albumin from plasma. However, the difficulty in assuring the reproducibility, purity, and safety accelerated the development of recombinant expression systems for protein production [4]. Up to now, mammalian, bacteria and yeasts are the main expression systems used to produce recombinant proteins, presenting several differences among them, especially in terms of posttranslational modifications. Glycosylation is the most frequent modification of recombinant proteins and has a huge impact on the biological activity. Almost 70% of all therapeutic recombinant proteins (membrane bound and secreted) are glycosylated [5]. Table 1 presents an overview of recent approved therapeutic recombinant proteins and the respective production platforms. As it can be observed in Table 1, the predominance of glycosylated proteins has been maintained.

2 Glycosylation and Its Impact on Protein Functionality and Immunogenicity

A major fraction of the currently available biopharmaceutical proteins is produced as secreted glycoproteins from mammalian cell cultures. The glycosylation profile of these proteins is one of the most important quality attributes as it can affect pharmacokinetics, secretion, solubility, receptor recognition, immunogenicity, and biological and clinical activity [6].

Glycosylation involves the enzymatic addition of monosaccharides or oligosaccharides (glycans) to specific amino acids in the polypeptide chain and can occur on virtually all proteins from mammalian cells [7]. Mammalian glycans consist in 10 common monosaccharides: glucose (Glc), fucose (Fuc), xylose (Xyl), mannose (Man), galactose (Gal), N-acetylglucosamine (GlcNAc), glucuronic acid (GlcA), iduronic acid (IdoA), N-acetylgalactosamine (GalNAc), sialic acid, or 5-N-acetylneuraminic acid (Neu5Ac) [8]. Two major glycosylation types are found in human proteins: N-linked and O-linked glycosylation [9]. The N-linked glycosylation process occurs in eukaryotes and archaea, but rarely occurs in bacteria.

The nature of the N-linked glycans attached to a glycoprotein is determined by the amino acid sequence of a protein and by the cell in which it is expressed. The N-glycosylation also varies between species and different species synthesize different types of N-linked glycans. The N-linked glycosylation occurs when a glycan is attached to nitrogen atom to an asparagine (Asn) in the sequon

Table 1
Overview of the therapeutic recombinant proteins approved over the last few years

Year	Product	Company	Clinical indication	Glycoprotein	Production platform
2016	Zinplava® (bezlotoxumab)	Sharp & Dohme Limited	To reduce the recurrence of Clostridium difficile infection in patients aged 18 years or older	Yes	CHO
	Lartruvo® (olaratumab)	Eli Lilly Nederland B.V.	Treatment of adult patients with soft tissue sarcoma (STS)	Yes	NS0
	Taltz® (ixekizumab)	Eli Lilly Nederland B.V.	Treatment of adults with moderate-to-severe plaque psoriasis	Yes	CHO
	Cinqair® (reslizumab)	Teva Pharmaceuticals Limited	Add-on to standard treatment of patients with severe asthma with an eosinophilic phenotype	Yes	NS0
	Zinbryta® (daclizumab)	Biogen Idec Ltd	Treatment of relapsing forms of multiple sclerosis (RMS)	Yes	NS0
	Tecentriq® (atezolizumab)	Genentech, Inc	Treatment of patients with urothelial carcinoma and metastatic non-small cell lung cancer	No	CHO
	Anthim® (obiltoxaximab)	Elusys Therapeutics, Inc	Treatment of inhalational anthrax due to *Bacillus anthracis*	Yes	NS0
2015	Empliciti® (elotuzumab)	Bristol-Myers Squibb	Patients with multiple myeloma who have received one to three prior therapies	Yes	NS0
	Portrazza® (necitumumab)	Eli Lilly Nederland B.V.	Patients with metastatic squamous non-small cell lung cancer	Yes	NS0
	Darzalex® (daratumumab)	*Janssen* Biotech, Inc.	Patients with multiple myeloma	Yes	CHO
	Nucala® (mepolizumab)	GlaxoSmithKline	Add-on to standard treatment of patients with severe asthma with an eosinophilic phenotype	Yes	CHO
	Strensiq® (asfotase alfa)	Alexion Europe SAS	Patients with perinatal/infantile- and juvenile-onset hypophosphatasia	Yes	CHO
	Praxbind® (idarucizumab)	Boehringer Ingelheim International GmbH	For emergency surgery/urgent procedures and/or in life-threatening or uncontrolled bleeding	No	CHO
	Tresiba® (insulin degludec)	Novo Nordisk A/S	Improve glycemic control in adults with diabetes mellitus	No	*Saccharomyces cerevisiae*
	Repatha® (evolocumab)	Amgen Europe B.V.	Treatment of certain patients with high cholesterol	Yes	CHO
	Praluent® (alirocumab)	Sanofi and Regeneron Pharmaceuticals	Treatment of certain patients with high cholesterol	Yes	CHO
	Unituxin® (dinutuximab)	United Therapeutics Europe Ltd	Treatment of pediatric patients with high-risk neuroblastoma	Yes	SP2/0
	Natpara® (parathyroid hormone)	NPS Pharmaceuticals	Control hypocalcemia in patients with hypoparathyroidism	No	*E. coli*
	Cosentyx® (secukinumab)	Novartis Europharm Ltd	Treatment of moderate to severe plaque psoriasis in adult patients who are candidates for systemic therapy or phototherapy	Yes	CHO
	Vonvendi® (Recombinant von Willebrand factor)	Baxalta US Inc	On-demand treatment and control of bleeding episodes in adults diagnosed with von Willebrand disease (VWD)	Yes	CHO
	Nuwiq® (Antihemophilic rFactor VIII)	Octapharma AB	Treatment and prophylaxis of bleeding in pediatric and adult patients with hemofilia A	Yes	HEK293F
	Ixinity®(Coagulation rFactor IX)	Cangene Corporation	Control and prevention of bleeding episodes and peri-operative management in patients with hemophilia B.	Yes	CHO

(continued)

Table 1 (continued)

Year	Product	Company	Clinical indication	Glycoprotein	Production platform
2014	Eloctate® (Antihemophilic rFactor VIII, FcFusion Protein)	Biogen Idec, Inc	Treatment and prophylaxis of bleeding in pediatric and adult patients with hemofilia A	Yes	HEK293
	Alprolix® (rFactor IX fused to a human IgG1 Fc domain)	Biogen Idec, Inc.	Control and prevention of bleeding episodes and peri-operative management in patients with hemophilia B.	Yes	HEK293
	Ruconest® (r C1 Esterase Inhibitor)	Salix Pharma-ceuticals, Inc.	Treatment of acute attacks in adult and adolescent patients with hereditary angioedema (HAE)	Yes	Milk of transgenic rabbits
	Pelegridy® (peginterferon beta-1a)	Biogen Idec Ltd	Indicated for relapsing forms of multiple sclerosis	Yes	CHO
	Myalept® (metreleptin)	Bristol-Myers Squibb	Indicated as an adjunct to diet as replacement therapy to treat the complications of leptin deficiency in patients with congenital or acquired generalized lipodystrophy	No	*E.coli*
	Afrezza® (r-insulin)	Mannkind	Improve glycemic control in adult patients with diabetes mellitus	No	*E. coli*
	Cyramza® (ramucirumab)	Eli Lilly Nederland B.V.	Treatment of patients with gastric cancer	Yes	NS0
	Entyvio® (vedolizumab)	Millenium Pharmaceuticals	Indicated for adults with moderately to severely active ulcerative colitis (UC) or Crohn's disease (CD)	Yes	CHO
	Sylvant® (siltuximab)	Janssen-Cilag International NV	Treatment of multicentric Castleman's disease	Yes	CHO

via a N-glycosidic bond. All N-linked glycans are based on the common core pentasaccharide (Man3GlcNAc2) and the sequon is an Asn-X-Ser or Asn-X-Thr sequence, where X is any amino acid except proline. In eukaryotes the core glycan will be further modified to result in a diverse range of N-glycan structures such as: N-acetylgalactosamine, galactose, neuraminic acid, N-acetylglucosamine, fucose, mannose, and other monosaccharides [6]. On the other hand the N-glycosilation in eubacteria and archaea result in a final glycan structure that is very similar to the common core without the modifications found in made in eukaryotes [10].

Although N-glycosylation is the most common and widely studied, O-glycosylation is a ubiquitous posttranslational modification and is present in approximately 80% of proteins trafficking through the secretory pathway [11, 12]. O-linked glycosylation is the attachment of a sugar molecule to an oxygen atom in an amino acid residue in a. This type of glycosylation occurs in the Golgi apparatus in eukaryotes. It also can occur in Archaea and bacteria [13]. The most common O-glycosylation involves the addition of a GalNAc, by the enzyme GalNAc transferase, to a serine or threonine residue in a fully folded protein. Even though there is no consensus sequence identified [12], glycosylation often occurs in a region of the protein with a high proportion of serine, threonine, and proline [14].

The capacity for glycosylation varies significantly among different eukaryotes, but also among different mammalian species [15, 16]. A study comparing the N-linked oligosaccharide structures of IgG, produced by different species, demonstrated huge differences, not only in glycan processing but also in isomers, suggesting remarkable changes in enzymatic specificities [16]. Moreover, the glycosylation profile of proteins produced in nonhuman mammalian cell lines may contain glycans that are immunogenic in humans [17]. Humans, unlike other mammals, have lost the ability to biosynthesize both the terminal glycans Galα1-3Galβ1-(3)4GlcNAc (alpha-Gal) and N-glycolylneuraminic acid (Neu5Gc) [18–20]. Hamster and murine cells used in industry express considerably higher levels of these epitopes; therefore, recombinant products derived from these cells have higher chance of presenting immunogenicity when compared to products derived from human cells [21, 22].

Besides the immunogenicity, N- and O-linked glycans have been shown to have also a large effect on the product efficacy, bioactivity, pharmacokinetics, and pharmacodynamics [14, 23, 24]. Erythropoietin (EPO), a glycoprotein hormone used for the treatment of anemia, has been documented to be very sensitive to its attached carbohydrates, therefore affecting the resulting pharmacokinetics [25–27]. Studies have shown that the sialylated EPO presented a half-life of about 3 h, in contrast to the 2 min half-life of the desialylated EPO [28]. Another study examined the differences in the glycosylation of 12 proteins of different sizes (9.5–52 kDa) produced by transient expression in two commonly used cell lines, HEK (Human Embryonic Kidney) and CHO (Chinese Hamster Ovary). For all recombinant proteins, significant differences in glycosylation were detected with higher sialylation in CHO cells, which can result in reduced antibody-dependent cell-mediated cytotoxicity (ADCC) activity of an antibody [15, 29, 30].

The complexity of the glycosylation network of metabolic reactions demands extensive care in the therapeutic recombinant glycoprotein manufacturing. The glycoform profile of a recombinant glycoprotein can be affected by a myriad of process conditions including the host cell line, the method of culture, the extracellular environment and the protein structure [6, 31, 32]. Biopharmaceutical regulatory agencies worldwide proposed several guidelines on requirements for the production of therapeutic glycoproteins. Manufacturers mainly refer to the International Conference on Harmonization (ICH) documents Q5E, specifying the comparability of site-specific glycan profile during scale-up and changes in manufacturing process, and Q6B, detailing the quality aspects for the characterization of such structures [32–35]. Hence, the variants of a glycoprotein must be carefully analyzed and controlled to guarantee product quality within defined acceptance limits [31, 36].

3 Recombinant Glycoprotein Production Platforms

Choosing the appropriate host for recombinant glycoproteins is a key step toward a successful functional protein production. A plethora of aspects needs to be considered, including host's main characteristics, biochemical environment, production costs, capacity to process and translate the RNA transcript, safety, efficacy, and stability of the product [37, 38]. The main characteristics of the production platforms (Table 2) that are being used for therapeutic recombinant glycoprotein production are addressed in this chapter. The production in whole plants and transgenic animals is not considered in this chapter.

3.1 Bacteria

The first biopharmaceutical approved (1982), a recombinant insulin (Eli Lilly & Co.'s Humulin®), was produced in bacterial expression systems (*E. coli*). Since then, it has been used to produce numerous commercially approved nonglycosylated therapeutic proteins, such as enzymes, cytokines and monoclonal antibodies [5], and has presented several limitations for recombinant glycoprotein production due to the lack of the enzymatic machinery required for mammalian-like glycosylation. For this reason, bacterial expression systems traditionally were not considered for the

Table 2
Comparison of the main characteristics of the production platforms that are being used for recombinant glycoprotein production

	Bacteria	Yeast	Plant cells	Insect cells	Mammalian cells	
					Mouse/murine	**Human**
Cell growth	Very fast	Fast	Moderate	Moderate	Slow	
Media requirement	Minimal and inexpensive media	Minimal and inexpensive media	Minimal and inexpensive media	Complex and expensive media	Complex and expensive media	
Process/product contamination	High risk (endotoxins)	Low risk	Low risk	High risk (viral)	High risk (viral)	
Genetic modification	Ease and well established	Ease and well established	Ease to Moderate	Moderate	Moderate and well established	
rP Yields	High	High	Low to Moderate	High	Moderate to High	
PTMs	Limited	Glycosylation Hypermannosylation	Glycosylation (possibility of plant-specific glycoforms) Do not perform sialylation	Glycosylation (possibility of insect cells-specific glycoforms Presence of high mannose or paucimannose residues Do not perform sialylation	Human-like glycosylation Presence of immunogenic nonhuman epitopes	Human glycosylation profile

production of complex and glycosylated proteins. This scenario may change over the next years due to the discovery of a protein N-glycosylation machinery in *Campylobacter jejuni* and its successful functional transfer into *E. coli* [39]. Although the significant progress made over recent years regarding *E. coli* glycoengineering, additional improvements are still needed to establish a cost-effective production process, since the yields currently achieved are insufficient for commercial purposes [40].

3.2 Yeast

Yeast-based expression system is considered very attractive and has been extensively used for the production of relevant proteins. Yeasts can be cultured in chemically defined media, grow rapidly achieving higher densities in a short period of time, present robust expression, have well characterized glycosylation machineries (N-glycosylation) and present the ability to scale up fermentation to industrial scale [4, 5]. Almost 20% of the biopharmaceuticals are being produced by *S. cerevisiae*, including insulin, hepatitis vaccines, and human serum albumin. Its intense use is also related to proper folding of many human proteins and relative ease of purification since the produced proteins are secreted to the extracellular medium [41]. Another yeast-based system is *Pichia pastoris*, considered a successful heterologous expression system with several recombinant proteins in clinical trials [42]. Yeast-based recombinant glycoproteins with marketing approval include: a Hepatitis B subunit vaccine using Hepatitis B surface antigen; recombinant human granulocyte macrophage-colony stimulating factor, Leukine®; recombinant human platelet derived growth factor, Regranex®; human papillomavirus subunit vaccine, Gardasil®; and a kallikrein inhibitor, Kalbitor® [43].

Although yeast has the potential to perform glycosylation, the hypermannosylated N-glycans, with more than 100 additional mannose residues, negatively affect the protein half-life upon administration and could provoke immune reactions in humans [5, 44]. Efforts have been made to genetically modify yeasts by eliminating the endogenous hypermannosylated glycans, followed by the introduction of elements required to generate human-like sialylated complex glycans [45]. Knockout of OCH1 (gene responsible for hypermannosylation triggering) in *Pichia pastoris* has been reported, resulting in a reduction in the number of mannoses (10–8) in glycans compared to a wild type strain [46]. In 2006, Merck acquired GlycoFi (*Pichia pastoris* glycoengineered for N-glycosylation) to synergize with its own expertise in biopharmaceutical production in *Saccharomyces cerevisiae* (Gardasil®) in order to produce enhanced products with costs of goods (COG) lower than obtained with mammalian cell lines [47].

3.3 Plant Cells

Similarly, to bacteria and yeast, plant cells can be easily cultured in basal culture medium and present a robust cell growth, being easily scaled up. Besides, they do not harbor human-trophic pathogens and endotoxins and are not subject to several disadvantages of recombinant glycoprotein produced in whole plants [48]. This platform has the ability to synthetize complex protein and also glycoproteins, showing greater similarity to humans in terms of N-glycan structure [49]. Accordingly, there is a recent and growing interest in using plant cells to produce biopharmaceuticals by the biotechnology companies around the world [48, 50].

Elelyso (Protalix BioTherapeutics), intended for the treatment of Gaucher's disease, was the first recombinant glycoprotein produced by plant cells that hit the market in 2012 [51]. It is produced in an engineered carrot plant root cell line, grown in a disposable bioreactor system (ProCellEx®) [3]. The proprietary ProCellEx® system has been also used to produce the recombinant glycoproteins Pegunigalsidase alfa (PRX-102) (alpha galactosidase-A; Alidornase alfa (PRX-110) (DNase I); and OPRX-106 (human tumor necrosis factor receptor II) that are in clinical trials phase. Another companies are involved in plant cell-derived biopharmaceuticals. Greenovation GmbH has developed a moss protonema-based platform for the production of Moss-Agal (alpha-galactosidase A) under clinical trial phase I investigation and Moss-FH (regulator factor H analogous) and Moss-GBA (glucocerebrosidase) in preclinical development. Biolex therapeutics developed the LEX platform (*Lemna*, a duckweed) which is focused on the difficulty in expressing proteins and optimized monoclonal antibodies. Nonetheless, the lead product under development (Locteron®, an interferon alpha) was the cause of company's bankruptcy filed in 2012. The LEX platform along with two biopharmaceuticals candidates (BLX-301, a humanized and glyco-optimized anti-CD20 antibody and BLX-155, a direct-acting thrombolytic) have been acquired by Synthon (Nijmegen, the Netherlands) [3]. This deal excludes Lacteron®.

Challenges remain over the yield of recombinant glycoproteins produced in plant cells, with reported yields for recombinant antibodies falling far short of those reported in mammalian (CHO) cell systems. This can be associated with incipient process optimization regarding media composition, fermenter design and cell line engineering [48, 52].

3.4 Insect Cells

Higher tolerance to osmolality and toxic metabolites, optimal protein expression at high levels, and no need for extensive culture handling are notable advantages related to the use of insect cells as a production platform for recombinant proteins [53]. Combined with the use of baculovirus expression vector (BEV), this platform can be used to rapidly produce high levels of recombinant proteins. *Spodoptera frugiperda* SF9 or SF21 and *Trichoplusia ni* BTI 5B1-4 (High Five™) are the main insect cell lines used for this purpose [17].

Although insect cells have the ability to perform most of the mammalian-like posttranslational modifications, this system is unable to perform sialylation, which is essential for serum half-life and biological activity of the protein [46]. Besides, recombinant proteins derived from insect cells also carry glycan structures that differ significantly from those present in mature human glycoproteins, including paucimannose-type glycans and immunogenic α-1,3-fucose attached to the innermost GlcNAc residue, in contrast to mammals, which exclusively synthesize core α-1,6-fucose residues [5]. Due to that, no recombinant glycoprotein is currently produced in this system as this would compromise in vivo bioactivity and potentially induce allergenic reactions [17]. Glycoengineering approaches to produce complex, human-like N-glycans are being developed to enable the human therapeutic use of recombinant glycoprotein production by BEV-insect cells platform [54]. Schneider-2 insect cells that produce fully complex type N-glycans, lacking only terminal sialic acids, also represent an alternative [17].

The insect cell-based platform has been successfully used to produce vaccine antigens and virus-like particles. Two FDA-licensed vaccines, Cervarix® (GlaxoSmithKline) and Flublok® (Protein Sciences) and one autologous prostate-cancer therapy product (Provenge® from Dendreon) have already been approved by regulatory agencies and many more are under clinical trial investigations [5, 55]. Cervarix® is a HPV vaccine composed of a truncated form of the major capsid L1 proteins of HPV types 16 and 18 expressed in *T. ni* High Five™ cells. Flublok® Influenza HA vaccine is based on trivalent recombinant hemagglutinin expressed in a proprietary Sf9-derived cell population (expressSF$^{+®}$). PROVENGE® is composed of a prostate surface antigen (PSA) produced in SF9 cells.

3.5 Mammalian Cells

Due to the limitations of producing glycosylate proteins in the production platforms addressed above, mammalian cell lines are the preferred host cell expression system to produce proteins with human-like glycosylation patterns. More than 50% of therapeutic recombinant proteins available on the market are produced using mammalian cells [56]. The fidelity of protein glycosylation is an essential aspect because protein ligand binding and stability could be affected, leading to rapid clearance and immunogenicity when administered into patients [57]. Chinese hamster ovary (CHO), baby hamster kidney (BHK-21) and murine myeloma cells (NS0 and Sp2/0) are the most used cell lines for therapeutic protein production [1, 56]. CHO is the most well-studied cell line, being used in the manufacture of more than 70% of therapeutic recombinant proteins at commercial scale. This host has an advantage of growing in single-cell suspension, facilitating the scale-up in bioreactors [15]. CHO cells allow gene amplification, have stronger expression units and advanced hosts, allow improved selection strategies and targeting the transgene to transcriptional hotspots, what lead to high protein yields. Yields of about 90 pg/cell/day and 5 g/L for an extended-batch culture process lasting up to three weeks have already been reported [58]. Besides, in view of the long track record of products accepted by FDA and EMA, the timeline required for approval can be reduced [1, 21]. Murine myeloma cells are the second cell type most frequent used by industries [1].

Although murine cell lines have the ability to produce similar human glycosylation pattern and correct protein folding, nonhuman cells contain epitopes highly immunogenic, namely, galactose-α-1,3-galactose (α-gal) and N-glycolylneuraminic acid (NGNA), affecting the therapeutic outcomes [59]. Taking this into consideration, human cell lines emerge as a powerful alternative for native human proteins production, increasing their efficacy and reducing the risk of possible immunoreactions. HEK 293 is a transformed cell line derived from human embryonic kidney and is widely employed for recombinant protein production by transient expression. This cell line is able to grow easily in suspension under xeno-free conditions, having, however, the tendency to form large aggregates. Four commercial products have been produced in this cell line: Eloctate® (B-deleted recombinant FVIII-Fc fusion protein (IgG1), Biogen Idec), Nuwiq® (B-deleted recombinant FVIII protein, Octapharma AG), Alprolix® (recombinant FIX-IgG1 Fc fusion protein, Biogen Idec) [60], and Trulicity® (recombinant glucagon-1-like peptide (GLP-1) Fc fusion protein, Eli Lily).

HT-1080 fibrosarcoma-derived cell is another human cell line that has been successfully used by the industry [1, 21, 61]. Employing gene activation technology, three commercially approved therapeutic glycoproteins are available on the market and have been produced by Shire: Elaprase® (Iduronate-2-sulfatase); Replagal® (Agalsidase alfa); and Vpriv® (velaglucerase alfa) [1, 62].

PER.C6 (human embryonic retina cells) from Crucell is probably one of the most advanced alternatives to CHO cell-based platforms for recombinant glycoprotein production [61]. This cell line has the ability for growth at extremely high cell densities (up to 10^8 cells/mL) and presents a high recombinant protein productivity (with low gene copy numbers) without requiring time-consuming amplification methods [21]. Up to now, more than 14 biopharmaceutical products utilizing the PER.C6 cell line are in Phase I/II clinical trials [63]. Besides PER.C6 cells, the human cell lines CAP (derived from human amniocytes) and HKB-11 (hybrid of HEK293-S and a human B-cell line) have been used for the production of products currently in preclinical development and represent promising human cell lines for commercial recombinant glycoprotein production [1, 21, 61, 62].

It seems clear that human cells lines have advantages that favor their use as recombinant protein producers and with continuous research, investment and technology advances, human cells lines will be considered the gold standard for protein production with applications in different therapeutic areas.

4 Concluding Remarks

A variety of different production platforms are currently available for therapeutic recombinant glycoprotein production purposes, the mammalian-based being the predominantly employed. It seems clear that besides cell growth capacities and recombinant protein yields, the potential to generate proteins with posttranslational modification profile (mainly glycosylation), similar to the nature counterparts, will dictate the choice of the production platform. New and improved production platforms are already available, and the results indicate that maybe the CHO cell-based platform can become less predominant in the coming years.

Acknowledgment

The authors acknowledge the São Paulo Research Foundation – FAPESP (Grants 2012/04629-8, 2013/08135-2 and 2016/02433-0), CAPES scholarship and the Center for Cell-based Therapy/CTC/FAPESP (Regional Blood Center of Ribeirão Preto) for financial support.

References

1. Dumont J, Euwart D, Mei B, Estes S, Kshirsagar R (2016) Human cell lines for biopharmaceutical manufacturing: history, status, and future perspectives. Crit Rev Biotechnol 36 (6):1110–1122
2. Leader B, Baca QJ, Golan DE (2008) Protein therapeutics: a summary and pharmacological classification. Nat Rev Drug Discov 7(1):21–39
3. Walsh G (2014) Biopharmaceutical benchmarks 2014. Nat Biotechnol 32 (10):992–1000
4. Sethuraman N, Stadheim TA (2006) Challenges in therapeutic glycoprotein production. Curr Opin Biotechnol 17(4):341–346
5. Ghaderi D, Zhang M, Hurtado-Ziola N, Varki A (2012) Production platforms for biotherapeutic glycoproteins. Occurrence, impact, and challenges of non-human sialylation. Biotechnol Genet Eng Rev 28:147–175
6. Butler M (2006) Optimisation of the cellular metabolism of glycosylation for recombinant proteins produced by mammalian cell systems. Cytotechnology 50(1–3):57–76
7. Stowell SR, Ju T, Cummings RD (2015) Protein glycosylation in cancer. Annu Rev Pathol 10:473–510
8. Cummings RD, Pierce JM (2014) The challenge and promise of glycomics. Chem Biol 21(1):1–15
9. Sola RJ, Griebenow K (2010) Glycosylation of therapeutic proteins: an effective strategy to optimize efficacy. BioDrugs 24(1):9–21
10. METaK D (2006) Introduction to glycobiology, 2nd edn. Oxford University Press, New York, NY
11. Steentoft C, Vakhrushev SY, Vester-Christensen MB, Schjoldager KT, Kong Y, Bennett EP, Mandel U, Wandall H, Levery SB, Clausen H (2011) Mining the O-glycoproteome using zinc-finger nuclease-glycoengineered SIMPLECELL lines. Nat Methods 8(11):977–982
12. Zielinska DF, Gnad F, Wisniewski JR, Mann M (2010) Precision mapping of an in vivo N-glycoproteome reveals rigid topological and sequence constraints. Cell 141(5):897–907
13. Flynne WG (2008) Biotechnology and bioengineering. Nova Publishers, Hauppauge, NY
14. Van den Steen P, Rudd PM, Dwek RA, Opdenakker G (1998) Concepts and principles of O-linked glycosylation. Crit Rev Biochem Mol Biol 33(3):151–208

15. Butler M, Spearman M (2014) The choice of mammalian cell host and possibilities for glycosylation engineering. Curr Opin Biotechnol 30:107–112
16. Raju TS, Briggs JB, Borge SM, Jones AJ (2000) Species-specific variation in glycosylation of IgG: evidence for the species-specific sialylation and branch-specific galactosylation and importance for engineering recombinant glycoprotein therapeutics. Glycobiology 10 (5):477–486
17. Durocher Y, Butler M (2009) Expression systems for therapeutic glycoprotein production. Curr Opin Biotechnol 20(6):700–707
18. Sandberg H, Kannicht C, Stenlund P, Dadaian M, Oswaldsson U, Cordula C, Walter O (2012) Functional characteristics of the novel, human-derived recombinant FVIII protein product, human-cl rhFVIII. Thromb Res 130 (5):808–817
19. Kannicht C, Ramstrom M, Kohla G, Tiemeyer M, Casademunt E, Walter O, Sandberg H (2013) Characterisation of the post-translational modifications of a novel, human cell line-derived recombinant human factor VIII. Thromb Res 131(1):78–88
20. Brooks SA (2004) Appropriate glycosylation of recombinant proteins for human use: implications of choice of expression system. Mol Biotechnol 28(3):241–255
21. Fliedl L, Grillari J, Grillari-Voglauer R (2015) Human cell lines for the production of recombinant proteins: on the horizon. New Biotechnol 32(6):673–679
22. Muchmore EA, Milewski M, Varki A, Diaz S (1989) Biosynthesis of N-glycolyneuraminic acid. The primary site of hydroxylation of N-acetylneuraminic acid is the cytosolic sugar nucleotide pool. J Biol Chem 264 (34):20216–20223
23. Lis H, Sharon N (1993) Protein glycosylation. Structural and functional aspects. Eur J Biochem 218(1):1–27
24. Lowe JB, Marth JD (2003) A genetic approach to mammalian glycan function. Annu Rev Biochem 72:643–691
25. Delorme E, Lorenzini T, Giffin J, Martin F, Jacobsen F, Boone T, Elliott S (1992) Role of glycosylation on the secretion and biological activity of erythropoietin. Biochemistry 31 (41):9871–9876
26. Walsh G, Jefferis R (2006) Post-translational modifications in the context of therapeutic proteins. Nat Biotechnol 24(10):1241–1252
27. Wasley LC, Timony G, Murtha P, Stoudemire J, Dorner AJ, Caro J, Krieger M, Kaufman RJ (1991) The importance of N- and O-linked oligosaccharides for the biosynthesis and in vitro and in vivo biologic activities of erythropoietin. Blood 77(12):2624–2632
28. Fukuda MN, Sasaki H, Lopez L, Fukuda M (1989) Survival of recombinant erythropoietin in the circulation: the role of carbohydrates. Blood 73(1):84–89
29. Croset A, Delafosse L, Gaudry JP, Arod C, Glez L, Losberger C, Begue D, Krstanovic A, Robert F, Vilbois F, Chevalet L, Antonsson B (2012) Differences in the glycosylation of recombinant proteins expressed in HEK and CHO cells. J Biotechnol 161(3):336–348
30. Scallon BJ, Tam SH, McCarthy SG, Cai AN, Raju TS (2007) Higher levels of sialylated Fc glycans in immunoglobulin G molecules can adversely impact functionality. Mol Immunol 44(7):1524–1534
31. Hossler P, Khattak SF, Li ZJ (2009) Optimal and consistent protein glycosylation in mammalian cell culture. Glycobiology 19 (9):936–949
32. Fournier J (2015) A review of glycan analysis requirements. Biopharm Int. http://www.biopharminternational.com/review-glycan-analysis-requirements
33. Lingg N, Zhang P, Song Z, Bardor M (2012) The sweet tooth of biopharmaceuticals: importance of recombinant protein glycosylation analysis. Biotechnol J 7(12):1462–1472
34. Administration FaD (1999) International conference on harmonisation; guidance on specifications: test procedures and acceptance criteria for biotechnological/biological products. Notice. Food and Drug Administration, HHS. Fed Regist 64(159):44928–44935
35. Administration FaD (2005) International conference on harmonisation; guidance on Q5E comparability of biotechnological/biological products subject to changes in their manufacturing process; availability. Notice. Fed Regist 70(125):37861–37862
36. Zhang L, Luo S, Zhang B (2016) Glycan analysis of therapeutic glycoproteins. MAbs 8 (2):205–215
37. Celik E, Calik P (2012) Production of recombinant proteins by yeast cells. Biotechnol Adv 30(5):1108–1118
38. Doran PM (2000) Foreign protein production in plant tissue cultures. Curr Opin Biotechnol 11(2):199–204
39. Schwarz F, Huang W, Li C, Schulz BL, Lizak C, Palumbo A, Numao S, Neri D, Aebi M, Wang L-X (2010) A combined method for producing homogeneous glycoproteins with eukaryotic N-glycosylation. Nat Chem Biol 6 (4):264–266

40. Jaffé SRP, Strutton B, Levarski Z, Pandhal J, Wright PC (2014) Escherichia coli as a glycoprotein production host: recent developments and challenges. Curr Opin Biotechnol 30:205–210
41. Nielsen J (2013) Production of biopharmaceutical proteins by yeast: advances through metabolic engineering. Bioengineered 4(4):207–211
42. Razaghi A, Tan E, Lua LH, Owens L, Karthikeyan OP, Heimann K (2017) Is Pichia pastoris a realistic platform for industrial production of recombinant human interferon gamma? Biologicals 45:52–60
43. Meehl MA, Stadheim TA (2014) Biopharmaceutical discovery and production in yeast. Curr Opin Biotechnol 30:120–127
44. Gemmill TR, Trimble RB (1999) Overview of N- and O-linked oligosaccharide structures found in various yeast species. Biochim Biophys Acta 1426(2):227–223
45. Hamilton SR, Gerngross TU (2007) Glycosylation engineering in yeast: the advent of fully humanized yeast. Curr Opin Biotechnol 18 (5):387–392
46. Khan AH, Bayat H, Rajabibazl M, Sabri S, Rahimpour A (2017) Humanizing glycosylation pathways in eukaryotic expression systems. W J Microbiol Biotechnol 33(1):4
47. Beck A, Reichert JM (2012) Marketing approval of mogamulizumab: a triumph for glyco-engineering. mAbs 4(4):419–425
48. Paul M, Ma JK (2011) Plant-made pharmaceuticals: leading products and production platforms. Biotechnol Appl Biochem 58(1):58–67
49. Hellwig S, Drossard J, Twyman RM, Fischer R (2004) Plant cell cultures for the production of recombinant proteins. Nat Biotechnol 22 (11):1415–1422
50. da Cunha NB, Vianna GR, da Almeida LT, Rech E (2014) Molecular farming of human cytokines and blood products from plants: challenges in biosynthesis and detection of plant-produced recombinant proteins. Biotechnol J 9(1):39–50
51. Fox JL (2012) First plant-made biologic approved. Nat Biotechnol 30(6):472–472
52. Holland T, Sack M, Rademacher T, Schmale K, Altmann F, Stadlmann J, Fischer R, Hellwig S (2010) Optimal nitrogen supply as a key to increased and sustained production of a monoclonal full-size antibody in BY-2 suspension culture. Biotechnol Bioeng 107(2):278–289
53. Andersen DC, Krummen L (2002) Recombinant protein expression for therapeutic applications. Curr Opin Biotechnol 13(2):117–123
54. Toth AM, Kuo CW, Khoo KH, Jarvis DL (2014) A new insect cell glycoengineering approach provides baculovirus-inducible glycogene expression and increases human-type glycosylation efficiency. J Biotechnol 182–183:19–29
55. Cox MM (2012) Recombinant protein vaccines produced in insect cells. Vaccine 30 (10):1759–1766
56. Zhu J (2012) Mammalian cell protein expression for biopharmaceutical production. Biotechnol Adv 30(5):1158–1170
57. Browne SM, Al-Rubeai M (2007) Selection methods for high-producing mammalian cell lines. Trends Biotechnol 25(9):425–432
58. Wurm FM (2004) Production of recombinant protein therapeutics in cultivated mammalian cells. Nat Biotechnol 22(11):1393–8.
59. Biaggio RT, Abreu-Neto MS, Covas DT, Swiech K (2015) Serum-free suspension culturing of human cells: adaptation, growth, and cryopreservation. Bioprocess Biosyst Eng 38(8):1495–1507
60. Swiech K, Picanco-Castro V, Covas DT (2017) Production of recombinant coagulation factors: are humans the best host cells? Bioengineered:1–9. doi:10.1080/21655979.2017.1279767
61. Picanco-Castro V, Biaggio RT, Cova DT, Swiech K (2013) Production of recombinant therapeutic proteins in human cells: current achievements and future perspectives. Protein Pept Lett 20(12):1373–1381
62. Swiech K, Picanco-Castro V, Covas DT (2012) Human cells: new platform for recombinant therapeutic protein production. Protein Expr Purif 84(1):147–153
63. PER.C6 Cell Lines. http://www.gmp-creativebiolabs.com/per-c6-cell-lines_74.htm. Accessed 23 Mar 2017

Chapter 2

Uncovering Innovation Features and Emerging Technologies in Molecular Biology through Patent Analysis

Cristiano Gonçalves Pereira and Geciane Silveira Porto

Abstract

Scientific research at universities has a crucial role in leveraging a country's innovative potential. Sectors that require greater investments in technology for the development of their research, such as biotechnology, need to be aware of the frontier state-of-the-art technology and the knowledge incrusted within it. Although the information available in scientific articles is well explored in academic environment, the patent literature, where much of the technological information is present, is still poorly accessed. This chapter is intended to instruct students and researchers at universities to look at patent document analysis as a source of scientific and technological information and explore its applications. Within this chapter, we use the technological area regarding immunoglobulins inventions (monoclonal and polyclonal antibodies) as example to provide directions on how to develop a patent landscape to get an overview of the inventions in a certain field; how to map a collaborative network of inventors/assignees to help the pursuit and identification of future partnerships; and lastly we describe the steps of how to set up a network of patent citations with the aim of forecasting emerging technologies. We strongly believe that incorporate data from patents in planning phase of research projects at academia, as well as to establish partnerships and join R&D efforts to invest on promising technologies, is of great relevance to leverage the growth of the biotechnology sector.

Key words Knowledge flows, Patent analysis, Innovation, Biotechnology, Technology forecasting

1 Introduction

Universities have long been recognized as one of the most important environment responsible to generate knowledge through its research and innovation activities. By doing so, the universities are considered a significant player in developing and stimulating the innovative capacity of a region or country. The academic research is a notable driver for uncovering new inventions in the industrial sector and particularly important in basic research, such as the molecular and cell biology studies. It also plays a critical role in the development and dissemination of new insights that have innovative and technological potential. Nevertheless, most of this knowledge has been typically published in the form of academic

Virgínia Picanço-Castro and Kamilla Swiech (eds.), *Recombinant Glycoprotein Production: Methods and Protocols*, Methods in Molecular Biology, vol. 1674, DOI 10.1007/978-1-4939-7312-5_2, © Springer Science+Business Media LLC 2018

textbooks or/and peer review standards of publication such as in scientific journals. When scientists from universities drive into this direction, they create a condition in which the need to produce and publish new discoveries quickly may lead the researchers to discount or overlook the intellectual property features of their scientific discoveries [1].

Despite all efforts, protecting the inventions generated in the academic environment by taking advantage of the intellectual property rights, such as patents, is still a practice which is consistently present only in a few research groups. Furthermore, most of the researches within the university still do not use patent databases as a source of technological information in the development of their projects, failing to stay updated about the state-of-art of technologies in their respective research areas.

The definition of patent per WIPO states that patent is an exclusive right granted by the government for an invention (product or a process) that provides, in general, a new way of doing something, or offers a new technical solution to a problem [2]. To get a patent, a detailed technical information about the invention, sufficiently for an expert in the field to be able to carry it out, must be disclosed to the public in a patent application. Because the patent process is costly and generally can take several years for granting, filing a patent generally means there is optimism in economic or technical contribution of the invention [3].

Technological information is a necessary condition for success in any process of research, industrial planning, development, manufacturing, commercialization, or management. Appropriate selection of sources of information is highly significant to a research project planning [4]. In this regard, patent literature plays a crucial position, in which knowledge disclosed is an important contributor to the base of science and technology upon which the Nations' economy is built. It is estimated that 70% of information disclosed in patents are never published in any other form, such as books, technical-scientific articles or any other medium of information [2].

The relevance of using patent as a source of technological information contributes either to avoid R&D investment duplications. European Patent Office estimated that approximately US$20 billion are wasted in the European Community every year because of duplicate inventions (http://www.epo.org/about-us/annual-reports-statistics/annual-report/2012.html). Another study related to vascular risk management comparing scientific papers and patent application, showed that international patent applications reached the public quicker through the corresponding peer review document if one was published, but more than two-thirds of the patent applications had no such companion paper in a scientific journal, supporting the uniqueness and remarkability of patent literature compared to peer reviewed literature [5]. Specifically, at the academic environment, Pereira et al. [6] assessed the importance of universities to include the patent literature as a source of scientific and technological information in their research projects, through verification of existing patent applications related to new insights contained in doctoral theses, by the own or unrelated research groups. The results showed that 54% of the theses had part or the whole knowledge disclosed in patent documents prior to the thesis defense, and some were specific about the topic and the results of the thesis, presuming that there was a likely duplicity in the development of the research.

Concerning patents being a reliable source of information, its document allows the scientists the use of data such as: information regarding inventors and assignees, classification by the technological area (IPC and CPC codes), abstract and patent claims, date of application and patent life time, geographical coverage, legal status, among others [7]. These data present a uniform worldwide format which is available electronically, usually freely accessible in most countries, and contain the technology description in a manner sufficiently clear for it to be carried out by a person skilled in the art.

All this easily accessible information provided through patents can be analyzed into many different approaches such as design a patenting landscape within a technological area, to study collaborations networks among assignees which results in patent application, as well as to forecast emerging technologies using patent citation analysis. In this chapter, we are going to describe those three methods of patent analysis and provide a brief example of its application in a technological area of molecular and cell biology.

1.1 Patent Landscape Analysis

Recently, the use of landscapes by industry, government, and academia to map scientific and technological trends has become very common. A landscape normally seeks to answer specific policy or practical questions and to represent the complex information about this activity in a clear and accessible manner for audiences with different background [8]. As countries within the OECD (Organization for Economic Co-operation and Development) strive to integrate science policy and innovation strategy [9], they place greater reliance on landscapes to track trends and support the coordination of activities [10], actors and institutions. Patent is considered the most commonly used indicator in designing landscapes. According to WIPO, a patent landscape analysis starts with a state-of-the-art search for the relevant technology in selected patent databases and then analyze the search results to answer specific questions such as patterns of patenting activity or features of innovation. The patent landscape outputs are disclosed in a visually manner to assist the understanding, conclusions or recommendations based on the empirical evidence which are provided [11].

Knowledge-intensive sectors such as molecular biotechnology have faced and increase in patent applications over the past decade and, consequently, the pool of patent data available for analysis has become even bigger. Recently, many studies have driven into patent landscapes to better describe an overview of patent activity regarding the potential technologies for therapeutic applications. And a study published in 2014 [12] described the patent landscape of induced pluripotent stem cell technologies (iPSC), 7 years after its discovery in 2007. The area is of great interest for patenting, along with its long-term therapeutic potential, the iPSC technologies are considered as well as valuable tools for drug discovery, toxicology testing and disease modeling. Another study published in 2016 [13] uses the patent landscape approach to verify patters of patent activity of a powerful tool in basic biological research with a wide range of potential commercial applications, the CRISPR-Cas gene editing technology. Those studies are of huge relevance, since it answers many questions related the technological area such as: How do new inventions evolve over the years? Which are the main companies that own the technologies and how do they collaborate with each other? Who are the main inventors and where they are located? In what technological areas are the inventions protected? Where the technologies are protected? The patent landscapes contribute to a high-level understanding of a specific sector, answering those and many other questions.

1.2 Collaboration Networks

According to the OECD, the innovative potential relies on how knowledge transits and the system of structure connections. In this respect, the importance of establishing partnerships between companies, government, and universities with the aim of enhancing technological innovation has been discussed and emphasized over the years in studies dealing with innovation systems [14, 15], the triple helix [16], open innovation [17] and, more recently, the joint-product orientation [18].

A very important indicator to measure the activities of innovation and interaction of knowledge between actors, is the number of publications and patents in common [19].The use of Social Network Analysis (SNA) to study cooperation among actors is a reliable and appropriate tool for understand and visualize the connections. Because the innovative process is complex, how the interaction among actors occurs can define the outcome of the effort. In this context, the flow of knowledge arising from cooperative structures can be interpreted by social network metrics and graphs. Researchers such as Cantner and Graf [20] applied the network analysis to describe the evolution of the innovation network in Jena, Germany, between 1995 and 2001, others such as Owen-Smith et al. [21] developed a work focused on R&D cooperation comparing the organization of scientific research in the US and Europe through network analysis. The use of network analysis was also applied to study the relationship between common academic inventors in patenting processes, at different technological areas [22]. The studies on collaboration networks seek to explain the performance of individual actors using the characteristics of the network in which they are inserted, to describe the structures of the collaboration network, or even to explain the development and dynamics of the network [23].

The collaborations networks can provide useful information regarding not only on how the dynamic of connections from companies or inventors occurs, it provides also information about what are the companies/inventors willing to establish cooperation, as well as to guide the scientists to others researcher groups or institution specialized on a certain technological in which a solution to a problem is pursued.

1.3 Patent Citation Analysis

Different from scientific papers citation, which the list of references is based on an author choice, the patent citation is an obligation of the patent applicant, who should comply with the legal requirement to disclose which prior scientific studies and other patents contributed to the development of the technology, in addition to disclose a complete description of the state of the art in the field of the invention as the technological background [24, 25]. Such information provides, with greater wealth of reliability, how the development of such technology has occurred. By interconnecting these patents through the citation relationship between them, it is possible to have, with relative accuracy, a line of knowledge development of the technology in question [26, 27].

The patent citation associates patents that have a certain correlation of content, providing simple links of unidirectional transfer of knowledge by creating networks. The patent citation networks then consist of related patent groups, in which the exchange of citations between them evolves over time and reveals technological trends. Network visualization provides a better understanding of the development and diffusion of technologies, and their analyses can be made from critical nodes, relevant trajectories, or the network as a whole [28].

The analysis of patent citation networks can indicate emerging technologies by tracking its most relevant trajectories. The SNA tools and the studies of trajectories of patent citations networks such as those described by Verspagen in 2007 [29] prove to be reliable tool to leverage the technological forecast studies and identify promising technologies. The result of applying these techniques aims to direct the technological efforts of the R&D areas such as biotechnology, for investment in technological trends for future exploration.

The chapter hereby is intended to provide useful instructions regarding the use of patent literature of source of scientific and technological information, which universities and researchers can take advantage of. The factors that motivated us to write this chapter were the significance of incorporate data from patents in planning phase of research projects at academia, as well as the need to establish partnerships and R&D efforts to invest on promising technologies needed to leverage the growth of the sector. Here, we describe the tools and methods of three different and complementary approaches of patent analysis applied at a molecular biology field by designing patent landscapes, mapping collaboration networks among patent assignees and identifying future-oriented technologies through patent citation network analysis.

2 Tools and Materials

Patent analysis is largely used in innovation studies to assess and monitor technological changes in different sectors. Several applications can be considered when performing patent analysis. Here, we describe the tools required to develop patent landscapes, collaborations networks, and the network analysis of patent citation.

2.1 Patent Database

The major patent database for patent search are described.

2.1.1 Open Source

- *Patentscope (www.wipo.int/patentscope)*: The WIPO database provides access to updated international Patent Cooperation Treaty (PCT) applications in full text format, as well as to patent documents of participating national and regional patent offices. Searching for keywords, names of applicants, international patent classification, and many other search criteria are available in multiple languages.
- *Espacenet (www.espacenet.com)*: The database provided by EPO, the European Commission, give you access to nearly 90 million patent documents worldwide, with information regarding patent families, legal status and citations. It has the option of smart search, advanced and classification search. Epacenet is not intended for bulk data retrieval and users are limited to ten search-related actions per minute (robots are not allowed).
- *Google Patents (patents.google.com)*: Currently, the engine covers 17 offices including PCT applications, although they cannot guarantee complete coverage, showing the results in a Google fashion. Patents in non-English language are machine-translated to English and indexed, so patent search can be performed using only English keywords. You can download the top 1000 results in a "csv" format.

2.1.2 Fee-Based

- *Clarivate Analytics - Derwent Innovation - (formerly Thomson Innovation from Thomson Reuters® (info.thomsoninnovation.com)*: Clarivate Analytics provides comprehensive international patent, business, news, and scientific literature coverage and analysis tools. English translation (machine and manual) of patents from the major offices from Europa, Asia-Pacific is also available. It has the support of Derwent World Patent Index (DWPI), conferring its unique abstracts and titles (and the data curation skills of an editorial staff of more than 500 experts) that offer a strong differentiation of products. The themescape and text clustering analysis are valuable tools to identify the technology "core" and landscape of a patent group.

- *Questel-Orbit (www.orbit.com)*: Worldwide coverage and contains full text of patent publications from 22 patent offices. It provides analysis tools as well excellent charting and visualizations. It has unique family grouping, called FAMPAT (gives invention based and duplicate free results). It has evaluation modules for business intelligence, summary of invention, prior art resources, normalized assignee names, and concept normalization.
- *PatBase from Minesoft (patbase.com)*: The database has a worldwide coverage (from over 100 patent-issuing authorities) and a user-friendly interface. It presents search results based on patent families rather than individual patent documents. PatBase displays unified classification and assignee data for each patent family record.

2.2 Computer Software Program for Storing, Organizing, and Manipulating Data

Microsoft Excel (Microsoft), Google Sheets (Google), Numbers (Apple), Apache OpenOffice (OpenOffice), etc.

2.3 OpenRefine (http:/openrefine.org/): Requires a Web Browser with JAVA® Support

A stand-alone open source desktop application for data wrangling based on data cleanup, uniformization and transformation to other formats.

2.4 Gephi: The Open Graph Viz Platform [30] (gephi.org)

An open-source software for visualization and exploration of all kinds of graphs and networks (check system requirements for installation). Gephi is a tool for people that have to explore and understand graphs.

2.5 Statistics and Plugins Associated with Gephi (Available at Marketplace.gephi.org/plugins/)

- *Multimode Networks Transformations by Jaroslav Kuchar (2012)*: This plugin allows multimode networks projection such as project your bipartite (two-mode) graph to monopartite (one-mode) graph.
- *OpenOrd Layout by Mathieu Bastian* [31]: it is a force-directed layout algorithms ideal for large graphs.
- *Noverlap by Mathieu Jacomy (2012)*: prevents the round nodes from overlapping.

- *Betweenness Centrality (using Network Diameter statistics):* measure how often a node appears on shortest path between nodes in the network.

3 Methods

3.1 Patent Search

The first step in a patent analysis is to clearly define which technology you are looking for and which means you will use to perform the query. The search for a certain technology can be done by using (*see* **Note 1**):

- Keywords: Keywords should define the "core" of the technology. Usually the keywords are searched in the title, abstract and claims of a patent document. Several keywords can be managed using Boolean truncation or proximity. A good patent search by keywords must cover several possibilities of the subject's occurrence using synonymous and word's variation.

  Example: CTLA4 monoclonal antibodies in cancer—Search in Tittle
  Abstract and Claims: ((monoclonal ADJ antibody) AND (cancer)) AND CTLA4
- *Technological areas:* Technological areas of a patent are generally defined by international codes in which patents are classified as International Patent Code (IPC), Cooperative Patent Code (CPC), US CLASS, etc. The patent classification is intended to code and group documents according to the technical features of the invention. The codes enable searching of documents of different languages disclosing similar technical content.

  Example: IPC search: C07K 16/00 (Immunoglobulins [IGs], e.g., monoclonal or polyclonal antibodies).
- *Patent from companies/inventors*: You can search for patents of a specific company or researchers (as inventors) using the search field Assignee/patent owner (companies, research institutes, university, foundation, etc.) or Inventor, respectively (*see* **Note 2**) Example: Patents from Genentech (Assignee: Genentech).

3.2 Extract the Results Dataset

After running the search as described previously, you must select the patent documents prior to export the data. You can also select, which information you need to include in the results datasheet (Patent number, Tittle, Abstract, Claims, Priority country, Assignee, Inventor's name, cited Refs, legal status, and many others depending on your interest and the patent database availability) (*see* **Note 3**). After selecting the data, you can export the results in a spreadsheet format, usually the ".csv" is the best option to further manipulate the data.

3.3 Data Managing

Once you have the file including the results (list of patents) of your search, you can open it using a software for spreadsheet manipulation, such as Microsoft Excel®, to extract information according to the type of patent analysis you want to perform and prepare the data for further assessment.

3.4 Normalizing Data with OpenRefine

Patent database use a large amount of data, and in most cases the data is messy which needs to be normalized, such as the names of patent's assignee and inventors prior leading it to SNA. Among the data collected there are many ambiguities due to writing variations of the same name ranging from small inconsistencies to the use of different abbreviations. These variations result in duplications of the same "node" in the network, which drastically affects the statistics and the interpretation of the analysis. The activity of standardizing those names, if performed manually, will require a lot of time and effort [32]. OpenRefine is free, open source software that was made available to the public in October 2010 by Google. The tool has the purpose to explore, clean, and handle a large volume of data, as well as allowing connection to external data sources through the web, which allows to enrich the content of the databases. In order to clear names, it has clustering tools, which allow to group words with different values which represents alternatives names to the same thing (e.g., Celgene, Celgene Corporation, Celgene A, and Celgene Corp). To do so, the tool uses a select number of different agglomeration methods and algorithms that are effective and efficient for use over a wide range of data. For the names normalization, follow the steps below:

1. Install OpenRefine and upload your spreadsheet containing the list of patents and create the project.
2. Select which column do you want to normalized and click on it to select "facet" and then "TextFacet" option.
3. Now click the "Cluster" button on the left panel to bring up a new pop-up.
4. At the new window, OpenRefine will provide options about the way of all the terms will be clustered together given the first selected clustering algorithms. By default, the first one is the key collision method named fingerprint (*see* **Note 4**).
5. OpenRefine will show how many clusters were found to be pretty much the same values according to the algorithm used. Click the "Merge?" checkbox if you want to merge the values to the default name (the first shown) or to the name of your choice by editing the "New Cell Value" option.

6. Subsequently at the bottom right of the popup window, click the "Merge Selected & Close" button (you have the option to "Merge Selected & Re-Cluster" if you want to use another algorithm).
7. Once normalized the data you have the option to export and save the normalized excel file or a ".csv" file to your computer.

3.5 Patent Landscape Designing

For landscapes designing, from a group of patents, you can select information regarding the Patent publication and/or application dates (*see* **Note 5**), Assignee's name, Inventor's name, patent legal status, IPC or CPC class, country code, and many others to draw charts for better visualization.

Here is an example: Patent applications concerning the technological area comprising the IPC class code C07K 16/00 (antibodies) between 1995 and 2015. It was found 18.856 INPADOC records (using Clarivate Analytics). The patent landscape can be performed using the patent information as follows (*see* **Note 6**).

1. *Evolution chart of Patent Publication and Application over the years*: To do so, you need to select the Publication Year and Application Year columns from patents to draw a chart on excel (*see* **Note 7**).

   Example: Fig. 1 shows the evolution chart of Patent Publication between 1995 and 2015 of immunoglobulin inventions selected by the IPC C07K 16/00.
2. *Top Assignees and Inventors*: Select the assignee and/or Inventor's name column after normalization (*see* Subheading 3.4) to

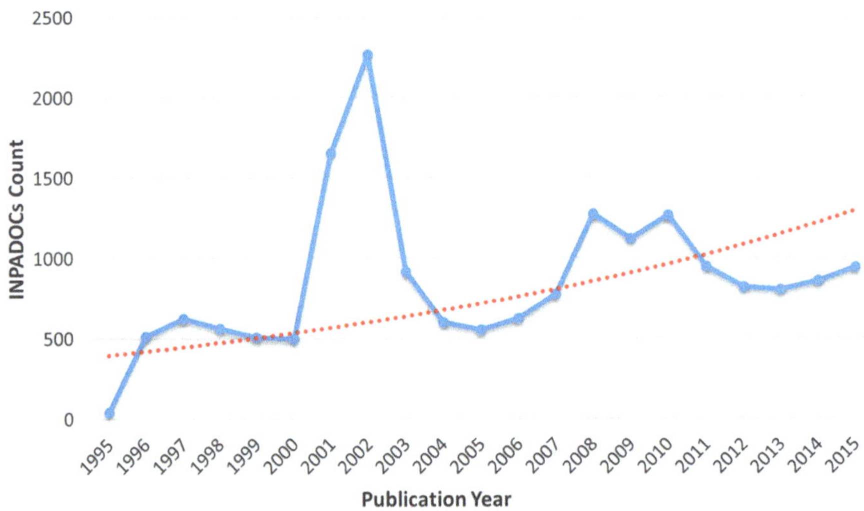

Fig. 1 Evolution of patent publication concerning the technological area selected by the IPC C07K 16/00 (immunoglobulins) filed between 1995 and 2015. The *red dotted line* represents the exponential tendency curve

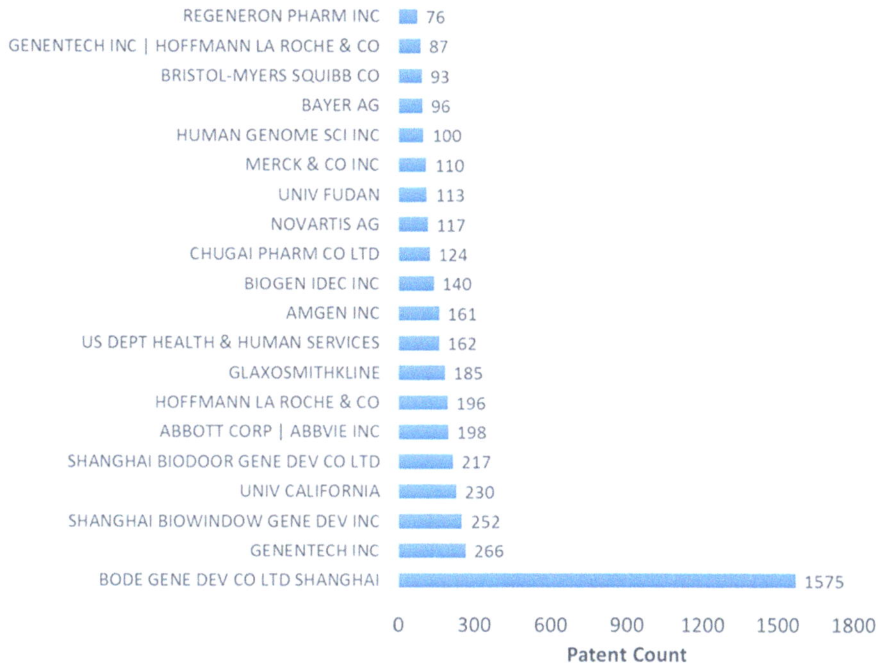

Fig. 2 The Top 20 normalized assignee/applicant of patents concerning immunoglobulins inventions filed between 1995 and 2015

draw a chart. You can select to show the top 5, 10, and 20 names; more than this will make visualization difficult.

Example: Fig. 2 shows the top 20 Assignees in INPADOCs count to the query of immunoglobulins patents (1995–2015).

3. *Top IPC and CPC class*: Regarding to patent class codes, you have two options to draw a chart according to the deep level of the code. The four-digit (C07K—Peptides) code gives a wide view of the technological area, while the full code (C07K 16/00—Immunoglobulins, e.g., monoclonal or polyclonal antibodies) represents the technological area more strictly.

4. *Top countries or organizations where the patent application was filed or granted*: The country was selected by the "Country Code" column. The code has two letters and can represent an organization as well, such as WO (World Intellectual Property Organization), EP (European Patent Office), and EA (Eurasian Patent Organization) (*see* **Note 8**). Depending on the patent database you are using, you can view the patent coverage in a world map. The Fig. 3, displays the patent coverage for immunoglobulins inventions (selected by the IPC C07K 16/00) in patents filed between 1995 and 2015 using Clarivate Analytics®.

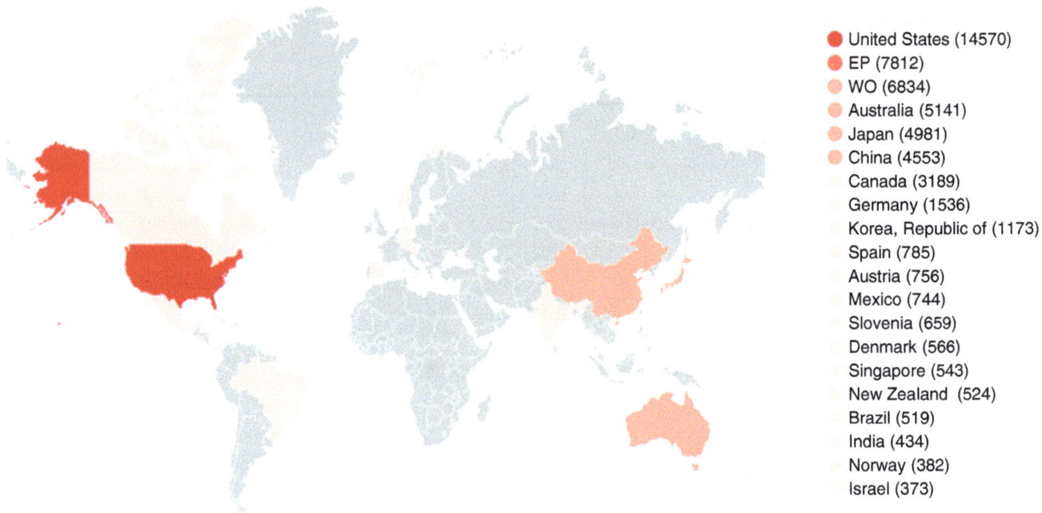

Fig. 3 The Thomson Innovation® world map of the coverage of patents regarding immunoglobulins. The data of "Country Code" of patent documents were disposed geographically and represented by colors (*dark colors*, elevated number of patents in in the respective countries, and *light colors*, the opposite)

The patent landscape can be explored using information other than those described here. Information regarding legal status of patents are useful to verify if patent application is still pending or if the patent has been withdrawn, rejected, granted, expired, lapsed or been revoked. In addition, you can use tools to display an overview of the topic of a group of patents based on assessment of text fields such as the title, abstracts and claims. To get the patent landscape based on patent contents, tools based on text mining and visualization provide the extraction of relevant information about the patent's "core" without the need to read an excessive pile of documents. There are many tools to facilitate the uncovering of relevant information about innovative technologies based on text mining and visualization tools such as VantagePoint (www.thevantagepoint.com), text clustering function provided by Clarivate Analytics®, ThemeScape (Clarivate Analytics®), and many others [33].

3.6 Collaboration Networks

The collaborations networks methods can be applied either to map the links among patent's assignees or patent's inventors (which will be named here as "actors"). The first step before generate the collaborations networks itself, is to prepare the data. This is a very time-consuming phase and can be carried out as follows:

1. Since we are working with collaboration among inventors/assignees which results in patent application, you should clean patents with just one value or with no information.

2. The next step is to split multivalued cells according to the separator between actors and enter the patent numbers.
3. Perform the normalization data described in Subheading 3.4.

Now that you have your normalized and cleaned datasheet, you can proceed to create the network. The collaboration network is a bipartite graph, since it has nodes with two different categories: the patent nodes and the assignee/inventor nodes. The steps to create the network are as follows:

1. Since it's a bipartite network, two datasheets must be prepared, one containing the nodes identification (nodes table) and another one with the links among patent and its assignees (edge table).
2. Both datasheets are uploaded into a "New Project" into Gephi software by "Importing Spreadsheet" within data laboratory section.
3. Now that you have uploaded both data, the Gephi software has created the bipartite network displaying the nodes (patents and actors) and edges (links) in the "Graph" tab within the "Overview" section (*see* **Note 9**).
4. To get a better visualization of the network, you can choose the layout algorithm OpenOrd and then the algorithm ForceAtlas 2 to separate the nodes (*see* **Note 10**).

To create the collaboration network containing links between only the actors, you need to first transform the network from a bipartite to a monopartite, eliminating the node with the category of "Patents." To do so, you need to apply the Multimode Networks Transformations Plugin to the bipartite network by selecting the attribute type (node category) and indicate the left matrix (Actor-Patent) and right matrix (Patent-Actor) to show the relationship among actors only. An actor can be an assignee or an inventor. Remember to click to the "Remove Nodes" and "Remove edges" checkbox. Once the network has been created, you can arrange the nodes according to different layouts, give name to nodes, resize the nodes per degree value, define communities using the modularity statistics [34], and use many others tools to deliver a better visualization of the relationships.

Example: Following the patent query of previous examples using Clarivate Analytics® database (Subheading 3.5, **steps 1**, **2**, and **4**) concerning the IPC search code C07K 16/00 (Immunoglobulins [IGs], e.g., monoclonal or polyclonal antibodies) on patent applications ranging from 1995 to 2015. Figure 4 shows the collaboration network among assignees which results in patent application related to the query mentioned above.

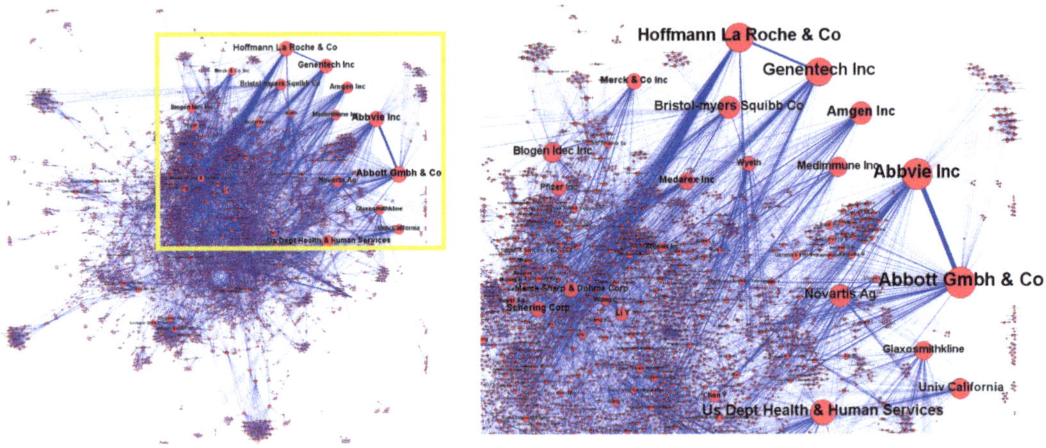

Fig. 4 Collaboration network among assignees/applicant which resulted in patent deposit concerning the technological area of immunoglobulins (filed between 1995 and 2015). The *left panel* shows an overview of the whole network where you can see the link (*blue*) of all the players (*red*). The *right panel* shows a zoom of the *yellow rectangle* on *left*. In the zooming version, you can see the thin edges and thick edges, where the edge thickness represents the weight of an edge according to the number of collaboration between players. The image was generated using Gephi software

3.7 Patent Citation Network Analysis

The patent citation network is created using the information of the "Cited Refs" where the references in this case are patents and not scientific papers. The patent citation network is monopartite and directed since the citation events are unidirectional. Follow the steps listed below to prepare the data before uploading it into Gephi and for network construction.

3.7.1 Datasheet Preparation

1. Identify and copy both the patent number column and the cited patent column and paste in a new datasheet.
2. Edit the column heading to "Source" for the patent number and "Target" for the cited patents.
3. Split multivalued cells according to the separator between actors (usually ";" or "| ") and enter the patent numbers.
4. Save the new file and open it in Gephi.

3.7.2 Creating the Patent Citation Network

1. Import the file containing the patent citation prepared previously into Gephi through Data Laboratory section as the "edge table" format.
2. At the import window mark the option "Create missing nodes" and check if the "directed" network is set (*see* **Note 11**).
3. Go to the Overview section and use the OpenOrd layout to get a better arrangement of nodes and edges (*see* **Note 10**).
4. To size the nodes according to the number of citations, go to the "attributes" in the overview section and click nodes > attributes > Select degree and choose the size button. You can set

an unique value for the degree or attribute the minimum and maximum node size.

5. Use different statistics metrics in the "statistics" window.
6. Identify communities for a network using the Modularity statistics [34] and give color to nodes according to the community ID (*see* **Note 9**).
7. You can also use filters to select a group of patents for further analysis, such as isolate the giant component, create an ego network, and separate nodes ranging for a certain degree.
8. To display the node ID in the Graph, go to Label Text Lettings and click in "ID" and then click the "Show Node labels" button (letter "T" button).
9. To save the Graph layout into your computer, go to the "Preview" section and edit the network appearance as you wish (node label size, edge format, edge thickness, etc.) and export the image by clicking the "export" button.

3.7.3 Forecasting Emerging Technologies Using Patent Citation Networks

Patent citation networks can be used as a tool for technological forecast and identify trends based on promising technologies. Many methods can be used to achieve this purpose; here, we describe one important approached based on Betweenness centrality of nodes and edges.

Betweenness centrality

The Betweenness Centrality can be used to identify critical patents in the patent citation network. A node with a high Betweenness centrality has the ability to facilitate or limit interaction between the nodes it links [35], in other words, it functions as a bridge which control the information flow of the network (*see* **Note 12**). Using Gephi it is possible to calculate the Betweenness Centrality of a node by running the statistics "Network Diameter." After clicking on it, a window will pop up for you to choose to apply the metric on a directed or undirected network and you have also the option to normalize the data. By clicking "OK," the metric will run and add each node score to a column in the "nodes" tab in "Data Laboratory" section. Use this information to identify the nodes with higher Betweenness Centrality (rank columns from higher to lower values). The node (patent) with higher Betweenness Centrality score has a greater importance to the network as a whole. You can explore the links (ups and downs) of this patent node to reach more technological trends based on recent patents linked to this node (*see* **Note 13**).

Example:

Goal: Identify promising technologies through Patent Citation Network Analysis related to immunoglobulins inventions (searched by IPC C07K 16/00 with application date ranging from 1995 to 2015). Analysis: The node with higher Betweenness Centrality score is the patent US20100076178A1 (Dual Variable Domain Immunoglobulins and Uses Thereof—2010) from Abbot Laboratories. The invention describes methods of making bispecific antibodies, a novel family of binding proteins capable of binding two or more antigens with high affinity, and their uses in the prevention, diagnosis, and/or treatment of disease. The high Edge Betweenness node has an "In-Degree" of five, which means that five more recent patents cited this invention (named "descendent" on the lineage of the analyzed patent). The Fig. 5a shows a network of knowledge flows considering only the patents which cited the US20100076178A1 (high Betweenness Centrality) patent. The

A

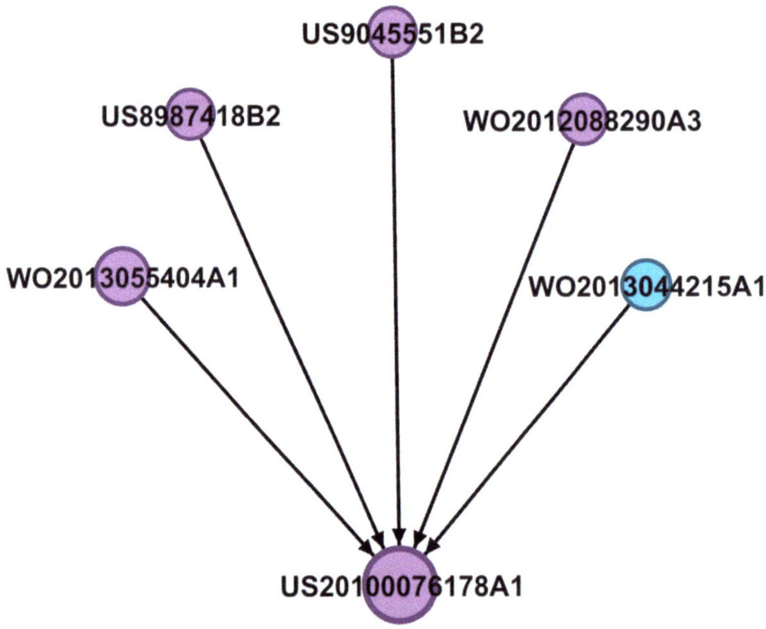

B

Patent	Condition	Title	Assignee	Pub. Year
US20100076178A1	Orign*	Dual Variable Domain Immumoglobulins and Uses Thereof	Abbott Laboratories	2010
WO2013055404A1	Descendent	Meditopes and meditope-binding antibodies and uses thereof	Beckman Research	2013
US8987418B2	Descendent	Dual specific binding proteins directed against IL-1β and/or IL-17	AbbVie	2015
US9045551B2	Descendent	Anti-DLL4/VEGF dual variable domain immunoglobulin and uses thereof	AbbVie	2015
WO2012088290A3	Descendent	Tri-variable domain binding proteins and uses thereof	Abbott Laboratories	2012
WO2013044215A1	Descendent	Vegf/dll4 binding agents and uses thereof	Oncomed	2013

Fig. 5 Knowledge flows among promising technologies based on patent citation. (**a**) The Citation Network with one level of depth of patents which cited the patent with higher Betweenness Centrality, US20100076178A1 (from the global citation network of patents concerning immunoglobulin inventions). (**b**) The *table* shows the information of patents in the condition of descendant on the US20100076178A1 lineage based on patent citation

table included on Fig. 5b brings the information regarding each of those patents which can be considered as promising technologies (*see* **Note 14**).

There are several tools to forecast emerging technologies such as developing technology roadmaps and scenario planning, performing tech mining, models for technology diffusion as the Fisher-Pry and others. Tools based on relevant paths is also convenient to track for future-oriented technologies, such as the methods described by Hummon and Dereian in 1989 [36] and implemented by Verspagen [29].

4 Notes

1. When performing a patent search, you can also use combinations of more than one field. Example: you can search for some keywords (on Tittle and Abstract fields) to find patents of on a specific company (assignee field) filed between 2003 and 2009 (Application date field) concerning a technological area of immunoglobulins (IPC or CPC fields).
2. If your intention is to do a broaden search for a company's patent ownership, you must include the corporation tree including its alternative names, subsidiaries, companies which were merged/acquired, etc. The Clarivate Analytics - Derwent Innovation® database provides a "Corporate tree" function to assess this information from an Applicant/Assignee (US and EP grants and EP and WO applications only).
3. Some databases do not provide a complete information in the exported data. Check on the database before starting collecting the data.
4. OpenRefine has six algorithms which are mentioned in order, from strictest to loosest: key collision: fingerprint; key collision: ngram-fingerprint; key collision: metaphone3; key collision: cologne-phonetic; nearest neighbor: levenshtein; and nearest neighbor: PPM.
5. Remember of the confidential period gap of patents, generally patent applications are published 18 months after the earliest priority date of the application.
6. You can choose to draw the data once analyzed into different charts (pie, bars, bubble, and many others).
7. Most of the patent database provide a graphical version of patent publication and application dates.
8. Besides the country which the patent was filed or granted, you can also select the information of "Priority Country" which means the country where the patent is first filed before being (possibly) extended to other countries.

9. In the "Overview" section of Gephi, go to the appearance tab to edit the nodes and edges as you wish to get a better representation of the network. You can set the node size (you can highlight the nodes which performed higher citations or were more cited by out-degree or in-degree score respectively) and color (you can color the nodes based on community ID after running the Modularity statistics) according to different parameters (some of them are available only after running its statistics).
10. Use the Noverlap layout to avoid node overlapping after using OpenOrd. If you use ForceAtlas 2, check the layout option "prevent overlap" in the settings.
11. The directed graph is arrow-oriented. The arrow orientation determines the citation flow, the arrow that goes to directed to a node adds an "in-degree" score to that node, whereas arrows that exit a node, adds an "out-degree" score to that node. In patent citation network, nodes that have elevated "in-degree" score means patents which are highly cited, whereas node that have high "out-degree" score represent patents which cited an elevated number of patents.
12. The edge-betweenness is based on the idea of "betweenness" of links in the network and it's defined as "the number of shortest paths between pairs of vertices that run along it" [37]. The edge betweenness value is higher on edges that connects nodes from different communities, which means that if the edge is removed, more likely it is that the network will disintegrate. So the Edge Betweenness represents crucial links on the networks as a whole. Recently the use of the Edge-Betweenness algorithm on patent citation has extensively being explored on citation networks [38–40].
13. One option to explore the ups and downs links of a specific node in a directed network, as the patent citation network, is to create ego networks. Ego networks consist of a focal node named "ego" and the nodes to whom ego is directly connected to. T isolate an ego network in Gephi, go to the tab "filters" in overview section and click to the folder "Topology" and then "Ego Network." Once the filter is selected, you need to specify the node ID and the depth level before running the filter.
14. For this type of analysis, you can also include patents with conditions of descendent and ascendant in more than one level of depth, to track the knowledge flows using the node with the higher Betweenness Centrality as reference.

Acknowledgments

This work was supported by the Fundação de Amparo à Pesquisa do Estado de São Paulo (FAPESP) grant number 2014/22500-8; 2012/22686-9 and 2015/13816-4 (which provided us the resources for a Clarivate Analytics - Derwent Innovation® subscription.)

References

1. Patino R (2009) Intellectual property rights and research disclosure in the university environment: preserving the commercialization option and optimizing market interest. J Am Assoc Lab Anim Sci 48:138–143
2. W.I.P.O. (WIPO) Patents. http://www.wipo.int/patents/en/
3. Basberg BL (1987) Patents and the measurement of technological change: a survey of the literature. Res Policy 16:131–141
4. WIPO (1995) WIPO Asian regional seminar on the use of patent information by industry : New Delhi, India, March 16 to 18, 1994. World Intellectual Property Organization, Geneva
5. HAM M (2011) Relating patenting and peer-review publications: an extended perspective on the vascular health and risk management literature. Vasc Health Risk Manag 7:265–272
6. Pereira CG, da Silva RR, Porto GS (2015) The scientific information provided through patents and its limited use in scientific research at universities. Braz J Sci Technol 2:2
7. Yoon B, Park Y (2004) A text-mining-based patent network: Analytical tool for high-technology trend. J High Technol Managem Res 15:37–50
8. Trippe A (2015) Guidelines for preparing patent landscape reports. WIPO, Geneva
9. OECD (2010) The OECD innovation strategy: getting a head start on tomorrow. OECD Publishing, Paris
10. OECD (2011) Collaborative mechanisms for intellectual property management in the life sciences. OECD, Paris
11. WIPO. Patent landscape reports. http://www.wipo.int/patentscope/en/programs/patent_landscapes/
12. Roberts M, Wall IB, Bingham I et al (2014) The global intellectual property landscape of induced pluripotent stem cell technologies. Nat Biotechnol 32:742–748
13. Egelie KJ, Graff GD, Strand SP et al (2016) The emerging patent landscape of CRISPR-Cas gene editing technology. Nat Biotechnol 34:1025–1031
14. Lundvall B-A (1992) National systems of innovation: towards a theory of innovation and interactive learning. IDEAS, London
15. Freeman C (1995) The â€™ national system of innovation â€™ in historical perspective. Camb J Econ 19:5–24
16. Leydesdorff L, Etzkowitz H (1996) Emergence of a triple helix of university-industry-government relations. Sci Public Policy 23:279–286
17. Chesbrough H (2003) The logic of open innovation: managing intellectual property. Calif Manag Rev 45:33–58
18. Foray D, Lissoni F (2010) University research and public-private interaction. Handb Econ Innov 1:275–314
19. Inzelt A (2004) The evolution of university-industry-government relationships during transition. Res Policy 33(6–7):975–995
20. Cantner U, Graf H (2006) The network of innovators in Jena: an application of social network analysis. Res Policy 35:463–480
21. Owen R, Macnaghten P, Stilgoe J (2012) Responsible research and innovation: from science in society to science for society, with society. Sci Public Policy 39:751–760
22. Balconi M, Breschi S, Lissoni F (2004) Networks of inventors and the role of academia: an exploration of Italian patent data. Res Policy 33:127–145
23. Van Der Valk T, Gijsbers G (2010) The use of social network analysis in innovation studies: mapping actors and technologies. Innov Manag Policy Pract 12:5–17
24. Criscuolo P, Verspagen B (2008) Does it matter where patent citations come from? Inventor vs. examiner citations in European patents. Res Policy 37:1892–1908
25. Jaffe A, Trajtenberg M, Fogarty M (2000) Knowledge spillovers and patent citations: evidence from a survey of inventors. Econ Rev 90:215–218

26. Carpenter MP, Narin F, Woolf P (1981) Citation rates to technologically important patents. World Patent Inf 3:160–163
27. Michel J, Bettels B (2001) Patent citation analysis: a closer look at the basic input data from patent search reports. Scientometrics 51:185–201
28. Cho T-S, Shih H-Y (2011) Patent citation network analysis of core and emerging technologies in Taiwan: 1997–2008. Scientometrics 89:795–811
29. Verspagen B (2007) Mapping technological trajectories as patent citation networks: a study on the history of fuel cell research. Adv Complex Syst 10:93–115
30. Bastian M,Heymann S, Jacomy M (2009) Gephi: an open source software for exploring and manipulating networks, third international AAAI conference on weblogs and social media, pp 361–362
31. Martin S, Brown WM, Klavans R et al (2011) OpenOrd: an open-source toolbox for large graph layout. Proc SPIE 7868:786806–786811
32. Wong CK, Ho SS, Saini B et al (2015) Standardisation of the FAERS database: a systematic approach to manually recoding drug name variants. Pharmacoepidemiol Drug Saf 24:731–737
33. Yang Y, Akers L, Klose T et al (2008) Text mining and visualization tools – impressions of emerging capabilities. World Patent Inf 30:280–293
34. Blondel VD, Guillaume J-L, Lambiotte R et al (2008) Fast unfolding of communities in large networks. J Stat Mech Theor Exp 10008:6
35. Hung SW, Wang AP (2010) Examining the small world phenomenon in the patent citation network: a case study of the radio frequency identification (RFID) network. Scientometrics 82:121–134
36. Hummon NP, Dereian P (1989) Connectivity in a citation network: the development of DNA theory. Soc Networks 11:39–63
37. Girvan M, Newman MEJ (2002) Community structure in social and biological networks. Proc Natl Acad Sci U S A 99:7821–7826
38. Bommarito MJ, Katz DM, Zelner JL et al (2010) Distance measures for dynamic citation networks. Physics A 389:4201–4208
39. Hsiao CH, Tang K-Y, Liu JS (2015) Citation-based analysis of literature: a case study of technology acceptance research. Scientometrics 105:1091–1110
40. Lu LYY, Liu JS (2016) A novel approach to identify the major research themes and development trajectory: the case of patenting research. Technol Forecast Soc Chang 103:71–82

Part I

Upstream Processes

Chapter 3

Production of Full-Length Antibody by *Pichia pastoris*

Adam Nylen and Ming-Tang Chen

Abstract

The methylotrophic yeast *Pichia pastoris* has become an increasingly popular host for recombinant protein expression in recent times. MRL pioneered a glycoengineered humanized *P. pastoris* expression system that could produce glycoproteins with glycosylation profiles similar to mammalian systems. Therapeutic glycoproteins produced by the humanized *P. pastoris* platform have shown comparable folding, stability, and in vitro and in vivo efficacies in preclinical models to their counterparts produced from the CHO cells. *P. pastoris* offers a cost and time efficient alternative platform for therapeutic protein production. This chapter describes a protocol for using *P. pastoris* to produce full-length monoclonal antibodies. It covers a broad spectrum of antibody expression technologies in *P. pastoris*, including expression vector construction, yeast transformation, high-throughput strain selection, fermentation, and antibody purification.

Key words *Pichia pastoris*, Monoclonal antibody, High-throughput screening, Fermentation, Purification

1 Introduction

Monoclonal antibodies (mAbs) have been used as treatments for a wide range of diseases, including cancers, autoimmune and infectious diseases, and organ transplantation. At present, mammalian cells like NS0 cells and in particular CHO cells are used as the expression platform for therapeutic mAb production. However, alternative mAb platforms including plant, insect cells, and microbial expression systems are being explored as cost and time efficient alternative expression systems [1–3]. Yeast *P. pastoris* is regarded as a compelling alternative to mammalian cell culture for the production of recombinant antibodies and antibody fragments. *P. pastoris* can grow in methanol to very high cell densities in bioreactors, exceeding 450 g/L wet cell weight. *P. pastoris* has the capability to perform posttranslational modifications such as glycosylation, disulfide isomerization, proteolytic processing, and secretes correctly folded protein into culture media. Other benefits of the *P. pastoris* system include ease of genetic manipulation, stable expression, rapid cell growth, low-cost scalable fermentation

Virgínia Picanço-Castro and Kamilla Swiech (eds.), *Recombinant Glycoprotein Production: Methods and Protocols*, Methods in Molecular Biology, vol. 1674, DOI 10.1007/978-1-4939-7312-5_3, © Springer Science+Business Media LLC 2018

processes, and little to no risk of human pathogenic virus contamination. Kalbitor® (ecallantide), a 60-amino acid polypeptide inhibitor of the protein kallikrein, was the first FDA approved *P. pastoris* derived product. There are currently several *P. pastoris*-produced full-length IgG molecules in clinical trials. ALD403, a humanized mAb that targets calcitonin gene-related peptide (CGRP) and contains a mutation at Fc N297 residue that abolishes glycosylation as a benign blocker [4], has successfully completed Phase 2 and is currently in Phase 3.

When mAbs are recombinantly expressed in yeasts, abnormal O-mannosylation often occurs. The key to achieve high yield and high quality mAb expression in yeasts is to suppress yeast-specific protein O-mannosylation [5]. Protein O-mannosylation is initiated at the endoplasmic reticulum by protein-O-mannosyltransferases (PMT's) that catalyze the addition of mannose residues to serine or threonine residues of target proteins. While there appear to be five *pmt.* genes encoding PMT homologues in *P. pastoris*, O-mannosylation of secreted heterologous proteins in *P. pastoris* is primarily dependent on the gene encoding Pmt2p [6, 7]. The methods described here use either a *pmt2* knockout strain or chemical PMT enzyme inhibitors (PMTi) to reduce protein O-mannosylation. This chapter covers a broad spectrum of techniques for producing full-length mAbs in *P. pastoris.* It includes expression cassette construction, DNA transformation, and clonal selection using high-throughput screening to identify high expression cell lines. In addition, protocols of mAb production including strain fermentation and antibody purification from the supernatant are also part of this chapter.

2 Materials

2.1 Construction of Expression Cassette, P. pastoris Transformation, and High Producing Antibody Strain Selection

1. Restriction enzymes EcoRI, FseI, PstI, and SalI.
2. One Shot® TOP10/P3 chemically competent *Escherichia coli* (Invitrogen).
3. S.O.C. medium: 2% tryptone, 0.5% yeast extract, 10 mM NaCl, 2.5 mM KCl, 10 mM $MgCl_2$, 10 mM $MgSO_4$, and 20 mM glucose.
4. LB-ampicillin plates: 1% Bacto tryptone, 0.5% yeast extract, 1% NaCl, and 1.5% agar; adjust pH to 7.5 using NaOH; add ampicillin to a final concentration of 50–100 μg/mL before pouring plates.
5. Corning® 96 Well Clear Round Bottom 1 mL Polypropylene Deep Well Plate (Corning, Inc.).
6. Yeast extract, Soytone, dextrose (YSD) plates containing 1% yeast extract, 2% Soytone, 2% glucose (dextrose), and 1.5% agar.

Table 1
Sequences of primers used in this study

Primer name	Primer sequence
AOX1p-FW	GGTTCCAATTGACAAGCTTTTGATTTTAACG
IgG-Fc-RV	AAACAGTAACTGGCTCTGGGAAGTAG
Kappa-Lc-RV	CTCTCTTGGGTAGAAGTTGTTCAACAA

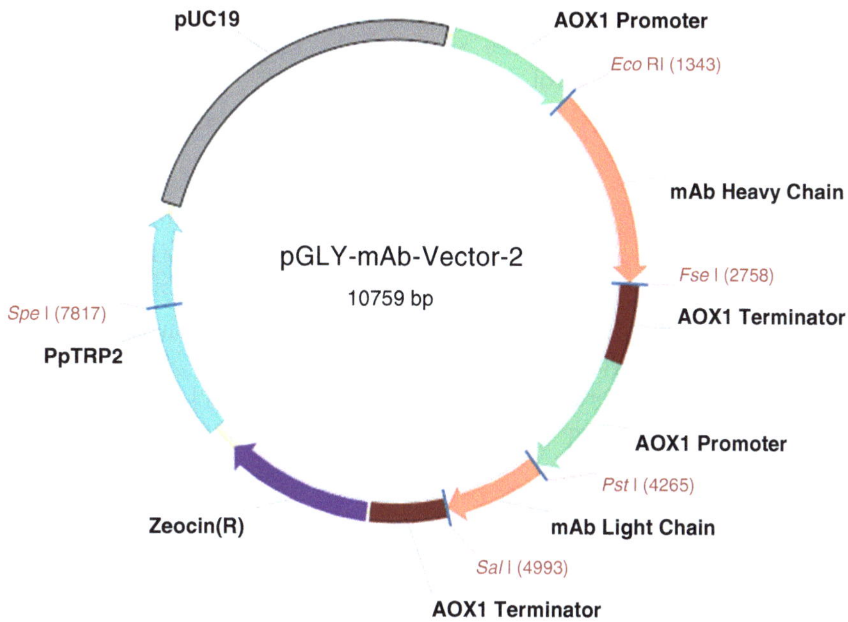

Fig. 1 *P. pastoris* depicts antibody expression vector pGLY-mAb-vector-2. The mAb expression vector pGLY-mAb-vector-2 is a TRP2::ZeoR roll-in plasmid that is used to introduce a sequence into *P. pastoris* TRP2 locus while duplicating the TRP2 target site by linearizing the plasmid within the TRP2 gene prior to transformation and using the ZeoR, zeocin resistance cassette, as a dominant selectable marker. This plasmid also contains dual AOX1-promoter driven cassettes of both the heavy chain and light chain genes of an anti-HER2 antibody. The plasmid also contains pUC19 sequence for maintenance in *E. coli*

7. 1 M D-sorbitol sterile and ice cold.
8. Zeocin.
9. Oligonucleotides were synthesized by Integrated DNA Technologies, Inc. (Coralville, Iowa).
10. Primers and their sequences are listed in Table 1.
11. Monoclonal antibody expression pGLY-mAb-vector-2 is constructed according to [8]. The plasmid map is illustrated in Fig. 1. The vector is derived from pUC19 which contains ampicillin resistant gene for *E. coli* transformation selection. Zeocin resistant cassette is included in the vector for *P. pastoris*

transformation selection. Restriction sites of EcoRI and FseI, or PstI and SalI, are engineered for heavy- and light-chain cloning. Heavy and light chains are codon optimized using *P. pastoris*-preferred codon usage. *Saccharomyces cerevisiae* alpha mating factor predomain and Kozak sequence are fused at the 5′end of heavy and light chain, respectively, and DNA is synthesized by Genewiz LLC.

12. *P. pastoris* strains NRRL-Y11430 (wild type) and YGLY35032 (*pmt2*) are used for construction of antibody producing yeast strains.
13. Transformation recovery media containing YSD (1% yeast extract, 2% Soytone, 2% dextrose (D-glucose)), and 1 M sorbitol.
14. Gene Pulser MXcell Electroporation System (Bio-Rad Laboratories).

2.2 Antibody Expressing Strain Selection

1. BSGY is composed of 100 mm potassium phosphate, 10 g/L yeast extract, 20 g/L Soytone, 40 g/L glycerol, 18.2 g/L sorbitol, 13.4 g/L YNB, and 4 mg/L biotin.
2. BMMY is composed of 100 mm potassium phosphate, 10 g/L yeast extract, 20 g/L Soytone, 20 mL/L methanol, 18.2 g/L sorbitol, 13.4 g/L YNB, and 4 mg/L biotin.
3. PMT inhibitor (PMTi) is a rhodanine-3-acetic acid derivative originally identified as an inhibitor of fungal protein mannosyltransferase [9].
4. Glycerol feed solution: 50% w/w glycerol, 5 mg/L biotin, and 12.5 mL/L PTM Salts (6.5 g/L $FeSO_4{\cdot}7H_2O$, 2.0 g/L $ZnCl_2$, 9 g/L H_2SO_4, 0.6 g/L $CuSO_4{\cdot}5H_2O$, 5 g/L H_2SO_4, 3 g/L $MnSO_4{\cdot}7H_2O$, 0.5 g/L $CoCl_2{\cdot}6H_2O$, 0.2 g/L $NaMoO_4{\cdot}2H_2O$, 0.2 g/L Biotin, 80 mg/L NaI, 20 mg/L H_3BO_4).
5. Methanol feed solution: 100% w/w methanol, 5 mg/L biotin, and 12.5 mL/L PTM Salts (6.5 g/L $FeSO_4{\cdot}7H_2O$, 2.0 g/L $ZnCl_2$, 9 g/L H_2SO_4, 0.6 g/L $CuSO_4{\cdot}5H_2O$, 5 g/L H_2SO_4, 3 g/L $MnSO_4{\cdot}7H_2O$, 0.5 g/L $CoCl_2{\cdot}6H_2O$, 0.2 g/L $NaMoO_4{\cdot}2H_2O$, 0.2 g/L Biotin, 80 mg/L NaI, 20 mg/L H_3BO_4).
6. ThermoFisher polyclonal goat anti-human kappa chain (Waltham, MA), Costar 96-well white polystyrene high binding assay plate (Corning, Corning, NY).
7. Bio-Stack microtiter plate washer was made by BioTek Instruments (Winooski, VT).
8. Rabbit anti-human IgG (Fc specific)-Alkaline Phosphatase antibody (Sigma Aldrich).

9. Wash buffer: TBS (50 mM Tris–HCl, 150 mM NaCl, pH 7.4), 0.2% Tween 20.
10. 1 L Scale DASGIP Parallel Bioreactor Systems (Eppendorf).
11. Beckman Allegra centrifuges (Beckman Coulter).
12. Sorvall Evolution RC Superspeed Centrifuge (Thermo Scientific).

2.3 Antibody Purification from Fermentation Supernatant

1. STREAMLINE rProtein A (80–165 mm particle size), SP Sepharose High Performance (SP HP) (34 mm particle size), and Tricorn 10/200 columns (GE Healthcare).
2. Polyethersulfone membrane filters (0.2 mm pore size) (Nalgene).
3. SARTOPORE 2 (0.8 + 0.45 mm) (Sartorius, Göttingen, Germany).
4. 4–20% Tris–HCl Ready Gels and prestained SDS–PAGE standards (broad range) (Bio-Rad Laboratories).
5. XK50/30 column (GE Healthcare).
6. HiTrap SP HP cation exchange column (GE Healthcare).
7. AKTA explorer 100 (GE Healthcare).

3 Method

3.1 Expression Plasmid Construction, Yeast Transformation, and Strain Selection

1. The first step is assembly of the antibody heavy chain and light chain amino acid sequences with *S. cerevisiae* alpha mating factor predomain signal sequences. The example shown in this chapter is an anti-Her2 mAb which has identical amino acid sequence to trastuzumab (*see* **Note 1**).
2. The heavy and light chains with signal sequences at their N-termini are reversely translated to DNA using *P. pastoris*-preferred codon usage by Genewiz LLC. Kozak sequence (GAAACG) is added before start codon “ATG” for both heavy and light chains. Two consecutive stop codons (TAA-TAG) are followed at the end of the coding regions. EcoRI and FseI sites are inserted to the 5′UTR and 3′ UTR regions of heavy chain, respectively, and PstI and SalI sites are added to the 5′UTR and 3′UTR regions of the light chain, respectively to facilitate cloning (*see* **Note 2**).
3. DNA of the heavy and light chains is synthesized by Genewiz LLC and delivered in pUC57 vector backbone.
4. Four pieces of DNA fragments, including heavy and light chains and two vector fragments from *P. pastoris* mAb expression pGLY-mAb-vector-2, are prepared using restriction enzymes digestion. Restriction enzymes digestion is carried

out using 1 μg pGLY-mAb-vector-2 vector DNA and 1 μg heavy chain or light chain plasmids synthesized by Genewiz. Four pieces of the fragments listed below are recovered from 1% agarose gel and purified by QIAquick Gel Extraction Kit (Qiagen) after enzymatic cutting: (a) pGLY-mAb-vector-2 cut with EcoRI and SalI, the larger piece, (b) pGLY-mAb-vector-2 cut with FseI and PstI, the smaller piece, (c) Genewiz synthesized heavy chain plasmid cut with EcoRI and FseI, and (d) Genewiz synthesized light chain plasmid cut with PstI and SalI.

5. These four pieces of DNA fragments are ligated together by Quick Ligation Kit from New England Biolabs following the provided protocol.
6. One microliter of ligation reaction is transformed into Top10 chemical competent cells by heat shock.
7. *E. coli* transformation selection is performed on LB agar plates which contain 50–100 μg/mL ampicillin.
8. *E. coli* transformants are screened by colony PCR using two primer pairs: (a) AOX1p-FW and IgG-Fc-RV for the heavy chain and (b) AOX1p-FW and Kappa-Lc-RV for the light chain (*see* **Note 3**).
9. *E. coli* plasmids isolated from colony PCR-positive transformants are digested with restriction enzymes (a) EcoRI and FseI to confirm the heavy chain insertion and (b) PstI and SalI to confirm correction light chain insertion.
10. Plasmid which harbored both heavy and light chain inserts and proper vector size are selected for *P. pastoris* transformation.
11. *P. pastoris* transformation is conducted as follows: 10 μg of pGLY-mAb-vector-2 derived mAb expressing plasmid is digested with SpeI for 3 h. The linearized DNA plasmid DNA is then precipitated in ice-cold isopropanol. DNA is pelleted by centrifugation and subsequently washed twice with 70% ethanol. The DNA pellet is dried under a vacuum and resuspended in 10 μL of sterile 1 M sorbitol.
12. Preparation of *P. pastoris*-competent cells: NRRL-Y11430 (wild-type *P. pastoris* strain) and YGLY35032 (*pmt2*) are inoculated in 50 mL BMGY media the day before transformation (*see* **Note 4**).
13. Overnight cultures of cells (OD_{600} in between 5 and 10) are centrifuged using 50 mL conical tubes to collect the cells. Cell pellet is washed twice with ice-cold sterile 1 M Sorbitol. Cells are suspended in 5–10 mL of 1 M ice-cold sorbitol and stored on ice for transformation.
14. Electroporation of mAb expression plasmids into *P. pastoris* cells is conducted as follows: linearized mAb expression plasmid (5–10 μL) is mixed with 100 μL of cold competent *P.*

pastoris cells (prepared in **step 13**) and incubated on ice for 5 min. The plasmid and *P. pastoris* cell resuspension is then transferred to electroporation cuvette (*see* **Note 5**). The mAb expressing plasmid is electrotransformed into *P. pastoris* NRRL-Y11430 and YGLY35032 using Bio-Rad Gene Pulser built-in *P. pastoris* program (25 μF, 200 ohm, 2000 V). Eight hundred microliters of recovery media YSD is immediately added into the cuvette.

15. Two-hundred microliters of the transformation mixture is plated out on YSD containing Zeocin plates to select *P. pastoris* transformants (*see* **Note 6**).

3.2 mAb Expression in 96-Deep-Well Plates and Isolation of High Producing Strains

1. Individual colonies transformed with the expression plasmid from selective agar plates are transferred into individual wells of a 96-well deep-well titer block containing 600 mL per well BSGY.
2. This plate, called the "seed plate," is then covered with a microporous rayon film and incubated for 48 h under standard growth conditions: 24 °C, 90% humidity, in a shaking incubator at a speed of 3000 × *g* with a 3-mm throw.
3. Multiple "expansion plates" are made by collecting 50 mL from the seed culture and inoculating into 600 mL of fresh BSGY (*see* **Note** 7).
4. Following 48 h under standard growth conditions, expansion plates are consolidated into one or two plates and then subjected to recombinant protein induction with 600 mL fresh BMMY medium with 3 mM PMTi per well.
5. Induction proceeds under standard growth conditions for 48 h. The supernatant is collected after centrifugation at 2000 × *g* for 5 min.
6. Screening clones for mAb titer using ELISA is then performed. Each well of a Costar 96-well white polystyrene high-binding assay plate is coated with 0.1 mg ThermoFisher polyclonal goat anti-human kappa chain for 1 h at room temperature.
7. The plate is then washed on a microtiter plate washer three times with 200 mL wash buffer.
8. Diluted sample (100 mL) is added followed by incubation for 1 h at room temperature. The plate is then washed as described above.
9. One hundred microliters per well of Rabbit anti-human IgG (Fc specific)-Alkaline Phosphatase antibody (diluted 1:10,000) is then added. The plates are incubated for additional 1 h at room temperature and washed.

10. After the preceding wash step, 100 mL 4-Methylumbelliferyl phosphate (4-MUP) is added to each well and allowed to incubate for 30 min in the dark.
11. Fluorescence is read using a microplate reader at 350/465 nm.
12. Expression titers are calculated by referencing experimental values against an immunoglobulin G standard.

3.3 Fermentation of Selected mAb-Producing Strains

1. Fermentations of *P. pastoris* antibody producing strains are carried out in 1 L glass bioreactors (Dasgips) under the described conditions.
2. An inoculum is prepared when cells from a patch plate are added to 100 mL of 4% BSGY medium in 500 mL baffled flask. The culture is then incubated for 48 ± 1 h at 24 °C, while shaking on an orbital shaker at 180 rpm with a 2.5 cm throw. The cell density of the inoculums is determined by optical density OD_{600} (after a 1:100 dilution in PBS) in a standard spectrophotometer.
3. Dasgip Control 4.x Scripting Software is used to control glycerol and methanol feeds, pH, temperature, and dissolved oxygen. The pH is controlled at 6.5 ± 0.1 using 15% NH_4OH and the pH is allowed to reach the set point before inoculation. Dissolved oxygen concentration is maintained by a steady airflow velocity of 0.7 vvm and cascading agitation rate from 600 to 1200 rpm.
4. The fermentation is carried out with the parameters presented in Table 2.
5. The fermentation is executed with the initiation of a batch phase with the inoculation of fermenter with the contents of the shake flask (v:v = 50:500). Batch phase continues until the initial glycerol charge (40 g/L) is consumed (typically 18–24 h).

Table 2
Fermentation parameters

Parameter	Unit	Value
Starting volume	[L]	0.5
Temp.	[°C]	24 ± 0.1
DO	[mg/L]	1.4 ± 0.1
pH	[]	6.5 ± 0.1
50% Glycerol feed	[g]	30
MeOH feed rate	[g/L/h]	2.0
Airflow rate	[vvm]	0.7

6. A fed batch phase is then initiated with the addition of 3 mM PMTi. A 50% w/w glycerol feed solution containing 12.5 mL/L PTM Salts and 12.5 mL/L of Biotin (40 mg/L) is fed to the fermenter continuously for 8 h.
7. At the end of the fed batch phase, the glycerol feed is terminated and a starvation period initiated. The starvation period is maintained for approximately 30 min. The decrease in metabolic activity is then detected by a sudden increase in dissolved oxygen concentration. At the end of the starvation period a dose of 3 mM PMTi is added.
8. An induction phase is initiated with the addition of a 100% methanol containing 12.5 mL/L PTM Salts and 12.5 mL/L of biotin (40 mg/L) fed continuously at 2 g/L/h. The induction phase continues for 72 ± 6 h.
9. At the end of induction phase, the entire volume is removed from the reactor and transferred into 1000 mL centrifuge bottles and centrifuged at 4 °C for 30 min at 15,000 × *g*.
10. The cell mass is discarded and the supernatant is retained for purification and analysis.

3.4 Purification of Recombinant mAb from Glycoengineered *P. pastoris* Fermentation Supernatant

The following protocol was modified according to [10] including the first step of protein A capture and the second step of polishing by HP SP chromatography.

1. Supernatant is filtered with SARTOBRAN P Sterile MidiCap, 0.65 + 0.45 mm, to further clarify the supernatant.
2. Protein A capture is performed with AKTA pilot at room temperature at a flow rate of 100 mL/min.
3. A column equilibration is performed with 3 column volumes of 20 mM Tris pH 7.0 and washed with 2 CV 20 mM Tris pH 7.0 and 3 column volumes of 20 mM Tris pH 7.0.
4. Target protein is eluted with 6 column volume of linear gradient from 0 to 100% 50 mM sodium citrate buffer, pH 3.0.
5. mAb fractions are pooled, filter-sterilize using 0.2 mm membrane, and keep at 4 °C.
6. Pooled mAb fractions are run on 4–20% SDS-PAGE gels using nonreducing (containing 100 mM NEM) and reducing sample buffers.
7. mAb concentrations are measured by a Bradford assay using commercial antibody as standard.
8. A second purification step is performed to polish by SP HP exchange chromatography.
9. 60 mL of SP HP is packed into XK16/30 column.

10. A 5× dilution of the protein A capture pool is created with H_2O (conductivity: <~10 mS/cm) and adjust to pH 5.0.
11. SP HP chromatography is performed on AKTA explorer 100 at room temperature at a flow rate of 5 mL/min.
12. A column equilibration is performed with 5 column volumes of 25 mM sodium acetate, pH 5.0 and loaded with the 5× diluted protein A capture pool. The column is then washed with 3 column volumes of 25 mM sodium acetate, pH 5.0.
13. Protein elution is performed with a linear gradient of 0–500 mM NaCl in 25 mM sodium acetate buffer, pH 5.0.
14. Fractions are run on 4–20% SDS-PAGE gels using nonreducing (containing 100 mM NEM) and reducing sample buffers.
15. Figure 2 shows a Coomassie-stained SDS-PAGE gel of protein A purified antibody expressed by clones transformed with an anti-HER2 mAb containing plasmid and cultivated in 1 Liter DasGip fermentor, from parental strains YGLY35032 and NRRL-Y11430.

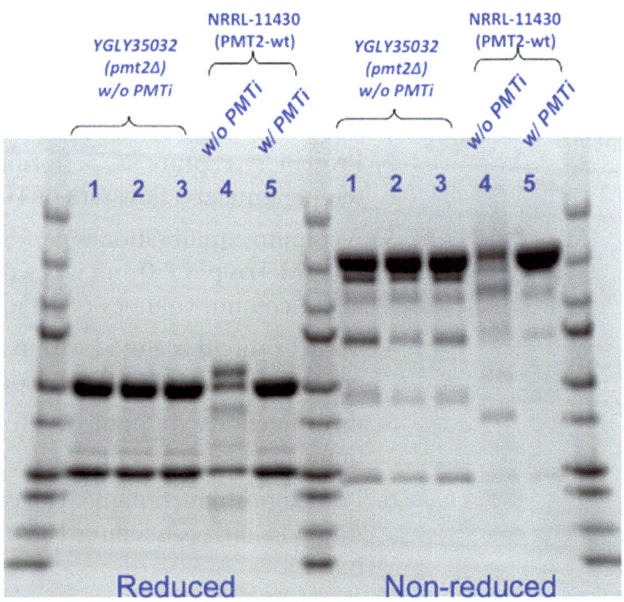

Fig. 2 Coomassie-stained SDS-PAGE gel of protein A purified antibody expressed by clones that are transformed with an anti-HER2 mAb containing plasmid and cultivated in 1 L DasGip fermentor, from parental strains YGLY35032 (*pmt2*) (*Lanes 1–3*) and NRRL-Y11430 (*Lanes 4 and 5*). All *pmt2* strains were cultivated in the absence of PMTi inhibitor. *Lane 4* shows NRRL-Y11430 strains cultivated without PMTi inhibitor. *Lane 5* shows NRRL-Y11430 strains cultivated with PMTi inhibitor

4 Notes

1. The anti-Her2 amino acid sequence used in this chapter is identical to trastuzumab (Herceptin®) sequence listed on the DrugBank database: https://www.drugbank.ca/drugs/DB00072.

   The amino acid sequences of the anti-Her2 heavy chain with *S. cevevisiae* alpha mating factor predomain (underlined) as signal sequence is listed below:

   MRFPSIFTAVLFAASSALAEVQLVESGGGLVQPGGSLRL
   SCAASGFNIKDTYIHWVRQAPGKGLEWVARIYPTN-
   GYTRYADSVKGRFTISADTSKNTAYLQMNSLRAED-
   TAVYYCSRWGGDGFYAMDYWGQGTLVTVS-
   SASTKGPSVFPLAPSSKSTSGGTAALGCLVKDYF-
   PEPVTVSWNSGALTSGVHTFPAVLQSS-
   GLYSLSSVVTVPSSSLGTQTYICNVNHKPSN
   TKVDKKVEPKSCDKTHTCPPCPA-
   PELLGGPSVFLFPPKPKDTLMISRTPEVTCVVVDV-
   SHEDPEVKFNWYVDGVEVHNAKTKPREEQYN-
   STYRVVSVLTVLHQDWLNGKEYKCKVSNKALPA-
   PIEKTISKAKGQPREPQVYTLPPSREEMTKN
   QVSLTCLVKGFYPSDIAVEWESNGQPEN-
   NYKTTPPVLDSDGSF-
   FLYSKLTVDKSRWQQGNVFSCSVMHEALH
   NHYTQKSLSLSPGK

   The amino acid sequences of the anti-Her2 light chain with *S. cevevisiae* alpha mating factor predomain as signal sequence (underlined) is listed below:

   MRFPSIFTAVLFAASSALADIQMTQSPSSLSASVGDRV-
   TITCRASQDVNTAVAWYQQKPGKAPKLLIYSAS-
   FLYSGVPSRFSGSRSGTDFTLTISSLQPEDFA-
   TYYCQQHYTTPPTFGQGTKVEIKRTVAAPSVFIFPPS-
   DEQLKSGTASVVCLLNNFYPREAKVQWKVD-
   NALQSGNSQESVTEQDSKDSTYSLSSTLT
   LSKADYEKHKVYACEVTHQGLSSPVTKSFNRGEC

2. During heavy chain and light chain codon optimization processes, potential EcoRI, FseI, PstI, and SalI restriction sites are avoided. The SpeI site, which was used to linearizing the plasmid, is also avoided in the coding region.
3. The same amino acid sequences could yield various codon optimized DNA sequences depended on the codon optimization algorithms used. The primers IgG-Fc-RV and kappa-Lc-RV used in this chapter were based on the DNA coding sequences of IgG1 and kappa light chain constant regions the

author used. Individual lab should redesign its own primers based on the actual DNA sequence used for heavy- and light-chain coding regions.

4. The growth rate of YGLY35032 (*pmt2*) strain is significantly slower than NRRL-Y11430 wild-type *P. pastoris* strain. To achieve the best transformation efficiency, pilot study with various inoculation cell densities is needed. OD_{600} value higher than 10 normally has lower transformation efficiency.
5. The cuvette should be kept on ice in order to achieve maximum transformation efficiency.
6. The sensitivity to Zeocin varies in different P. pastoris strains, YGLY35032 (*pmt2*) strain is more sensitive to Zeocin comparing with NRRL-Y11430 wild-type strain. Zeocin resistance also depends on the DNA copy numbers integrated into yeast chromosome.
7. Two to eight times of expansion plates are created for every seed plate depending upon the amount of material required for subsequent analysis and expression level of the antibody of interest. This two-step batch growth phase method allows the biomass to grow to saturation twice and therefore serves to normalize OD_{600} in microtiter plates.

References

1. Grohs BM, Niu Y, Veldhuis LJ, Trabelsi S, Garabagi F, Hassell JA, McLean MD, Hall JC (2011) J Agric Food Chem 58:10056–10063
2. Komarova TV, Kosorukov VS, Frolova OY, Petrunia IV, Skrypnik KA, Gleba YY, Dorokhov YL (2010) PLoS One 6:e17541
3. Li H, Sethuraman N, Stadheim TA, Zha D, Prinz B, Ballew N, Bobrowicz P, Choi BK, Cook WJ, Cukan M, Houston-Cummings NR, Davidson R, Gong B, Hamilton SR, Hoopes JP, Jiang Y, Kim N, Mansfield R, Nett JH, Rios S, Strawbridge R, Wildt S, Gerngross TU (2006) Nat Biotechnol 24:210–215
4. Dodick DW, Goadsby PJ, Silberstein SD, Lipton RB, Olesen J, Ashina M, Wilks K, Kudrow D, Kroll R, Kohrman B, Bargar R, Hirman J, Smith J (2014) Lancet Neurol 13:1100–1107
5. Kuroda K, Kobayashi K, Kitagawa Y, Nakagawa T, Tsumura H, Komeda T, Shinmi D, Mori E, Motoki K, Fuju K, Sakai T, Nonaka K, Suzuki T, Ichikawa K, Chiba Y, Jigami Y (2008) Appl Environ Microbiol 74:446–453
6. Nett JH, Cook WJ, Chen MT, Davidson RC, Bobrowicz P, Kett W, Brevnova E, Potgieter TI, Mellon MT, Prinz B, Choi BK, Zha D, Burnina I, Bukowski JT, Du M, Wildt S, Hamilton SR (2013) PLoS One 8:e68325
7. Chen M, Choi B, Davidson R (2013) In: United States Patent Application US 2015/0337274 A1
8. Zhang N, Liu L, Dumitru CD, Cummings NR, Cukan M, Jiang Y, Li Y, Li F, Mitchell T, Mallem MR, Ou Y, Patel RN, Vo K, Wang H, Burnina I, Choi BK, Huber HE, Stadheim TA, Zha D (2011) Glycoengineered Pichia produced anti-HER2 is comparable to trastuzumab in preclinical study. MAbs 3:289–298
9. Orchard M, Neuss J, Galley C, Carr A, Porter D, Smith P, Scopes D, Haydon D, Vousden K, Stubberfield C, Young K, Page M (2004) Bioorg Med Chem Lett 14:3975–3978
10. Jiang Y, Li F, Button M, Cukan M, Moore R, Sharkey N, Li H (2011) Protein Expr Purif 74:9–15

Chapter 4

Human Cells as Platform to Produce Gamma-Carboxylated Proteins

Aline de Sousa Bomfim, Marcela Cristina Corrêa de Freitas, Dimas Tadeu Covas, and Elisa Maria de Sousa Russo

Abstract

The gamma-carboxylated proteins belong to a family of proteins that depend on vitamin K for normal biosynthesis. The major representative gamma-carboxylated proteins are the coagulation system proteins, for example, factor VII, factor IX, factor X, prothrombin, and proteins C, S, and Z. These molecules have harbored posttranslational modifications, such as glycosylation and gamma-carboxylation, and for this reason they need to be produced in mammalian cell lines. Human cells lines have emerged as the most promising alternative to the production of gamma-carboxylated proteins. In this chapter, the methods to generate human cells as a platform to produce gamma-carboxylated proteins, for example the coagulation factors VII and IX, are presented. From the cell line modification up to the vitamin K adaptation of the produced cells is described in the protocols presented in this chapter.

Key words Gamma-carboxylated proteins, Vitamin K cycle, Human cell lines, Coagulation factors, Recombinant proteins

1 Introduction

Recombinant biopharmaceuticals have to meet high quality standards to ensure the best possible effects and safety of the patient, controlled by regulatory authorities. Thus, the functionality and quality of the product are a major concern to be tackled during recombinant production, especially with increasing complexity of the protein. The proper function of a protein crucially depends on correct posttranslational modifications including glycosylation, gamma-carboxylation, phosphorylation, and formation of disulfide bonds [1].

The gamma-carboxylated proteins belong to a family of proteins that depend on vitamin K for normal biosynthesis. These proteins have a 45-amino acid residue long domain in the N-terminal end called Gla (gamma-carboxyglutamic acid) domain, which contains from 9 to 12 Gla residues [2–4]. The Gla domain

Virgínia Picanço-Castro and Kamilla Swiech (eds.), *Recombinant Glycoprotein Production: Methods and Protocols*, Methods in Molecular Biology, vol. 1674, DOI 10.1007/978-1-4939-7312-5_4, © Springer Science+Business Media LLC 2018

is found in proteins of coagulation system, for example, factor VII (FVII), factor IX (FIX), factor X (FX), prothrombin, and proteins C, S [5, 6], and Z [7]. Other Gla domain-containing proteins include a Gas6, a homolog of protein S that appears to be a regulator of apoptosis [8]. There are also four Gla domain-containing transmembrane proteins termed proline-rich Gla proteins (PRGP) 1 and 2, and transmembrane Gla proteins (TMG) 3 and 4. Most of the soluble Gla domain-containing proteins are almost exclusively synthesized in the liver. In addition to the Gla domain-containing proteins, there are two Gla proteins called matrix Gla protein (five Gla residues) and bone Gla protein (osteocalcin; with three Gla residues) that regulates tissue mineralization [9, 10]. The clotting factors which are the main class of gamma-carboxylated proteins will be discussed in the following sections.

1.1 Domain Organization of Coagulation Proteins

Prothrombin and clotting FVII, FIX, and FX as well as proteins C and S, require vitamin K for proper biosynthesis. These proteins contain several distinct domains starting with a 45-amino acid residue long N-terminal domain that includes 9–12 Gla residues [3]. In FVII, FIX, FX, and protein C, the Gla domain is followed by two epidermal growth factor (EGF)-like domains, and a C-terminal serine protease domain. Prothrombin differs from these proteins once contains two kringle domains (KR) in place of the two EGF-like domains. Moreover, prothrombin contains a tetradecapeptide loop structure between the Gla domain and the first KR. In protein S, the Gla domain is followed by a thrombin-sensitive region, four EGF-like domains, and a sex hormone-binding globulin-like domain (SHBG)-like domain [11]. Figure 1 shows a schematic representation of the coagulation factors FVII and FIX which are addressed in this chapter.

1.2 Vitamin K-Dependent Gamma-Carboxylation System

The vitamin K-dependent gamma-carboxylation system involves several proteins that are embedded in the membrane of the endoplasmic reticulum [12]. The system modifies vitamin K-dependent proteins by converting glutamic acid (Glu) residues in the proteins to calcium-binding gamma-carboxyglutamic acid (Gla) residues. Calcium binding by the Gla residues is crucial to the proteins to perform their biological functions [13]. FVII, FIX, and FX; prothrombin; protein S; protein C; and protein Z, that are produced by the liver, are dependent on Gla modification. The vitamin K-dependent gamma-carboxylation system (Fig. 2) consists of (1) the vitamin K-dependent gamma-carboxylase, which requires the reduced hydroquinone form of vitamin K_1 (vitamin K_1H_2) as a cofactor and (2) the warfarin-sensitive enzyme vitamin K_1 2,3-epoxide reductase (VKOR), which produces the cofactor. Concomitant with gamma-carboxylation, the hydroquinone is converted to the metabolite vitamin K_1 2,3 epoxide, which is reduced back to the

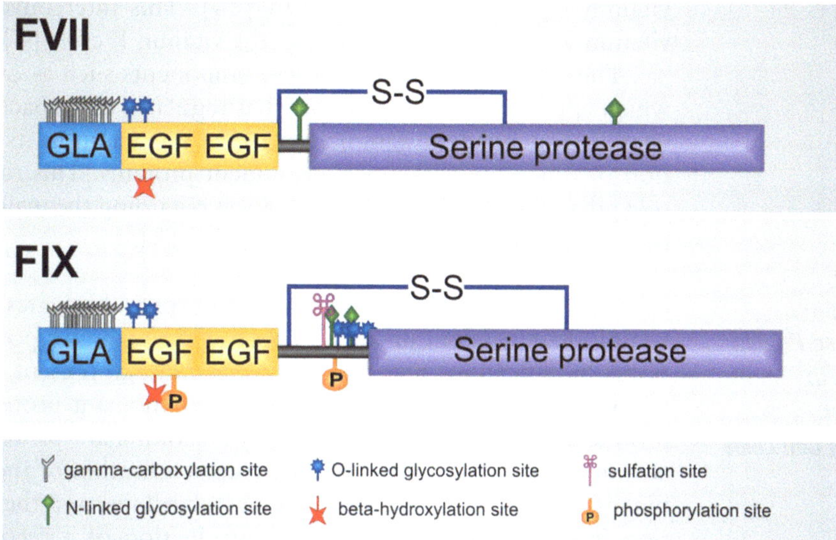

Fig. 1 Schematic representation of FVII and FIX coagulation proteins. Locations of posttranslational modifications (identified in the key) are denoted. Homologous domains are similarly colored. The disulfide bonds connecting the heavy and light chains are shown. Gla, gamma-carboxyglutamic acid domain; EGF, epidermal growth factor-like domain; serine protease

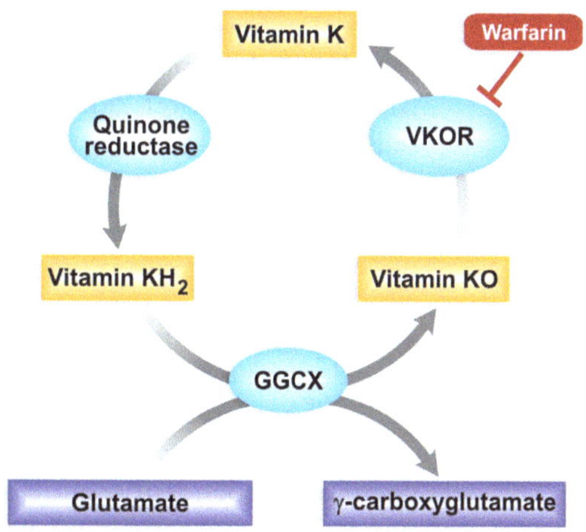

Fig. 2 The vitamin K-dependent gamma-carboxylation system. The gamma-carboxylase converts vitamin K-dependent proteins to gamma-carboxyglutamic acid (Gla) containing proteins by adding CO_2 to glutamic acid (Glu) residues in newly synthesized proteins. The gamma-carboxylase requires reduced vitamin K_1 (vitamin KH_2) as a cofactor for this posttranslational modification reaction. Concomitant with gamma-carboxylation, vitamin KH_2 is converted to vitamin K_1 2,3-epoxide (vitamin KO). The epoxide is reduced by the warfarin-sensitive enzyme vitamin K_1 2,3-epoxide reductase (VKOR) to the vitamin KH_2 cofactor. This cyclic interconversion of vitamin K metabolites constitutes the vitamin K cycle [6]. At high tissue concentrations, vitamin K_1 quinone (vitamin K_1) can be reduced to vitamin KH_2 by the alternative pathway of the cycle

vitamin K_1H_2 cofactor by VKOR [14]. This interconversion of vitamin K metabolites is known as the vitamin K cycle [4].

The vitamin K cycle has other components such as calumenin which appears to be a chaperone that regulates the capacity of the vitamin K-dependent gamma-carboxylation system in its production of functional vitamin K-dependent proteins. This regulatory role puts calumenin in a central position regarding the maintenance of a normal blood-coagulation system [4, 15].

1.3 Gamma-Carboxylated Protein Expression in Mammalian and Human Cell Lines

A variety of cell lines has been used to express biologically active vitamin K-dependent proteins [16–19]. The extensive posttranslational modifications that these proteins undergo present a unique set of challenges not found for other recombinant proteins. The production of an accurately processed recombinant protein is not always routine, and more extensive characterization of the recombinant product is required. The choice of cell line and the production level impact significantly the production of a recombinant protein, identical to that of its natural counterpart [20].

Mammalian expression system is the major choice for the production of proteins with harbor posttranslational modifications due to its ability to synthesize proteins that are similar to those naturally occurring in human [21]. Mammalian cell lines have the basic machinery to express and secrete recombinant protein, and a huge number of cell lines from various tissues and species with suitable growth properties are available. Two hamster cell lines, BHK (Baby hamster kidney cells) and Chinese hamster cell ovary cells (CHO) and two mouse cell lines, SP2-0 (hybridoma) and NS0 (myeloma), supply most of the biopharmaceutical industry. Research and clinical studies have provided substantial safety information about the use of these cells [22]. Despite the safety and efficacy of recombinant protein expression in murine cells, the gamma-carboxylation pattern is of profound importance when considering host cell potential as expression systems for the biosynthesis of human recombinant gamma-carboxylated proteins [23].

In the last decades, the major obstacle to the production of vitamin K-dependent coagulation factors has been the low expression of active recombinant proteins for pharmaceutical purposes [20, 24–26]. Studies have demonstrated that the low specific activity was due to a relatively low Gla content of the gamma-carboxylated proteins from murine cell lines, showing that these cells have a limited gamma-carboxylation. The limitation has been reported for a variety of host cell lines such as CHO [24], BHK [27], COS [28], and mouse fibroblast LM cells [29]. An ideal cell line specifically for the production of vitamin K dependent proteins would possess a high capacity for gamma-carboxylation as well as meet several other criteria such as adaptability to serum free culture, high protein synthetic capacity, ability to maintain high viability, robust growth, and, importantly, a track record of safety [26].

In this context, human cell lines emerge as a trend toward the production of gamma-carboxylated proteins. Academic and industrial laboratories have demonstrated successful expression of vitamin K dependent protein in active form in several mammalian cell lines such as HEK293, HepG2, HuH-7, and SK-Hepl [20, 30–32].

The human cell line derived from human embryonic kidney cells, called Hek 293 transformed with sheared fragments of adenovirus type 5 DNA, was established almost 35 years ago [33], but only in the last few years, it has been widely used in the production of recombinant proteins and viral vectors. Several studies have reported that this cell line produced fully γ-carboxylated factor IX [20, 34] and protein S [32]. Owing to its human origin, this cell line is preferred for the advantage that the clotting factors produced would possess human posttranslational modifications and therefore potentially reduce immunogenicity.

In the human body, the clotting factors are proteins synthesized in the liver. In these circumstances, HepG2 and HuH-7 cell lines, derived from hepatocellular carcinoma, have the appropriate machinery to perform the production of gamma-carboxylated proteins and their posttranslational modifications. Studies have reported that both cell lines may represent an effective cellular system for the production of recombinant FIX exhibiting posttranslational modifications similar to plasma-derived FIX [30, 31]. SK-Hep-1 is another human cell line that emerges as a promising option for the production of gamma-carboxylated proteins. These cells are derived from hepatic adenocarcinoma and have shown great results in the production of biologically active factors IX and VII [19], showing that these cells are capable of performing the correct posttranslational modifications.

Human cells lines have emerged as the most promising alternative to the production of gamma-carboxylated proteins. In this chapter, the methods to generate human cells as a platform to produce gamma-carboxylated proteins, for example the coagulation factors VII and IX, are presented. From the cell line modification up to the vitamin K adaptation of the produced cells is described in the protocols presented below. The methods for the gamma-carboxylation analysis are not included.

2 Materials

1. Adherent HEK 293T cells for virus production.
2. Adherent SK-Hep-1, HuH-7 and HKB-11 cells for gamma-carboxylated protein production.
3. Vectors: pCMVR8.91 (viral capsid containing gag, pol, rev and tat proteins from HIV-1), pMD2 VSV-G (envelope cassette is

derived from the vesicular stomatitis virus (VSV-G) instead of HIV-1 envelope) and an expression vector derived from HIV-1 containing the target gene (*see* **Note 1**).

4. Dulbecco's Modified Eagle's Medium supplemented with 10% (v/v) fetal bovine serum (FBS - complete medium)—medium for adherent HEK 293T, SK-Hep-1 and HuH-7 cells.
5. Dulbecco's Modified Eagle Medium: Nutrient Mixture F-12 supplemented with 10% (v/v) fetal bovine serum (complete medium)—medium for adherent HKB-11 cells. The media should be sterilized by filtration (0.22 μm) and stored at 4 °C.
6. Phosphate-buffered saline (PBS).
7. 10× trypsin solution 0.5% trypsin, 0.2% EDTA.
8. Disposable sterile pipettes.
9. Polybrene [1 mg/mL]: cationic polymer used for cell transduction.
10. Sterile Eppendorf® tubes.
11. Sterile tissue culture dishes (100 mm).
12. 6-well tissue culture plates.
13. T75 and T175 tissue culture flasks.
14. Sterile polypropylene centrifuge tubes (15 and 50 mL).
15. Sterile disposable syringes (10–50 mL).
16. Disposable syringes filters with PVDF membrane (pore size 0.45 μm).
17. Trypan blue vital stain: Mix 0.9% (w/v) NaCl and 1% (w/v) trypan blue in 4:1 ratio. Use within 24 h.
18. Polyethylenimine (PEI) [1 mg/mL]: a stable cationic polymer used for cell transfection (*see* **Note 2**).
19. Hemocytometer.
20. Inverted microscope.
21. Centrifuge.
22. Tube agitator.
23. Humidified CO_2 incubator.
24. Trypsin–EDTA 0.25%.
25. Vitamin K 10 mg/mL intravenous.

3 Methods

3.1 General Maintenance of Cell Lines

The human adherent cell lines should be expanded as rapidly as possible and stocks (at least ten vials) frozen at a low passage number (*see* **Note 3**).

1. The cell vial should be thawed at 37°C in a water bath.
2. Transfer the thawed cells into sterile polypropylene tube (15 mL) with 5 mL of appropriate medium.
3. Centrifuge the cells at 300 × *g* for 10 min.
4. Discard the cell supernatant and resuspend the cell pellet with 5 mL of appropriate medium.
5. Transfer the cell to a T75 flask with 10 mL of appropriate medium.
6. Incubate under appropriate conditions. For most cell lines it will at 37°C in 5% CO_2 in a humidified incubator.
7. Replace the cell medium every 2–3 days.
8. When the cell culture reaches 90–100% confluency, it is necessary to subculture the cells.
9. Remove the cell medium from the flask and discard it.
10. Add 5–10 mL of PBS (depending of the flask size) and wash over the cells.
11. Remove the PBS from the flask and discard it.
12. Add 3–5 mL of trypsin–EDTA 0.25% (depending of the flask size) and incubate at 37°C in 5% CO_2 in a humidified incubator for 3–5 min.
13. Inactivate the trypsin–EDTA 0.25% by the addition of the same volume of medium supplemented with 10% of fetal bovine serum.
14. Harvest the cells into a sterile polypropylene tube (15 mL).
15. Centrifuge the cells at 300 × *g* for 10 min.
16. Discard the cell supernatant and resuspend the cell pellet with 5 mL of appropriate medium.
17. Dilute a sample of the cell suspension into trypan blue vital stain and count live (cells that exclude trypan blue) and dead (cells that stain with trypan blue) cells using a hemocytometer.
18. Calculate the cell viability and plate the desired amount of cells.

3.2 Transfection for Virus Production

HEK293T cell line is widely used as a lentiviral package system. This cell line carries the SV40 large T-antigen that induces unwanted DNA replication at the SV40 ori contained within the integrated lentiviral expression vector. In this method it is necessary to use a vector containing the transgene and two auxiliary vectors having the SV40 replication origin. Cotransfection increases transgene transcription and the production of viral proteins, and, thereafter, more viral particles will be secreted in the cell culture supernatant.

1. Twenty-four hours before the transfection, seed 1×10^7 cells in a T175 flask. It should result in a cell monolayer with 60–80% confluency.
2. In a sterile polypropylene tube (50 mL), add 6 mL of DMEM without FBS supplementation.
3. Add the DNAs: 19.5 μg of pCMVR8.91, 10.5 μg of pMD2 VSV-G, and 30 μg of transgene vector, a total of 60 μg of DNA.
4. Mix gently in the tube agitator.
5. Add PEI dropwise under agitation at a ratio of 1 DNA:3 PEI (*see* **Note 4**).
6. Allow the mixture to stand for 15 min in room temperature.
7. Add to the mixture in the T175 flask 19 mL of DMEM supplemented with FBS.
8. Incubate under appropriate conditions (37°C, 5% CO_2) for 16 h.
9. After 16 h, replace culture medium.
10. Incubate under appropriate conditions (37°C, 5% CO_2) for 48 h.

3.3 Virus Harvesting and Storage

1. After 48 h, harvest cell supernatant in a sterile polypropylene tube (50 mL).
2. Centrifuge the supernatant at $300 \times g$ for 10 min.
3. Use a sterile disposable syringe and a 0.45-μm filter unit to filter the centrifuged cell supernatant.
4. Distribute 1 mL aliquot fractions of supernatant into sterile Eppendorf® tubes.
5. Freeze the aliquots at −80°C (*see* **Note 5**).

3.4 Virus Titering by Flow Cytometry

1. Perform the experiment in duplicate.
2. Twenty-four hours before the virus titering, seed 2×10^5 HEK 293T cells in a 6-well plate.
3. Dilute the frozen virus in three different ratios, e.g., 1:1, 1:2, and 1:3, in cell culture medium in a final volume of 2 mL (*see* **Note 6**).
4. Add 2 μL of Polybrene to the mixture in each well.
5. Incubate under appropriate conditions for 16 h.
6. After 16 h replace culture medium.
7. Incubate under appropriate conditions for 48 h.
8. After 48 h, trypsinize and resuspend cells as described in Subheading 3.1.

9. Aliquot 200 μL of cell suspension into flow cytometry tube and perform the analysis of the percentage of EGFP (enhanced green fluorescent protein)-positive cells (*see* **Note 7**).
10. To determine the virus titer, the following equation can be used. The value obtained in the equation represents infectious virus per mL of cell-free supernatant.

$$T = (F \times C/V) \times D.$$

F = percentage of EGFP-positive cells (% EGFP/100).

C = number of cells seeded per well.
V = volume (mL) per well.
D = virus dilution.

3.5 Transduction of Cells

Transduction performed by lentiviral vectors enables rapid generation of stable mammalian cell lines producing recombinant proteins. This system is flexible and allows for the insertion of large genes in the host chromatin. Transduction of cells with lentivirus is simple, and cell lines are able to produce proteins for an undefined time due to stable integration of the proviral genome into the host genome. High levels of protein production result primarily from two points of lentiviral transduction. First, it is the ability to achieve high transgene copy number through multiple rounds of transduction. The second is the genomic context into which lentiviral vectors integrate, showing a distinct preference for integration in gene coding elements. Due to their characteristics, lentiviral vectors can infect different types of stationary cells or cells in division with high efficiency [35].

3.5.1 Transduction of Adherent Cells

1. Seed in a 6-well plate a culture of the cells to be transduced and grow it up until 80% confluency.
2. Remove the growth medium.
3. Add the viral supernatant using a multiplicity of infection (MOI) of 1 (*see* **Note 8**).
4. Add 1 μL of Polybrene per each mL of viral supernatant.
5. Incubate under appropriate conditions for 8–16 h.
6. Remove the culture medium and add normal growth medium.
7. Expand the culture and use a suitable assay or select for transduced cells, for example: G418 or flow cytometer when the vector expresses EGFP (*see* **Note 9**).

3.5.2 Spinoculation

Spinoculation, also known as centrifugal inoculation, has been widely used to augment the in vitro infectivity of a variety of viruses in adhesion and suspension cultures [36]. This approach increases the deposition of virions on the cell surface [37].

1. Seed in a 6-well plate a culture of the cells to be transduced and grow it up to 80% confluency.
2. Remove the growth medium.
3. Add viral supernatant using a multiplicity of infection (MOI) of 1.
4. Add 1 μL of Polybrene per each mL of viral supernatant.
5. Centrifuge the plate at 450 × *g* for 50 min at 20°C.
6. Incubate under appropriate conditions for 8–16 h.
7. Remove the culture medium and add normal growth medium.

3.5.3 Vitamin K Cell Adaptation

The gamma-carboxylated proteins are vitamin K dependants. Therefore, it is mandatory to supplement the culture medium with vitamin K.

1. Seed 1×10^6 modified cell line in a T75 flask.
2. Add 15 mL of the appropriate medium supplemented with 5 μg/mL of vitamin K (*see* **Notes 10** and **11**).
3. Maintain the vitamin K supplementation during the protein production process (*see* **Notes 12**).

3.6 Final Considerations

1. Studies have shown that cotransduction of genes related to vitamin K cycle as VKORC1 and gamma-carboxylase increased the protein production [4, 38]. Thus, these approaches can be considered an alternative to increase gamma-carboxylated protein production.
2. To evaluate the gamma-carboxylated profile of proteins secreted by human cell lines there are various analytical methods. One of the most used methods is the liquid chromatography–mass spectrometry (LC-MS). This method enables relative quantification of gamma-carboxylation and can be performed according to the purpose of the study.

4 Notes

1. The expression vector used in this protocol has a gene reporter that codes for an EGFP which allows for the transduced cell selection by flow cytometry.
2. Many alternative transfection agents have been used for vector production including cationic lipid-based reagents such as Lipofectamine™ (ThermoFisher Scientific) [39] and calcium-phosphate (CaP) coprecipitation. Both approaches have proven to be as effective as PEI for the generation of lentiviral vectors, however, lipid-based agents become very costly for the use in large-scale production [40]. PEI, on the other hand, is a

relatively inexpensive approach, and is suitable for large-scale lentiviral vectors production. PEI does not require tight regulation of transfection conditions (e.g., pH), and it is also substantially less toxic than CaP, meaning that media change following transfection is not necessary. Moreover, PEI has proven to be effective for the transfection of both adherent and suspension cultures, and can be used in the presence or absence of serum [41, 42].

3. It is advisable to grow cells in the absence of antibiotics because their use might mask persistent low-grade infections.
4. The PEI concentration cannot exceed 10% of total incubation medium used at the transfection.
5. The frozen viruses are stable for 6 months at −80°C.
6. In the virus titer the percentage of EGFP (enhanced green fluorescent protein)-positive cells should not exceed 20%, to avoid overestimation of virus titer.
7. Without fluorescent genes, it is necessary to estimate expression at the mRNA or DNA level by measuring the number of copies of stably integrated lentiviruses. Therefore, real-time PCR methods allow for rapid and precise quantification of lentiviral transgene mRNA expression in transduced target cells.
8. Target cells can be transduced with an MOI of 1–10 viruses per cell and many transduction cycles can be performed.
9. This protocol uses adherent human cell lines for the production of gamma-carboxylated proteins. As an alternative, these cell lines can be adapted for suspension cell culture that could be scaled up for the production of proteins requiring gamma-carboxylation in high yields [43].
10. Vitamin K can be used in a range concentration of 5–10 μg/mL depending on the cell type.
11. Each procedure using vitamin K has to be done in the absence of light.
12. Before the first analysis of the secreted protein, maintain the cells with the vitamin K supplementation for at least 10 days.

Acknowledgments

The authors acknowledge São Paulo Research Foundation—FAPESP (2015/19017-6), Conselho Nacional de Pesquisa—CNPq (142406/2016-3), Centro de Pesquisa, Inovação e Difusão (CEPID), and National Institute of Science and Technology in Stem Cell and Cell Therapy—INCTC for financial support and Sandra Navarro for drawing the figures.

References

1. Fliedl L, Grillari J, Grillari-Voglauer R (2015) Human cell lines for the production of recombinant proteins: on the horizon. New Biotechnol 32:673–679. doi:10.1016/j.nbt.2014.11.005
2. Stenflo J, Fernlund P, Egan W, Roepstorff P (1974) Vitamin K dependent modifications of glutamic acid residues in prothrombin. Proc Natl Acad Sci U S A 71:2730–2733. doi:10.1073/pnas.71.7.2730
3. Stenflo J (1999) Contributions of Gla and EGF-like domains to the function of vitamin K-dependent coagulation factors. Crit Rev Eukaryot Gene Expr 9:59–88
4. Wallin R, Hutson SM (2004) Warfarin and the vitamin K-dependent γ-carboxylation system. Trends Mol Med 10:299–302. doi:10.1016/j.molmed.2004.05.003
5. Dahlbäck B, Villoutreix BO (2003) Molecular recognition in the protein C anticoagulant pathway. J Thromb Haemost 1:1525–1534
6. Aktimur A, Gabriel MA, Gailani D, Toomey JR (2003) The factor IX gamma-carboxyglutamic acid (Gla) domain is involved in interactions between factor IX and factor XIa. J Biol Chem 278:7981–7987. doi:10.1074/jbc.M212748200
7. Broze GJ (2001) Protein Z-dependent regulation of coagulation. Thromb Haemost 86:8–13
8. Melaragno MG, Fridell YW, Berk BC (1999) The Gas6/Axl system: a novel regulator of vascular cell function. Trends Cardiovasc Med 9:250–253
9. Hoang QQ, Sicheri F, Howard AJ, Yang DSC (2003) Bone recognition mechanism of porcine osteocalcin from crystal structure. Nature 425:977–980. doi:10.1038/nature02079
10. Luo G, Ducy P, McKee MD et al (1997) Spontaneous calcification of arteries and cartilage in mice lacking matrix GLA protein. Nature 386:78–81. doi:10.1038/386078a0
11. HANSSON K, STENFLO J (2005) Post-translational modifications in proteins involved in blood coagulation. J Thromb Haemost 3:2633–2648. doi:10.1111/j.1538-7836.2005.01478.x
12. Wallin R (2001) A molecular mechanism for genetic warfarin resistance in the rat. FASEB J 15:2542–2544. doi:10.1096/fj.01-0337fje
13. Furie B, Bouchard BA, Furie BC (1999) Vitamin K-dependent biosynthesis of γ-carboxyglutamic acid. Blood 93:1798–1808
14. Cain D, Hutson SM, Wallin R (1998) Warfarin resistance is associated with a protein component of vitamin K 2, 3-epoxide reductase enzyme complex in rat liver. Thromb Haemost 80:128–133
15. Vatandoost J, Pakdaman SF (2016) The effects of influencing factors on γ-carboxylation and expression of recombinant vitamin K dependent coagulation factors. J Biomed. doi:10.17795/jmb-6077
16. Xiao W, Li CQ, Xiao XP, Lin FZ (2013) Expression and fast preparation of biologically active recombinant human coagulation factor VII in CHO-K1 cells. Genet Mol Res 12:6813–6824. doi:10.4238/2013.December.16.7
17. do Amaral RLF, de Sousa Bomfim A, de Abreu-Neto MS et al (2016) Approaches for recombinant human factor IX production in serum-free suspension cultures. Biotechnol Lett 38:385–394. doi:10.1007/s10529-015-1991-1
18. Dadehbeigi N, Ostad SN, Faramarzi MA, Ghahremani MH (2008) Sex hormones affect the production of recombinant factor IX in CHO and HEK-293 cell lines. Biotechnol Lett 30:1909–1912. doi:10.1007/s10529-008-9774-6
19. de Sousa Bomfim A, Cristina Corrêa de Freitas M, Picanço-Castro V et al (2016) Human cell lines: a promising alternative for recombinant FIX production. Protein Expr Purif 121:149–156. doi:10.1016/j.pep.2015.11.023
20. Berkner KL (1993) Expression of recombinant vitamin K-dependent proteins in mammalian cells: factors IX and VII. Methods Enzymol 222:450–477
21. Zhu J (2012) Mammalian cell protein expression for biopharmaceutical production. Biotechnol Adv 30:1158–1170. doi:10.1016/j.biotechadv.2011.08.022
22. Rose T, Winkler K, Brundke E et al (2008) Alternative strategies and new cell lines for high-level production of biopharmaceuticals. In: Knäblein J (ed) Modern biopharmaceuticals: design, development and optimization. Wiley, Weinheim, pp 761–777
23. Swiech K, Picanço-Castro V, Covas DT (2012) Human cells: new platform for recombinant therapeutic protein production. Protein Expr Purif 84:147–153. doi:10.1016/j.pep.2012.04.023
24. Kaufman RJ, Wasley LC, Furie BC et al (1986) Expression, purification, and characterization of recombinant gamma-carboxylated factor IX synthesized in Chinese hamster ovary cells. J Biol Chem 261:9622–9628

25. Wajih N, Hutson SM, Owen J, Wallin R (2005) Increased production of functional recombinant human clotting factor IX by baby hamster kidney cells engineered to overexpress VKORC1, the vitamin K 2,3-epoxide-reducing enzyme of the vitamin K cycle. J Biol Chem 280:31603–31607. doi:10.1074/jbc.M505373200
26. Kumar SR (2015) Industrial production of clotting factors: challenges of expression, and choice of host cells. Biotechnol J 10:995–1004. doi:10.1002/biot.201400666
27. Busby S, Kumar A, Joseph M et al (1985) Expression of active human factor IX in transfected cells. Nature 316:271–273. doi:10.1038/316271a0
28. Messier TL, Pittman DD, Long GL et al (1991) Cloning and expression in COS-1 cells of a full-length cDNA encoding human coagulation factor X. Gene. doi:10.1016/0378-1119(91)90141-W
29. De La Salle H, Altenburger W, Elkaim R et al (1985) Active gamma-carboxylated factor IX expressed using recombinant DNA techniques. Nature 316:268–270
30. de Castilho Fernandes A, Fontes A, Gonsales N et al (2011) Stable and high-level production of recombinant factor IX in human hepatic cell line. Biotechnol Appl Biochem 58:243–249. doi:10.1002/bab.32
31. Enjolras N, Dargaud Y, Pérot E et al (2012) Human hepatoma cell line HuH-7 is an effective cellular system to produce recombinant factor IX with improved post-translational modi fi cations. Thromb Res 130:266–273
32. Grinnell BW, Walls JD, Marks C et al (1990) Gamma-carboxylated isoforms of recombinant human protein S with different biologic properties. Blood 76:2546–2554
33. Graham FL, Smiley J, Russell WC, Nairn R (1977) Characteristics of a human cell line transformed by DNA from human adenovirus type 5. J Gen Virol 36:59–72. doi:10.1099/0022-1317-36-1-59
34. Blostein M, Cuerquis J, Landry S, Galipeau J (2008) The carboxylation efficiency of the vitamin K-dependent clotting factors: studies with factor IX. Haemophilia 14:1063–1068. doi:10.1111/j.1365-2516.2008.01828.x
35. Spencer HT, Denning G, Gautney RE et al (2011) Lentiviral vector platform for production of bioengineered recombinant coagulation factor VIII. Mol Ther 19:302–309. doi:10.1038/mt.2010.239
36. Osborn JE, Walker DL (1968) Enhancement of infectivity of murine cytomegalovirus in vitro by centrifugal inoculation. J Virol 2:853–858
37. Yan R, Zhang Y, Cai D et al (2015) Spinoculation enhances HBV infection in NTCP-reconstituted hepatocytes. PLoS One 10: e0129889–e0129889. doi:10.1371/journal.pone.0129889
38. Wajih N, Owen J, Wallin R (2008) Enhanced functional recombinant factor VII production by HEK 293 cells stably transfected with VKORC1 where the gamma-carboxylase inhibitor calumenin is stably suppressed by shRNA transfection. Thromb Res 122:405–410. doi:10.1016/j.thromres.2007.11.002
39. Cribbs AP, Kennedy A, Gregory B, Brennan FM (2013) Simplified production and concentration of lentiviral vectors to achieve high transduction in primary human T cells. BMC Biotechnol 13:98. doi:10.1186/1472-6750-13-98
40. Segura MM, Mangion M, Gaillet B, Garnier A (2013) New developments in lentiviral vector design, production and purification. Expert Opin Biol Ther 13:987–1011. doi:10.1517/14712598.2013.779249
41. Reed SE, Staley EM, Mayginnes JP et al (2006) Transfection of mammalian cells using linear polyethylenimine is a simple and effective means of producing recombinant adeno-associated virus vectors. J Virol Methods 138:85–98. doi:10.1016/j.jviromet.2006.07.024
42. Segura MM, Garnier A, Durocher Y et al (2010) New protocol for lentiviral vector mass production. Methods Mol Biol 614:39–52
43. Biaggio RT, Abreu-Neto MS, Covas DT, Swiech K (2015) Serum-free suspension culturing of human cells: adaptation, growth, and cryopreservation. Bioprocess Biosyst Eng 38:1495–1507. doi:10.1007/s00449-015-1392-9

Chapter 5

Production of Recombinant Factor VIII in Human Cell Lines

Daianne Maciely Carvalho Fantacini and Virgínia Picanço-Castro

Abstract

Human cell lines can produce recombinant proteins much more similar to their natural counterpart, compared to other mammalian cell lines, reducing potential immunogenic reactions. Recombinant proteins produced in nonhuman cells can have in its structure glycan epitopes, such as Galα1,3-Gal (alpha-Gal) and N-glycolylneuraminic acid (Neu5Gc) residues, that are antigenic to humans and can potentially affect the efficacy of the recombinant product. Therefore, the production of recombinant factor VIII (rFVIII) in human cell lines is a new approach to avoid nonhuman glycosylation. Here, we describe a protocol to produce rFVIII in the human cell line SK-HEP, using a lentiviral vector to produce high quantities of the recombinant protein.

Key words Recombinant FVIII, Human cell line, SK-HEP, Lentiviral vector, Transduction

1 Introduction

1.1 Hemophilia A: Treatments

Hemophilia A is a congenital bleeding disorder caused by a deficiency or absence of factor VIII (FVIII). Its prevalence is 1 in 5000 male live births [1]. The level of severity depends on the amount of FVIII in the circulating plasma. Hemophilia A patients are classified as severe with FVIII levels less than 1%, moderate with FVIII levels of 2–5% and mild with FVIII levels of 6–40% [2]. Currently, the main treatment for hemophilia is factor VIII replacement therapy. This therapy is performed with infusions of coagulation factor VIII (FVIII derived from plasma or recombinant FVIII) concentrates, and these infusions help to stop bleeding before any damage is done.

By the early 1970s, hemophiliacs were treated with lyophilized plasma concentrates of coagulation factors. However, in the 1980s, 60–70% of people with severe disease became infected with the human immunodeficiency virus (HIV) resulting from the transmission of blood-borne viruses [1].

As an alternative, after the description of the structure of FVIII and the cloning of the gene, recombinant FVIII (rFVIII) products

Virgínia Picanço-Castro and Kamilla Swiech (eds.), *Recombinant Glycoprotein Production: Methods and Protocols*, Methods in Molecular Biology, vol. 1674, DOI 10.1007/978-1-4939-7312-5_5, © Springer Science+Business Media LLC 2018

were developed [3]. The first product was approved to be marketed in 1992–1993 and up to 2015, all rFVIII commercially available has been produced in either Chinese hamster ovary (CHO) or baby hamster Kidney cells (BHK) [4, 5]. Despite these products have replaced effectively the plasma concentrates; they display a nonhuman pattern of post-translational modifications (PTMs). Therefore, an incorrect reproduction of these PTMs in nonhuman cells may trigger immune reactions and lead to the formation of inhibitors against FVIII [6].

1.2 Human Cell Lines Used to Manufacture rFVIII

Our research group has shown that human hepatic cell lines, such as SK-HEP and HepG2 cells, can be used as a rFVIII production platform [7]. The use of hepatic cell lines for the expression of rFVIII is motivated by the fact that physiologically most of the secreted FVIII is produced in the liver [8]. SK-HEP cells are particularly interesting because of their ability to naturally express von Willebrand factor, a natural stabilizer of rFVIII [9].

In order to produce rFVIII with a human-like glycosylation profile, this chapter will show step by step the production of recombinant rFVIII in SK-HEP cell line, using lentiviral vectors.

2 Materials

2.1 Cell Culture

1. Adherent HEK 293FT cells and SK-HEP cells.
2. *Adherent HEK 293FT cell complete medium*: Dulbecco's modified Eagle's medium high glucose supplemented with 10% (v/v) fetal bovine serum, 0.1 mM MEM nonessential amino acids (NEAA), 200 mM solution GlutaMAX, and 1 mM MEM sodium pyruvate. The medium base (i.e., without FBS) should be filtered and sterilized (0.2 μm) into sterile containers and stored at 4 °C.
3. *SK HEP cell medium*: Dulbecco's modified Eagle's medium high glucose supplemented with 10% (v/v) fetal bovine serum.
4. Phosphate-buffered saline (PBS).
5. Trypsin solution 0.05%.
6. Disposable sterile pipets.
7. Polypropylene centrifuge tubes (15 and 50 mL).
8. Sterile Eppendorf Tubes.
9. Trypan blue.
10. Hemocytometer.
11. Water bath.
12. Centrifuge.
13. Inverted microscope.
14. BD FACSCALIBUR.

2.2 Transient Transfection

1. PEI powder (WM 25,000).
2. Vectors: pCMVR8.91 (viral proteins gag, pol, rev, and tat from HIV-1), pMD2 VSV-G (envelop) and cPPT-C(FVIIIDB) IGWS [7].
3. 20% sucrose solution: Dissolve 20 g of UltraPure sucrose, 100 mM NaCl, 20 mM HEPES (pH 7.4), and 1 mM EDTA. Adjust the volume to 100 mL by adding H_2O and filter-sterilize. This solution can be stored at 4 °C for at least 6 months.

2.3 Real-Time PCR

1. Universal PCR Master Mix.
2. MicroAmp® Optical 96-well reaction plate.
3. Nuclease-free water.
4. 7500 Fast Real-Time PCR System.
5. Sequences: 5-ACCTGAAAGCGAAAGGGAAAC-3 sense and 5-CACCCATCTCTCTCCTTCTAGCC-3 antisense primers and a probe LV2, 5-FAM-AGCTCTCTCGACGCAG-GACTCGGC-MGB3 (Applied Biosystems).
6. Human β-actin is provided by Applied Biosystems (Hs03023880_gl).

3 Methods

3.1 Transient Transfection of HEK 293FT for Lentiviral Production

1. The transfection protocol can be done using a cationic liposome reagent as Lipofectamine or a positively charged polymer as polyethyleneimine (PEI). Here, we described the PEI-based transfection, which is a very efficient and low cost method.
2. Prepare the PEI stock solution dissolving PEI powder (WM 25,000) in water preheated to 80 °C at a concentration of 1 mg/mL and adjust the pH to 7.0. Filtrate through 0.22 μm membrane (*see* **Note 1**). Make aliquots and store at -80 °C.
3. *Day 1*: Plate 3.5×10^6 HEK 293FT cells in a 100 mm^2 dish in complete DMEM medium (6 mL), supplemented with 10% fetal bovine serum, 24 h prior to transfection (*see* **Note 2**). Cells should reach 70–80% confluence on the day of transfection.
4. *Day 2*: Calculate the volume of vectors needed for the experiment following these quantities: 2 μg of factor VIII vector, 1.5 μg of 8.91, and 0.5 μg of VSV-G.
5. Add 0.5 mL of DMEM serum-free medium in a polypropylene tube of 15 mL and then add volume of DNA calculated in the previous step.

6. Add PEI (PEI volume is calculated based on the total DNA, added at 1:3 ratio). In this example, total DNA used for a 100 mm^2 dish is 4 μg; thus, 12 μg of PEI should be used.
7. Mix by pipetting up-and-down gently.
8. Incubate for 10 min at room temperature.
9. Remove medium from plate and wash with 5 mL PBS.
10. Add 5 mL of fresh medium to cover the plate.
11. Add DNA–PEI mixture to cells.
12. Incubate for 8–12 h.
13. *Day 3*: Prewarm 8 mL of complete medium to 37 °C to be used for feeding the cells. At 8–12 h post-transfection, remove medium from transfected cells and replace with the volume of DMEM complete medium (8 mL). Incubate cells in incubator (37 °C and 5% CO_2) for an additional 48 h.
14. *Day 5* (First harvest of viral particles): Collect the supernatant containing the viral particles carefully and place it in a collection tube (*see* **Note 3**). Store at 4 °C (until the next day).
15. Add a volume 5 mL of fresh prewarmed DMEM complete medium to cells and incubate for an additional 24 h.
16. *Day 6* (Second harvest of viral particles): Collect viral particles by carefully removing the media. Pool it with first harvest (day 5).
17. Centrifuge at 300 × *g* for 5 min at 4 °C to remove debris.
18. Filter the pool of viral particles through a 0.45 μm pore (*see* **Note 4**).
19. This protocol (summarized in Fig. 1) can be scaled up relative to the surface area of the culture dish/flask to be used. Recommended cell seeding densities, amount of vector required, and volume of medium for different plate formats are outlined in Table 1.

3.2 Ultracentrifugation for Lentivirus Concentration

The concentration of the viral particles requires ultracentrifugation to remove impurities and to ensure a high infection rate.

1. Sterilize six Ultra-clear SW28 centrifuge tubes by autoclaving.
2. Transfer viral supernatant to Beckman ultracentrifugation tubes and add 4 mL of 20% sucrose solution to the bottom of each tube carefully (*see* **Note 5**).
3. Spin for 2 h at 112,500 × *g* and 4 °C using SW28Ti rotor.
4. Carefully remove tubes from the rotor, pour off supernatant and leave tubes on a paper towel in an inverted position for 10 min to allow the residual liquid to drip away from the pellet (*see* **Note 6**).

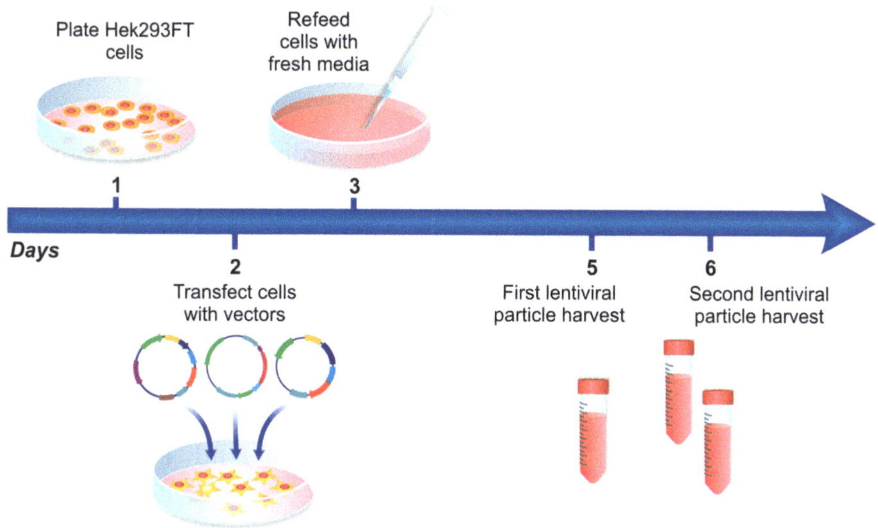

Fig. 1 Procedure overview

Table 1
Recommended cell densities, amount of vector, and volume of DMEM medium required for each cell culture format

Format	Initial number of cells plated	Plasmid: FVIII, 8.91, VSV-G (μg)	Volume of medium serum free, mL	Volume of DMEM complete medium, mL	Volume of DMEM complete medium (second), mL
T75 flask	6×10^6 cells/flask	10, 6.5, 3.5	2	12	10
T175 flask	1×10^7 cells/flask	15, 10, 5	4	20	15

5. Add 100 μL of PBS to each centrifuge tube and store them at 4 °C. It may take from 30 min to overnight for the pellet to be completely dissolved. Do not incubate longer than 24 h.
6. Pipette all viral concentrates up/down several times to ensure that the viral pellet is completely dissociated. Combine liquid from all resuspended pellets in a single SW28 ultracentrifuge tube. Store virus in aliquots of 100 μL at −80 °C. Do not refreeze (*see* **Note 7**).

3.3 Quantification of Viral Titer

There are several methods used to measure the concentration of infectious virus in a sample. In this chapter we will show two methods: (1) quantitative PCR to detected lentiviral LTR integrated region and (2) flow cytometric (FACS) to detected GFP.

3.3.1 Quantitative PCR Method

This method is based on determining the number of vector copies integrated in genomic DNA extracted from transduced cells.

1. Plate HEK 293FT cells at a density of 1×10^5 cells in a 6-well plate 1 day before transduction.
2. The next day, remove the medium and replace with 1 mL/well of fresh medium DMEM.
3. Transduce the cells by adding 12.5-, 25-, and 50 μL of concentrated vector aliquot per well (*see* **Note 8**).
4. For unconcentrated vector, use the following volumes: 250 μL, 500 μL, and 1 mL (*see* **Note 9**).
5. The next day, replace the medium for fresh medium (2 mL) and incubate cells at standard cell culture conditions.
6. After 48 h, wash the cells with 1 mL PBS, detach them by adding 500 μL of trypsin per well and incubate for 3 min at 37 °C.
7. Inactivate the trypsin by adding 1 mL of medium fresh and centrifuge the cells for 5 min at $500 \times g$.
8. Resuspend the pellet in 1 mL PBS and centrifuge again.
9. Isolate the genomic DNA from transduced cells using a DNeasy Kit, according to the manufacturer's instructions.
10. Determine DNA concentration (*see* **Note 10**).
11. The concentration of lentivirus integrated in cell genome can be assessed with two pairs of primers: a lentiviral-specific transgene (LTR gene) and a single copy gene-specific reference gene (beta actin). The number of lentiviral vector copies will be calculated based on standard curves for transgene (LTR) and estimation of absolute DNA titers will be achieved by comparing crossing point values derived from DNA samples to those obtained from a standard curve of known concentrations of plasmid lentiviral DNA.
    (a) *Standard curve for genomic DNA*: Prepare serial dilutions of human genomic DNA resulting in curve from 10^6 to 10^2 copies of human β-actin gene, a gene that is present as one copy per human cell. To prepare the standard curve using genomic DNA, the first step is to determine the mass of DNA in which a gene of interest is present at 1000,000 copies. This is will be the first point of the curve; the other points will be achieved by serial dilution. First we need to calculate the cell mass. The average weight of a DNA base pair is 650 and 1 Da equals 1.67×10^{-24} g. It is considered that the human genome (haploid) is 3.3×10^9 bp in length. Therefore, total weight of the human genome (haploid) is 2.15×10^{12} Da ($3.3 \times 10^9 \times 650$) or 3.59×10^{-12} g

Table 2
Calculations to determine the final concentration of genomic DNA

Copies number	Mass of genome (pg)	Volume to be pipetted in qPCR reaction, μL	Concentration of gDNA (pg/μL)
10^6	3.59×10^6	5	7.18×10^5
10^5	3.59×10^5	5	7.18×10^4
10^4	3.59×10^4	5	7.18×10^3
10^3	3.59×10^3	5	7.18×10^2
10^2	3.59×10^2	5	7.18×10^1

($2.15 \times 10^{12} \times 1.67 \times 10^{-24}$). Due to B-actin exists as a single copy gene per haploid genome, 3.59×10^{-12} g (or 3.59 pg) of human genomic DNA contains one copy of the B-actin gene.

12. To calculate the mass of genomic DNA containing the copies of interest, use the formula: *copy of interest* × *mass of haploid genome* = *mass of genomic DNA required*. For example, the first point of the curve is 1000,000 copies. Therefore, the mass of DNA required is 3,590,000 pg.
13. Table 2 shows the values relative to the other points of the curve.
14. The standard curve is then generated by serial tenfold dilution until 100 molecules. For the dilutions, it should be used the formula $C_1 V_1 = C_2 V_2$. For example, the stock concentration of genomic DNA is 1 μg/μL (or 1000,000 pg/μL) and each dilution has a final volume of 50 μL. Therefore, $C_1 = 1000{,}000$; $C_2 = 718{,}000$ (final concentration relative to 10^6 copies); $V_2 = 50$ μL and V_1 is unknown. After the calculation, V_1 is 35.9 μL and it should be added 10.5 μL of nuclease-free water to achieve the final volume of 50 μL.
15. Standard curve for lentiviral vector: to create a standard curve using a plasmid DNA template, the first step is to calculate the mass of a single plasmid molecule (*see* **Note 11**). The plasmid size is 12,300 bp, and it is accepted that DNA base pair is 650 and 1 Da equals 1.67×10^{-24} g. Thus, the mass of one plasmid molecule is 1.33×10^{-17} g ($12{,}300 \times 650 \times 1.67 \times 10^{-24}$) (*see* **Note 12**). Use the formula *copy of interest* × *mass of one plasmid molecule* = *mass of plasmid DNA required* to achieve the copy numbers of interest.
16. A dilution of the vector has to be performed to obtain the true curve. In this case, we use a curve of 10^7–10^3 plasmid molecules.

17. Considering that the first point of the curve is 10,000,000 copies, the concentration of plasmid DNA required is 2.66×10^{-11} g/μL (10,000,000 × 1.33 × 10^{-17}/5). In this example, a plasmid DNA stock of 100 ng/μL (or 1×10^{-7} g/μ L) is considered.
18. Using the formula $C_1V_1 = C_2V_2$ it is possible to calculate the volume required for 10,000,000 copies, where C_1 is 1×10^{-7}; C_2 is 2.66×10^{-11}; V_2 is 50 (final volume); and V_1 is unknown. After the formula application, V_1 is 13.3×10^{-3} μL. However, this is not a workable volume, and, thus, it is necessary to prepare three predilutions (each 1:10) for reducing the concentration. Once the plasmid is at a workable volume, 13.3 μL (predilution 3) is added with 36.7 μL of nuclease-free H_2O to achieve the final volume of 50 μL. This is the equivalent point to 10,000,000 copies. After that, the standard curve is generated by serial 10-fold dilution until 1000 copies (Table 3).
19. Amplification *Reaction*: Prepare Master mix on ice as described in Table 4. To amplify both standard curves. The mix should be prepared to LTR and β-actin probes separately and pipet in MicroAmp® Optical 96-well reaction plate.
20. Centrifuge the plate for 1 min at 200 × *g* to bring all liquid to the bottom of the wells.
21. Real-time PCR is performed with an ABI PRISM 7500 Sequence Detector. The amplification cycles used are: one cycle of 95 °C for 10 min followed by 40 cycles of 95 °C for 15 s and 60 °C for 1 min.

Table 3
Example of a serial dilution to create a plasmid standard curve

Dilutions	Initial concentration (g/μL)	Volume of plasmid DNA (μL)	Volume of nuclease-free H_2O	Final concentration (g/μL)	Copy number	
Predilution 1	1×10^{-7}	5	45	1×10^{-8}	–	
Predilution 2	1×10^{-8}	5	45	1×10^{-9}	–	
Predilution 3	1×10^{-9}	5	45	1×10^{-10}	–	
Dilution 1	1×10^{-10}	13.3	36.7	2.66×10^{-11}	10,000,000	
Dilution 2	2.66×10^{-11}	5	45	2.66×10^{-12}	1000,000	Curve
Dilution 3	2.66×10^{-12}	5	45	2.66×10^{-13}	100,000	
Dilution 4	2.66×10^{-13}	5	45	2.66×10^{-14}	10,000	
Dilution 5	2.66×10^{-14}	5	45	2.66×10^{-15}	1000	

Table 4
Real-time PCR reaction

2× taqMan master Mix	10 μL × *n*
Probe[a]	1 μL × *n*
H_2O	4 μL × *n*
Sample	5 μL
Final volume reaction	20 μL

[a]Concentration required to probe and primers are 250 and 500 nM, respectively

22. The points of the curve and transduced cell DNA should be tested in duplicate and it is import to include in all PCR reaction a sample with no template as a negative control.
23. The integrating copy number represents the vector titer expressed as infectious units per milliliter and is determined as follows: Infectious units/mL = ($Vector_{copy}$/(β-$actin_{copy}$/$2_{copy/cell}$) × number of target cells (count at day before transduction) × dilution factor [10].

3.3.2 Flow Cytometric (FACS) Method

This method can only be used to titer vectors that carry a transgene that is monitored by flow cytometry, such as GFP.

1. Carry out transduction as described in **steps 1–7** in the previous methodology.
2. Resuspend the pellet in 500 μL PBS and transfer the cells into a 5 mL FACS tube.
3. Analyze cells for GFP expression using FACS. Use a well that has between 1% and 20% of cells expressing GFP to determine titer (*see* **Note 13**).
4. Calculate infectious units according to the following formula: Infectious units/mL = 100,000 × (% of GFP positive cells/100) × dilution factor.

3.4 Transduction of SK HEP Cell Lines

The day before transduction, plate 1 × 10^6 SK HEP cells in a six wells plate.

1. Calculate the volume needed of viral particles to perform MOI 3. Multiplicity of infection (MOI) means the number of viral particles per cell. You should calculate using the formula: (total number of cell seeded × desired MOI) /infectious units/mL.
2. Add polybrene (final concentration 8 μg/mL) to volume of viral particles and mix gently (*see* **Note 14**).
3. Remove medium from plate and add 1 mL of fresh medium.
4. Add the mixture of **step 2** (polybrene plus viral particles) to plate and gently swirl it to mix. Incubate for 18–24 h at 37 °C (*see* **Note 15**).

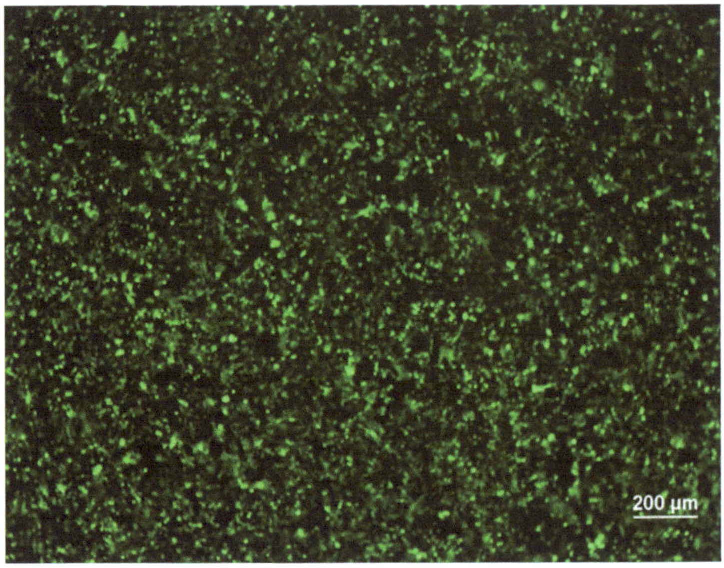

Fig. 2 SK HEP cells were transduced with CMV-F8-GFP vector. The cells were visualized by Olympus Fluorescence microscope 6 days post transduction

5. Replace the medium for fresh medium (2 mL) and incubate cells at standard cell culture conditions.
6. Four to six days after transduction, fluorescence of stable transduced cells may be detected by fluorescence microscopy (Fig. 2).
7. Analyze cells for GFP expression using FACS (*see* **Note 16**).

3.5 rFVIII Detection Method

Biological activity of FVIII should be measured either using One-stage clotting assay or Chromogenic assay. It is suggested using Chromogenic assay.

1. Plate 4×10^6 cells in a T75 flask.
2. Twenty-four hours after seeding, remove the medium and replace with 10 mL of fresh medium. Incubate cells for 48 h.
3. Collect the supernatant and centrifuge at $200 \times g$ for 6 min to remove the debris.
4. Use the supernatant to perform a chromogenic assay according to the manufacturer's instructions and determine the concentration of active factor VIII.

4 Notes

1. It is critically important to filtrate PEI solution because the presence of undissolved PEI particles will precipitate DNA.

2. The use of antibiotic is optional. To better efficiency of transfection opt for a culture without antibiotic and make sure if the cells are mycoplasm free. *Important*: Use alcohol 70% to sterilize the materials.
3. The first harvest of viral particles can be stored at 2–8 °C. Avoid storing at 4 °C for more than 48 h. If it is not possible, freeze at −80 °C. However, multiple freeze–thaw cycles of viral particles may reduce the infectious viral titer by approximately 30% per cycle.
4. If desired, the viral particles may be used directly if a high titer is not necessary or may be concentrated by ultracentrifugation.
5. It is extremely important to use volumes of supernatant above 25 mL in order to not deform the tube.
6. Sterilize the paper towel by UV during 30 min.
7. Viral stocks are stable at −80 °C for approximately 6 months, after that they should be retitered.
8. For each volume added you should reduce from the total fresh medium (for example, 50 μL of vector aliquot in 950 μL of fresh medium).
9. For the volume of 1 mL of vector aliquot, is not necessary to add fresh medium. For each volume added, you should reduce from the total fresh medium (for example, 500 μL of vector aliquot in 500 μL of fresh medium).
10. The DNA concentration should be 20 ng/μL.
11. It should be made an enzymatic digestion in order to linearize the plasmid because it will impact on PCR efficiency. After the linearization, the product should be purified by a commercial Kit.
12. To calculate the mass of a single plasmid should be used the size of the entire plasmid (plasmid + insert).
13. Below 1%, the FACS may not be accurate enough to reliably determine the number of GFP-positive cells. Above 20%, the chance for each GFP-positive target cell to be transduced twice significantly increases, resulting in underestimation of the number of transducing particles.
14. Hexadimethrine bromide is used to enhance the efficiency of transduction. The final concentration 8 μg/mL is referred to volume added in the following step. The viral particles should be thawed on ice. Avoid repeated thawing and freezing cycles. Furthermore, use extra caution when using viral particles because they are considered as Class of risk II.
15. Avoid more than 24 h incubation. Polybrene may be toxic.
16. If the percentage of positive GPF cells is less than 50%, perform another cycle of transduction.

Acknowledgments

We would like to thank Sandra Navarro for drawing the figure. This work was supported by Coordenação de Aperfeiçoamento de Pessoal de Nível Superior (CAPES) and Center for Cell-based Therapy of Ribeirão Preto. The authors acknowledge the São Paulo Research Foundation—FAPESP 2016/02433-0.

References

1. Mannucci PM, Tuddenham EG (2001) The hemophilias – from royal genes to gene therapy. N Engl J Med 344(5):1773–1779. doi:10.1056/NEJM200106073442307
2. White GC, Rosendaal F, Aledort LM et al (2001) Definitions in Hemophilia. Thromb Haemost 85:560
3. Vehar GA, Key B, Eaton D et al (1984) Structure of human FVIII. Nature 312:337–342. doi:10.1038/312337a0
4. White GC, McMillan CW, Kingdon HS, Shoemaker CB (1989) Use of recombinant antihemophilic factor in the treatment of two patients with classic haemophilia. N Engl J Med 320:166–170. doi:10.1056/NEJM198901193200307
5. Schwartz RS, Abildgaard CF, Aledort LM et al (1990) Human recombinant DNA-derived antihemophilic factor (factor VIII) in the treatment of haemophilia A. N Engl J Med 323:1800–1805. doi:10.1056/NEJM199012273232604
6. Ghaderi D, Taylor RE, Padler-Karavani V et al (2010) Implications of the presence of N-glycolylneuraminic acid in recombinant therapeutic glycoproteins. Nat Biotechnol 28:863–867. doi:10.1038/nbt.1651
7. Picanco V, Heinz S, Bott D, Behrmann M, Covas DT, Seifried E, Tonn T (2007) Recombinant expression of coagulation factor VIII in hepatic and non-hepatic cell lines stably transduced with third generation lentiviral vectors comprising the minimal factor VIII promoter. Cytotherapy 9:785–794. doi:10.1080/14653240701656053
8. Shahani T, Covens K, Lavend'homme R et al (2014) Human liver sinusoidal endothelial cells but not hepatocytes contain factor VIII. J Thromb Haemost 12(1):36–42. doi:10.1111/jth.12412
9. Da Rosa NG, Swiech K, Picanço-Castro V et al (2012) SK-HEP cells and lentiviral vector for production of human recombinant factor VIII. Biotechonol Lett 34:1435–1443. doi:10.1007/s10529-012-0925-4
10. Matsui H, Shibata M, Brown B et al (2007) Ex vivo gene therapy for hemophilia A that enhances safe delivery and sustained in vivo factor VIII expression from lentivirally engineered endothelial progenitors. Stem Cells 25:2660–2669. doi:10.1634/stemcells.2006-0699

Chapter 6

Strategies to Suspension Serum-Free Adaptation of Mammalian Cell Lines for Recombinant Glycoprotein Production

Angelo Luis Caron, Rafael Tagé Biaggio, and Kamilla Swiech

Abstract

Serum-free suspension cultures are preferably required for recombinant protein production due to its readiness in upstream/downstream processing and scale-up, therefore increasing process productivity and competitiveness. This type of culture replaces traditional cell culturing as the presence of animal-derived components may introduce lot-a-lot variability and adventitious pathogens to the process. However, adapting cells to serum-free conditions is challenging, time-consuming, and cell line and medium dependent. In this chapter, we present different approaches that can be used to adapt mammalian cell lines from an anchorage-dependent serum supplemented culture to a suspension serum-free culture.

Key words Adaptation, Cell culture, Human cell lines, Mammalian cell lines, Serum-free medium, Suspension culture

1 Introduction

Biopharmaceutical industry is always concerned in increasing productivity while decreasing costs. Therefore, a robust and reliable cell line capable of achieving high densities and high productivity is desired [1]. These highly productive systems involve chemically defined serum (and/or protein)-free media and suspension adapted cell lines [2–4]. Chemically defined serum-free or protein-free media offer lot-to-lot consistency, reduced production costs and have been used in recombinant protein production aiming bloodborne pathogens elimination and immunogenicity reduction. The choice of the correct cell culture medium is determinant for maintaining optimal cell growth, viability and productivity [5]. This medium should contain all essential nutrients required for the growth and maintenance of cells as growth factors, hormones, amino acids, carbohydrates, lipids, vitamins, trace elements, and others related to chemical and mechanical protection as

Virgínia Picanço-Castro and Kamilla Swiech (eds.), *Recombinant Glycoprotein Production: Methods and Protocols*, Methods in Molecular Biology, vol. 1674, DOI 10.1007/978-1-4939-7312-5_6,

antiproteases, antioxidants, and surfactants [6, 7]. A traditional way of cultivating cells is adding to a basal medium a nutrient supplement derived from human or animal sources. The most commonly used supplement is the fetal bovine serum (FBS). Despite its efficacy, its elimination as a supplement is highly recommended for regulatory, economical and ethical reasons: (1) animal derived sera is an ill-defined mixture of components and presents quantitative and qualitative variations among lots; (2) serum harvesting methods may cause unnecessary suffering for the bovine fetus; (3) serum can be a source of contamination with adventitious agents such as viruses, mycoplasma and prions; (4) serum may contain many different proteins and its presence imposes a high cost to the downstream purification process [3, 7, 8]. Additionally, suspension systems are preferable due to its readiness in upstream/downstream processing and scale-up, increasing therefore process productivity and competitiveness. Moreover, this system avoids mechanical or enzymatic dissociation steps [2, 9–11].

For the establishment of serum-free suspension cultures, adherent cells must be adapted to serum-free culture through serum withdrawal and cultured in an agitated system. However, this serum-free adaptation process is difficult, time-consuming, stressful for the cells, and should be done individually for each cell line and production process [3, 7, 9]. The speed, ease, and success rate to adapt cells to serum-free suspension conditions will depend on intrinsic factors related to each cell line, the similarity of medium compositions and how to fit each protocol during the adaptation process [4]. Nevertheless, genetically modified cells for the production of a recombinant protein may present reduced productivity after the adaptation process [12]. It is worth mentioning that not all cell lines are able to grow under serum-free conditions, especially lineages that have been already transformed. Therefore, adaptation protocols should be chosen and conducted carefully [13–16]. In this chapter, we present different approaches that can be used to adapt mammalian cell lines from an anchorage-dependent serum supplemented culture to a suspension serum-free culture, as presented in Fig. 1. The protocols described here were successfully employed in the suspension serum-free adaptation of the human cell lines SK-Hep-1, Huh-7, and HKB-11 [14].

2 Materials

1. Adherent cells for suspension serum-free adaptation.
2. Dulbecco's Modified Eagle's Medium supplemented with 10% (v/v) fetal bovine serum, 3.7 g/L sodium bicarbonate, and 2.4 g/L HEPES buffering agent (*see* **Note 1**). This medium is related to two adherent cell lines: SK-Hep-1 and HuH-7. The

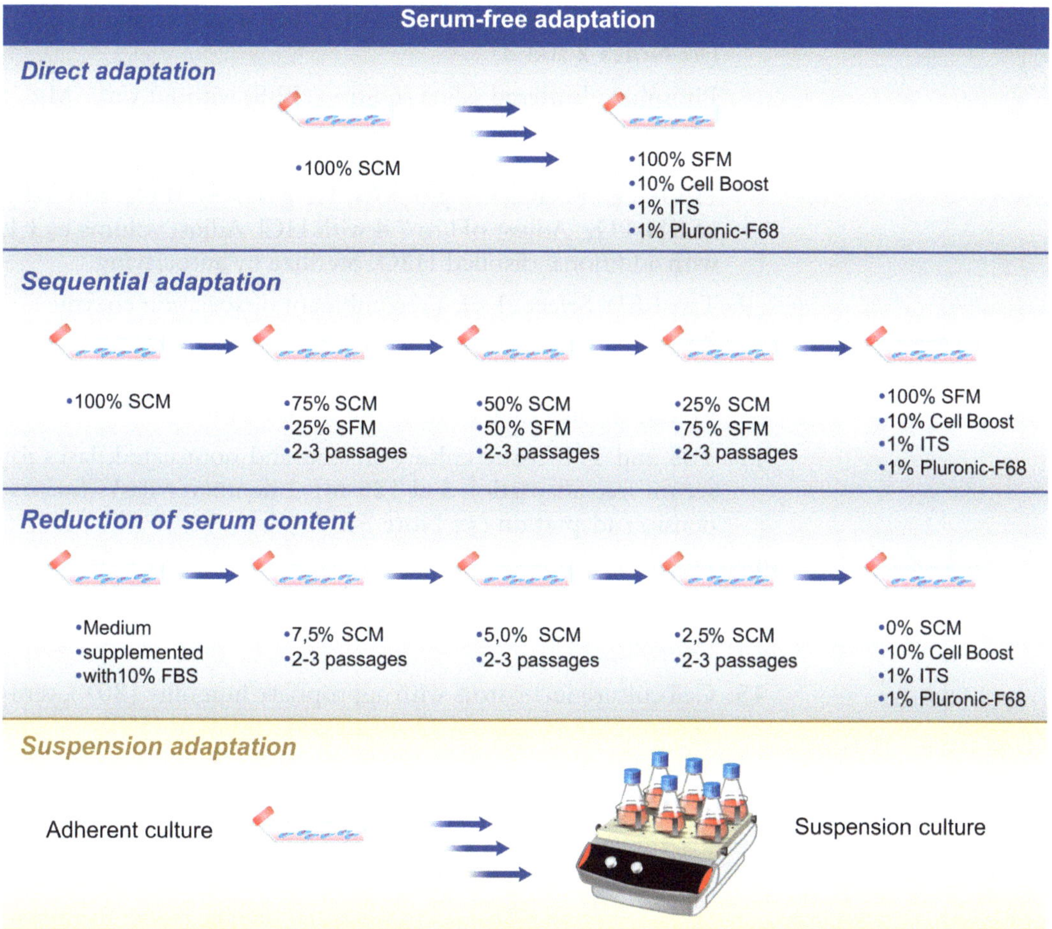

Fig. 1 Different approches for serum-free suspension adaptation

medium should be sterilized by filtration (0.22 μm) into sterile containers and stored at 4 °C.

3. Dulbecco's Modified Eagle's Medium: Nutrient Mixture F-12 supplemented with 10% (v/v) fetal bovine serum, 3.7 g/L sodium bicarbonate, and 2.4 g/L HEPES buffering agent. This medium is for adherent HKB-11 cells. The media should be filter-sterilized (0.22 μm) into sterile containers and stored at 4 °C.
4. FreeStyle™ 293 Expression Medium (Gibco) supplemented with 0.5% (v/v) Pluronic®F-68 nonionic surfactant 100× (*see* **Note 2**).
5. CDM4CHO (HyClone) supplemented with 6 mM GlutaMAX™-I, 2.2 g/L sodium bicarbonate and 0.1% (v/v) Pluronic®F-68 nonionic surfactant 100× (*see* **Notes 2** and **3**).
6. CD 293 AGT™ (Gibco) supplemented with 4 mM GlutaMAX™-I (*see* **Notes 2** and **3**).

7. 293 SFM II (Gibco) supplemented with 4 mM GlutaMAX™-I (*see* **Notes 2** and **3**).
8. Phosphate-buffered saline solution (PBS) without Ca^{2+}/Mg^{2+}: 137 mM NaCl, 2.7 mM KCl, 8.7 mM Na_2HPO_4, and 1.5 mM KH_2PO_4, pH 7.4. Dissolve the following in 800 mL distilled H_2O: 8 g of NaCl, 0.2 g of KCl, 1.2 g of Na_2HPO_4, and 0.2 g of KH_2PO_4. Adjust pH to 7.4 with HCl. Adjust volume to 1 L with additional distilled H2O. Sterilize by autoclaving.
9. TrypLE™ Select (1×), a recombinant dissociation enzyme (*see* **Note 4**).
10. Trypan Blue solution 0.4%. Mix 0.9% (w/v) NaCl and 1% (w/v) Trypan Blue in 4:1 ratio. Use within 24 h.
11. T25 and T75 tissue culture-treated and nontreated flasks for serum-free adaptation and 125-mL Erlenmeyer flasks for suspension adaptation (*see* **Note 5**).
12. Hemocytometer.
13. Inverted microscope.
14. Centrifuge.
15. Cell culture incubators with appropriate humidity (80%), temperature (37 °C), and CO_2 in air concentration (5%).
16. Orbital shaker.

3 Methods

For all the adaptation approaches described here, it is recommended that cells are in mid-logarithmic phase with viability higher than 90%. Before and after the adaptation it is important to characterize the growth behavior of the cells through the population doubling time (*see* **Note 6**). During adaptation, a decrease in cell viability is expected. If viable cells are below 80% or showing no growth, repeat the current step of the protocol until cells recover. New culture conditions may change the physiological stimuli that cells receive. Moreover, the new environment may change cell metabolism by induction or repression of protein synthesis in response to certain nutrient. Thus, it is common that cells change their metabolism, growth rate, morphology and productivity during the adaptation process and, in extreme circumstances, a subpopulation can be selected from large heterogeneous starting population. Therefore, it is important to characterize, before and after adaptation, all growth parameters as they may shift during the process. Also, we recommend to check productivity and glycosylation and/or phosphorylation patterns to ensure there was no changes in protein glycosylation profile, as it may interfere in biological activity [4, 6, 17].

After serum-free adaptation, it is important to establish a work and master cell bank of your adapted cells. Serum-free formulations must be used for cryopreservation to keep your process completely free of serum components (*see* **Note 7**).

3.1 Direct Adaptation

1. Cultivate normally your cells at 1–2 × 10^5 cells/mL (initial cell density) using a standard cell culture plasticware (T25 flask) and DMEM 10% FBS (*see* **Note 8**).
2. After 3–5 days, cells should be 80–90% confluent. Remove and discard all of the serum containing medium from the flask containing the cell monolayer.
3. Serum can inactivate the cell dissociation reagent (TrypLE™ Select), so it is important to wash the flask surface with 4 mL PBS (or with a sufficient volume to cover the cell monolayer) before adding your dissociation reagent.
4. Trypsinize your cells by pipetting 2 mL of the TrypLE™ Select into the flask and guarantee that all the monolayer area is covered with it. Incubate the flask at 37 °C for 2–10 min (*see* **Note 9**). Stop the reaction with a trypsin inhibitor as soybean trypsin inhibitor, an animal component free reagent (*see* **Note 10**). Add 1 mL trypsin inhibitor solution (1 mg/mL using serum-free media) for every mL of trypsin solution used for dissociation. Take a sample in an Eppendorf® tube to count the cells. Viability should be above 80%.
5. Pellet the cells by centrifugation at 150–300 × *g* for 5 min. Discard supernatant and resuspend the pellet in serum-free medium, such as FreeStyle™ 293 Expression Medium, to an initial concentration of 1–2 × 10^5 cells/mL. Incubate the T25 flask for 3–5 days.
6. Repeat **steps 2–4** for at least three passages in serum-free medium, and check if cells are growing sustainably and reproducibly. Ideally, the doubling time after adaptation should be similar to regular value before adaptation. Cell viability above 90% is desired (*see* **Note 11**). If this approach is too aggressive to the cells, we recommend the sequential protocol described below.

3.2 Sequential/Weaning Adaptation

1. Cultivate normally your cells at 1–2 × 10^5 cells/mL using a standard cell culture plasticware (T25 flask) and DMEM 10% FBS.
2. After 3–5 days, cells should be 80–90% confluent. Remove and discard all of the serum containing medium from the flask containing the cell monolayer.
3. Trypsinize the cells as described in Subheading 3.1, **steps 3** and **4**.

4. Resuspend the cells in a mix of 25% (v/v) serum-free medium and 75% (v/v) serum-containing medium to a final volume of 4 mL at cell density of 1–2 × 10^5 cells/mL. Incubate for 3–5 days until the cells reach confluence.
5. Repeat Subheading 3.1, **steps 3** and **4** to detach the cell monolayer. As some cells may detach from the treated surface of the flask with reduced serum content, transfer the conditioned medium into a polypropylene tube and mix the trypsinized cells to it. Afterward, proceed as described in Subheading 3.1, **step 5**.
6. Resuspend the cells in a mix of 50% of serum-free medium and 50% of serum-containing medium to a final volume of 4 mL at cell density of 1–2 × 10^5 cells/mL. Incubate for 3–5 days or until cells are confluent (*see* **Note 12**). From now on, it is interesting to keep a backup culture in the previous mixture to avoid starting over from the beginning.
7. Mix 75% of serum-free medium and 25% of serum-containing medium to a final volume of 4 mL at cell density of 1–2 × 10^5 cells/mL. Incubate for 3–5 days or until cells reach confluence (*see* **Note 13**). We recommend the cryopreservation of the adapted cells from this step to the end to avoid loss of adapted cells.
8. If the cell viability is higher than 80%, resuspend the cells with serum-free medium. To make sure the cells are adapted to serum-free conditions, it is recommended to check cell growth rates and cell densities for at least three passages in serum-free medium. Cell viabilities above 90% are desired. If cells are growing sustainably and reproducibly and growth parameters have returned to their original values, then the adaptation process was successful (*see* **Note 11**). Another weaning-based approach is described below if you intend to use only serum-free medium during the entire adaptation process.

3.3 Reduction of Serum Content

1. Cultivate normally your cells at 1–2 × 10^5 cells/mL using a standard cell culture plasticware (T25 flask) and DMEM 10% FBS.
2. After 3–5 days, cells should be 80–90% confluent. Remove and discard all of the serum containing medium from the flask containing the cell monolayer.
3. Trypsinize the cells as described in Subheading 3.1, **steps 3** and **4**.
4. Seed the cells at 1–2 × 10^5 cells/mL in the serum-free medium of your choice supplemented with 10% FBS. Incubate for 3–5 days and subculture when cells reach confluence. Repeat this step 2 or 3 times.

5. Check if cells are growing sustainably and reproducibly and if viability is above 90%. Repeat **steps 2** and **3**. Seed the cells at $1–2 \times 10^5$ cells/mL in serum-free medium supplemented with 7.5% of FBS (Final volume of 4 mL). Incubate for 3–5 days or until cells reach confluence.
6. Check if cells are growing well and viability is above 90%. Transfer the conditioned medium into a polypropylene tube as some cells may already begin to detach. Repeat **steps 2** and **3**. Seed the cells at $1–2 \times 10^5$ cells/mL in the serum-free medium of choice supplemented with 5.0% of FBS (Final volume of 4 mL). Incubate for 3–5 days and subculture cells when they reach confluence (*see* **Note 12**).
7. Repeat the previous steps seeding the cells at $1–2 \times 10^5$ cells/mL in serum-free medium supplemented with 2.5% of FBS. Incubate for 3–5 days. Cell viability may decrease during this step (*see* **Note 13**).
8. Repeat **step 2** and resuspend the cells in serum-free medium. To make sure cells are adapted to serum-free conditions, it is recommended to check cell growth rates and cell density for at least three passages in serum-free medium. Cell viability above 90% are desired. If cells are growing sustainably and reproducibly and growth parameters have returned to their original values, then the adaptation process was successful (*see* **Note 11**).

3.4 Adaptation to Suspension Culture

Some cell lines require separate serum-free and suspension adaptation steps. As anchorage-dependent cells may detach throughout serum removal, it is common to observe floating cells in culture before complete withdraw of serum. That is why we recommend starting serum-free adaptation before suspension adaptation. However, if you want to begin with the suspension adaptation or your cells are still associated with the treated surface, follow the protocol below.

1. Collect the adherent cells by trypsinization as disclosed in Subheading 3.1, **steps 3** and **4** and pellet cells at $150–300 \times g$ for 5 min.
2. Resuspend the pellet with FBS-supplemented growth medium at a cell density of approximately 0.5×10^6 cells/mL (*see* **Note 14**).
3. Transfer suspended cells into a 125-mL shake flask with a working volume of 20 mL.
4. Place the flask in an orbital shaker maintained inside the incubator. The rotation speed should be determined empirically. Typically, cells are shaken at 80–120 rpm.

5. Assess cell density and viability on a daily basis. When cell density reaches 1–3 × 10^6 cells/mL or after 2–3 days post-seeding, passage cells to an initial concentration of 0.5 × 10^6 cells/mL.
6. If the cells maintained in serum-free medium present a sustainable and reproducible growth for at least three passages (achieving 1–3 × 10^6 cells/mL and cell viability of 90% or higher), cells are considered adapted to suspension culture. At this point, initial cell seeding can be reduced to approximately 0.2–0.3 × 10^6 cells/mL.

3.5 Final Considerations

Different mammalian cell lines have already been adapted to suspension serum-free cultures [1, 13–16, 18–23]. However, a universal protocol for serum-free suspension adaptation is not possible as each cell line presents unique responses to the process [1, 15]. Therefore, it is advisable to test different protocols and serum-free media, always monitoring the response of your cell line in terms of growth and viability.

4 Notes

1. We do not recommend the use of antibiotics in serum-containing and serum-free media as it may mask the presence of bacterial contamination.
2. Nowadays, around 50 suppliers prepare serum-free media for different cell lines. Commercial serum-free formulations are usually a trade secret and we have to rely on the supplier's advice or screen a number of different media for the cell line of interest. Nutritional requirements may greatly differ from different cell lines and processes. Even for the same cell line, a medium formulated for high cell density is not necessarily the same for high recombinant protein production. Therefore, the medium used to adapt a cell line is not necessarily the one that works best in a transfection process or to obtain high cell density [14]. Some examples of serum-free formulations are listed in Table 1. If possible, check if the medium contains any kind of attachment factors. Its absence may facilitate the adaptation process [24].
3. The use of GlutaMAX™-I instead of L-glutamine is preferable as the first is a L-alanine-L-glutamine dipeptide, which makes it more stable in aqueous solution. Nevertheless, the gradual glutamine release from the dipeptide due the action of cellular peptidases can be compared to a fed-batch culture.
4. For cell detachment, we recommend the use of a commercially available recombinant trypsin as TrypLE™ Select (1×) (Gibco)

Table 1
Commercial serum-free formulations available for mammalian cell lines

Serum-free media	Company	Cell line	Designed for
EX-CELL™Vero SF	Sigma-Aldrich	CHO	Suspension
EX-CELL™293-S	Sigma-Aldrich	HEK293	Suspension
EX-CELL™302	Sigma-Aldrich	Vero cells	Microcarrier
EX-CELL™420	Sigma-Aldrich	Sf9 and Sf21	Suspension and virus production
FreeStyle™ 293 Expression Medium	Gibco	HEK293	Suspension, transfection and adaptation of HEK293 cells
ProCHO5	Lonza	CHO	Adaptation process
ProCHO5	Lonza	CHO	Suspension
CD293 AGT	Gibco	HEK293	Suspension
293 SFM II	Gibco	HEK293	Suspension
SFM4Transfx-293	HyClone	HEK293	Transfection of suspension cells

to avoid animal-origin adventitious agents from animal origin components as the Trypsin 0.5% solution.

5. It is possible to use also smaller suspension systems as the Corning® 50 mL Mini Bioreactor, a 50 mL polypropylene tube with vent caps.
6. Doubling time is calculated by the following formula:

$$D_t = t \times \log 2/[\log(N/\text{No})]$$

where t is cultivation time, N is the final number of cells, and No is the initial number of cells.

7. Cryopreservation medium for serum-free adapted cells can be obtain by using the serum-free medium in which the cells are adapted, supplemented with 10% (v/v) dimethyl sulfoxide. Commercial serum-free formulations are also available for cryopreservation purposes. Some examples: HyCryo® Cryopreservation Medium (HyClone), Synth-a-Freeze® Cryopreservation Medium (Gibco), Cell Freezing Medium-DMSO Serum-free (1×) (Sigma-Aldrich) among others.
8. Serum-free media contain less proteins than serum-containing medium, so adapted cells are more sensitive to pH, temperature, osmolality, mechanical forces, and enzyme treatments. Therefore, in some cases, it is recommended to use higher cell density.
9. Different cell lines have different surface glycoproteins, requiring different incubation times. HKB-11 cells can be detached

from the treated surface in 2–3 min at 37 °C. However, Huh-7 cells can take around 10 min to be detached.

10. If a trypsin inhibitor is not available, we recommend the use of PBS to dilute the trypsin containing solution in a ratio of 2:1. If necessary, centrifuge and repeat the process to remove any residual trypsin content.
11. The number of subcultures before complete serum-free adaptation will change according to the cell line. During SK-Hep-1 and HKB-11 serum-free adaptation for example, cells showed different behavior during passages. If cell growth declines during any point through the adaptation, we recommend returning to previous serum concentration to allow cell growth to stabilize before moving forward with the adaptation. Culture parameters as cell density and viability tend to increase while doubling time tends to decrease when cells are well adapted to serum-free conditions. If cells are showing low viability or high doubling times probably they have not been adapted yet.
12. Some cell lines support better the adaptation process. In some cases, it is possible to start the serum withdraw directly from 50% of serum-free medium (5% of serum supplementation).
13. It is expected that cells start to grow slower and with lower viability. The transition from passage with 25% serum-free medium (2.5% serum-containing medium) to completely serum-free medium may be too stressful for the cells. If the viability is less than 80%, we recommend additional passages before complete serum withdraw, as 1.25%, and 0.65% of serum-containing medium. If cells supported serum-free medium, but viability decreases or cell growth stops, it is possible to supplement the serum-free medium with commercial supplements such as amino acids, defined proteins, vitamins, trace elements, or serum substitute enriched with all of these components.
14. If you successfully adapted your cell line to serum-free medium in the static flasks, use the same medium without serum for suspension adaptation.

Acknowledgment

This work was supported by FAPESP grants 2012/04629-8 and 2012/02109-7 and CAPES scholarship.

References

1. Rodrigues ME, Costa AR, Henriques M et al (2013) Advances and drawbacks of the adaptation to serum-free culture of CHO-K1 cells for monoclonal antibody production. Appl Biochem Biotechnol 169:1279–1291
2. Wurm FM (2004) Production of recombinant protein therapeutics in cultivated mammalian cells. Nat Biotechnol 22:1393–1398
3. Grillberger L, Kreil TR, Nasr S, Reiter M (2009) Emerging trends in plasma-free manufacturing of recombinant protein therapeutics expressed in mammalian cells. Biotechnol J 4:186–201
4. Johnson T (2006) Promises and pitfalls of cell line adaptation. BioProcess Int 3:S52–S56
5. Costa AR, Withers J, Rodrigues ME et al (2013) The impact of cell adaptation to serum-free conditions on the glycosylation profile of a monoclonal antibody produced by Chinese hamster ovary cells. New Biotechnol 30:563–572
6. Gstraunthaler G (2003) Alternatives to the use of fetal bovine serum: serum-free cell culture. ALTEX 20:275–281
7. van der Valk J, Brunner D, De Smet K et al (2010) Optimization of chemically defined cell culture media–replacing fetal bovine serum in mammalian in vitro methods. Toxicol In Vitro 24:1053–1063
8. van der Valk J, Mellor D, Brands R et al (2004) The humane collection of fetal bovine serum and possibilities for serum-free cell and tissue culture. Toxicol In Vitro 18:1–12
9. Chico E, Rodríguez G, Figueredo A (2008) Biorreatores para Células Animais. In: Moraes Â, Augusto E, Castilho L (eds) Tecnol. Cultiv. Células Animais Biofármacos a Ter. Gênica, 1st edn. Roca, Barcelona, Spain, p 528
10. Freshney IR (2015) Culture of animal cells: a manual of basic technique and specialized applications. In: Culture of animal cells, 6th edn. John Wiley & Sons, Inc., Hoboken, NJ, USA, pp i–xxxi
11. Zhu J (2012) Mammalian cell protein expression for biopharmaceutical production. Biotechnol Adv 30:1158–1170
12. Costa A, Rodrigues M, Henriques M et al (2011) Strategies for adaptation of mAb-producing CHO cells to serum-free medium. BMC Proc 5:P112
13. do Amaral RL, de Sousa Bomfim A, de Abreu-Neto MS et al (2016) Approaches for recombinant human factor IX production in serum-free suspension cultures. Biotechnol Lett 38:385–394
14. Biaggio RT, Abreu-Neto MS, Covas DT, Swiech K (2015) Serum-free suspension culturing of human cells: adaptation, growth, and cryopreservation. Bioprocess Biosyst Eng 38(8):1495–1507
15. González Hernández Y, Fischer RW (2007) Serum-free culturing of mammalian cells–adaptation to and cryopreservation in fully defined media. ALTEX 24:110–116
16. Jaluria P, Konstantopoulos K, Betenbaugh M, Shiloach J (2008) Egr1 and Gas6 facilitate the adaptation of HEK-293 cells to serum-free media by conferring enhanced viability and higher growth rates. Biotechnol Bioeng 99:1443–1452
17. Ghaderi D, Zhang M, Hurtado-Ziola N, Varki A (2012) Production platforms for biotherapeutic glycoproteins. Occurrence, impact, and challenges of non-human sialylation. Biotechnol Genet Eng Rev 28:147–175
18. Côté J, Garnier A, Massie B, Kamen A (1998) Serum-free production of recombinant proteins and adenoviral vectors by 293SF-3F6 cells. Biotechnol Bioeng 59:567–575
19. Cho MS, Yee H, Chan S (2002) Establishment of a human somatic hybrid cell line for recombinant protein production. J Biomed Sci 9:631–638
20. Schiedner G, Hertel S, Bialek C et al (2008) Efficient and reproducible generation of high-expressing, stable human cell lines without need for antibiotic selection. BMC Biotechnol 8:13
21. Perrin P, Madhusudana S, Gontier-Jallet C et al (1995) An experimental rabies vaccine produced with a new BHK-21 suspension cell culture process: use of serum-free medium and perfusion-reactor system. Vaccine 13:1244–1250
22. Haldankar R, Kopchick JJ, Ridgway D (1999) Stable production of a human growth hormone antagonist from CHO cells adapted to serum-free suspension culture. Biotechnol Prog 15:336–346
23. Pau MG, Ophorst C, Koldijk MH et al (2001) The human cell line PER.C6 provides a new manufacturing system for the production of influenza vaccines. Vaccine 19:2716–2721
24. Corning (2012) Adaptation of cell cultures to a serum-free medium - SFM adaptation protocol. In: Invitrogen. http://cellgro.com/media/upload/file/techinfosheets/new/Adaptation of Cell Cultures to a Serum-free Medium.pdf. Accessed 9 Jan 2017

Chapter 7

Production of Recombinant Rabies Virus Glycoprotein by Insect Cells in a Single-Use Fixed-Bed Bioreactor

Daniella C. Ventini-Monteiro, Renato M. Astray, and Carlos A. Pereira

Abstract

A single-use fixed-bed bioreactor (iCELLis nano) can be used for cultivating non adherent insect cells, which can be then recovered for scaling up or for harvesting a membrane-associated viral glycoprotein with high quality in terms of preserved protein structure and biological function. Here, we describe the procedures for establishing genetically modified *Drosophila melanogaster* Schneider 2 (S2) cell cultures in the iCELLis nano bioreactor and for quantifying by ELISA the recombinant rabies virus glycoprotein (rRVGP) synthesized. By using the described protocol of production, the following performance can be regularly achieved: 1.7 ± 0.6 × 1E10 total cells; 2.4 ± 0.8 × 1E7 cells/mL and 1.2 ± 0.9 μg of rRVGP/1E7 cells; 1.5 ± 0.8 mg of total rRVGP.

Key words Insect cells, Single-use fixed-bed bioreactor, Rabies virus glycoprotein, ELISA

1 Introduction

Single-use bioreactors for animal cell culture are attractive tools for industries and research laboratories because of their ease of handling characteristics [1]. Cell adherence and secreted or cell associated products are main parameters and often limit the use of a given cell culture technology. Cell associated products usually requires cell recovering for product harvesting. Fixed-bed bioreactor provide the cell cultivation with good environment in view of a lower hydrodynamic stress due to immobilization and a medium perfusion for nutrients supply and metabolites removal [2]. It also leads to a production of higher quality protein [3].

Viral glycoproteins are among the most important proteins in view of their importance in virus infection and immunity. Their oligomerization leads to a suitable biological function and represent a challenge for recombinant expression, synthesis, purification and quantification [4]. The recombinant rabies virus glycoprotein (rRVGP), a cell membrane associated-product synthesized by stable transfected S2 cells [5] represent a valuable target for studies of

Virgínia Picanço-Castro and Kamilla Swiech (eds.), *Recombinant Glycoprotein Production: Methods and Protocols*, Methods in Molecular Biology, vol. 1674, DOI 10.1007/978-1-4939-7312-5_7, © Springer Science+Business Media LLC 2018

structure and function [6]. A large-scale production of rRVGP with preserved structure is expected for replacing the classical rabies vaccine constituted by inactivated virions produced by infected mammalian cell cultures [4].

Our aim in this chapter is to provide a detailed methodology for entrapping, growing and recovering recombinant S2 cells in a disposable fixed-bed bioreactor and for quantitatively evaluate the preserved rRVGP produced by ELISA.

2 Materials

2.1 Recombinant S2 Cell Culture in a Single-Use Fixed-Bed Bioreactor

1. iCELLis nano disposable bioreactor (PALL Life Science) of a total volume of 1 L, fixed-bed with microfiber carriers (MC) (200 mL, 4 m^2 of available surface to the cells and 3544 MC) containing a magnetic impeller.
2. pH and DO probe.
3. Reusable stainless steel lid.
4. GL45 cap.
5. Tubing and connections.
6. Dock station.
7. Controller.
8. Perfusion module.
9. SF-900 III SFM media (Thermo Fisher Scientific).
10. Cell lysis solution: 0.1% Crystal Violet, 0.1 M citric acid.
11. Stable transfected S2 cells for rRVGP expression under the control of $CuSO_4$ inducible metallothionein promoter [2, 3].

2.2 rRVGP Quantification by ELISA

1. ELISA kit for rRVGP (Institut Pasteur, Paris, France).
2. Suspension buffer: 25 mM Tris–HCl, 25 mM NaCl, 5 mM $MgCl_2$, and 1 mM PMSF (phenylmethylsulfonyl fluoride), pH 7.2.
3. Lysis buffer: 25 mM Tris–HCl, 25 mM NaCl, 5 mM $MgCl_2$, 1 mM PMSF (phenylmethylsulfonyl fluoride), and Igepal® 0.4%, pH 7.2.
4. ELISA plates (High binding type).
5. Carbonate buffer: 0.05 M Na_2CO_3 adjusted to pH 9.6 with 0.05 M $NaHCO_3$.
6. Blocking solution: carbonate buffer with 0.3% BSA (bovine serum albumin) and 5% sucrose.
7. Phosphate buffered saline 10× (PBS): 1.37 M NaCl, 0.1 M Na_2HPO_4, 27 mM KCl, 18 mM KH_2PO_4, pH 7.0.
8. PBS–Tween pH 7.0: PBS with 0.05% Tween 20.

9. PBS–Tween–BSA pH 7.0: PBS with 0.05% Tween 20, 0.5% BSA.
10. Citrate buffer: 40 mM $Na_3C_6H_5O_7, 2H_2O$ (trisodium citrate), 11 mM citric acid, pH 5.6.
11. Substrate–chromogen solution: 2 mg/mL o-phenylenediamine (OPD), hydrogen peroxide 30% (110 vol) 1/1000, in citrate buffer.
12. Stopping solution; 4 NH_2SO_4.
13. IgG D1A primary antibody (Pasteur Institute, Paris), diluted 1:2000 in carbonate buffer.
14. IgG D1A-PO peroxidase-conjugated antibody (Pasteur Institute, Paris), diluted 1:2000 in PBS–Tween–BSA buffer.
15. Reference antigen for quantification, Lyophilized inactivated and purified rabies virus (PV strain) containing 10 μg of RVGP.

3 Methods

3.1 Single-Use Fixed-Bed Bioreactor Preparation, Probe Calibration, and Autoclaving

1. To assemble the lid, place it on the bioreactor and tighten it. Do the same with the GL45 cap.
2. Connect the tubings to the correspondent pipes (NaOH addition to the short protruding pipe; sampling to the longest plunging pipe; recirculation loop (RL) IN to the other short pipe, RL OUT medium pipe; for Inlet gases and vent out, just connect tubing with 0.22 μm filters).
3. To calibrate the pH probe, connect the pH cable to the pH probe, put the probe in buffer 4.0, and once the response is stabilized, confirm the 4.0 calibration. Proceed in a similar way with buffer 7.0.
4. To calibrate the DO probe, connect the DO cable to the DO probe. Put the DO probe into the bioreactor, and introduce nitrogen to eliminate the air. Once the value is stabilized close to 0%, calibrate it at 0%. To calibrate at 100%, let DO probe in free contact with the surrounding air of the room, and once the value stabilizes close to 100%, calibrate it at 100%.
5. Insert the calibrated probes and screw into the lid manually.
6. Once the bioreactor is ready, autoclave at 121.5 °C for 30 min.

3.2 S2 Cell Seeding

1. Cultivate routinely recombinant S2 cells in T-flasks with serum-free SF-900 III medium (Thermo Fisher Scientific).
2. To prepare the cell seeding for the bioreactor experiments, inoculate 5 × 1E5 cells/mL into orbital shaker flasks. After 2–3 days, the culture reaches cell seeding values (7 × 1E8 cells) (*see* **Note 1**).

3.3 Starting the Cell Culture

The culture parameters for S2 cell cultivation and rRVGP production in iCELLis bioreactor (Fig. 1) are presented in Table 1 (*see* **Note 2**).

1. Under the laminar flow hood put sampling manifold and recirculation tubing. Outside the laminar flow hood, put the reactor into the dock.
2. Insert the temperature probe into the thermo well (together with some glycerin), connect the probes with the respective cables, and connect the gas tubing to the Vent IN.
3. Set the parameters, according to Table 1, on the controller.
4. Fill the reactor with 400 mL of prewarmed medium through the sampling fold (*see* **Note 3**).
5. Switch ON the temperature, agitation, DO and check the flow meters for desired values.
6. Add the inoculum through the sampling manifold, and, if necessary, add additional medium (*see* **Notes 4** and **5**).

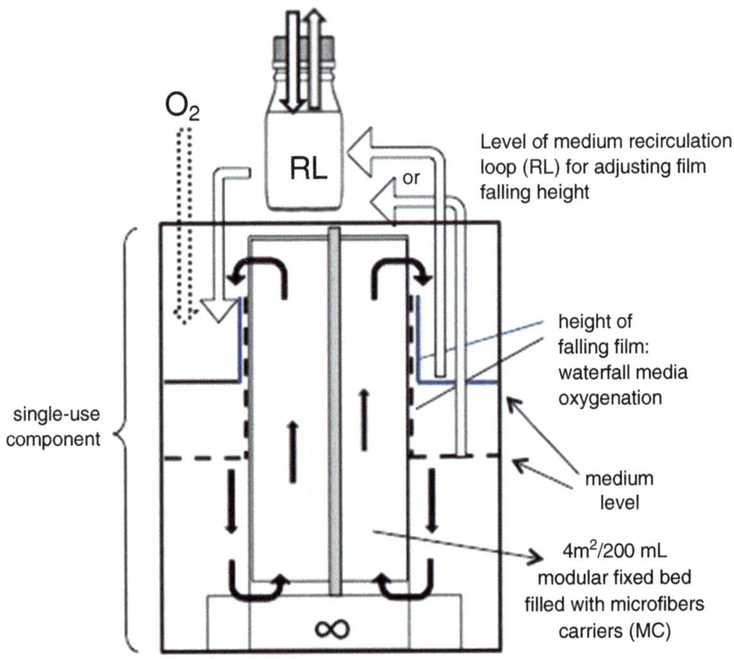

Fig. 1 The iCELLis nano bioreactor. A magnetic driver impeller leads the medium flows through fixed-bed from *bottom* to *top*. Then medium falls as a thin film down the outer wall where the O_2 transfer takes place. A recirculation loop (RL) can be installed for medium recirculation or perfusion. The medium level can then be adjusted to allow a higher waterfall falling film height and increase the O_2 transfer. Air or O_2 is supplied at the head of the disposable vessel (reproduced from reference 3 with permission from Elsevier)

Table 1
Parameters for iCELLis nano bioreactor culture of a S2 cell population expressing the rRVGP upon $CuSO_4$ induction of the metallothionein promoter.

Cell seeding	2 × 1E5 cells/MC 3.5 × 1E6 cells/mL of fixed bed 1E6 cells/mL (total medium inside bioreactor) 0.7 × 1E9 total cells
Fixed-bed compaction	1.5
Number of MC	3544
Fixed bed	4 m^2/200 mL
Total medium volume	1200 mL 700 mL bioreactor +500 mL RL
Falling film	6 cm
Cells	S2
Medium	SF-900 III
DO	30% (30 mL/min)
Temperature	28 °C
rpm growth (liquid speed)	610 (0.5 cm/s)
rpm entrapment (liquid speed)	755 (1 cm/s)
Entrapment time period	24 h
Medium exchange	50% of Recirculating Loop (RL) volume at exponential growth phase
pH	Measured/not controlled
Medium recirculation	15 time/day
Cell recovering	1200–1500 rpm with medium without cells

MC microfiber carriers present in the fixed bed, *RL* recirculation loop

3.4 Follow-Up of the Cell Culture in the Bioreactor

In order to have an adequate follow-up of the culture, MC sampling and supernatant sampling need to be performed regularly.

1. Transfer the reactor to the laminar flow hood.
2. Switch off temperature, DO, and agitation.
3. Open the clamp onto sampling manifold and using a vacuum pump transfer approximately 100 mL from the bioreactor to the sampling bottle. Make a sample with a pipet and pump air to transfer the media back to the bioreactor.
4. With the bioreactor still inside the laminar flow hood, unscrew the cap and sample some MC through the slots with sterile tweezers. Screw the cap back onto the lid.

5. Once the sampling is performed, put the bioreactor into the dock and switch ON the controllers.
6. For the cells attached onto MCs, the sampled MCs have to be treated with a lysis solution in order to free the nuclei (*see* **Note 6**).
7. When the cell culture achieves the exponential growth phase, induce the cells for protein production (*see* **Note** 7), and start with the recirculation loop by adding 500 mL of fresh media into the corresponding bottle and tuning ON the pumps.
8. To recover the cells from the fixed bed, set the agitation at maximum speed (1500 rpm). Repeat this procedure 2 or 3 times with PBS (300 mL). Almost 90% of the cells can be recovered from the fixed bed at high viability rate (*see* **Note 8**).
9. Turn all controllers OFF, disconnect filters, and close gas inlets.
10. Autoclave the bioreactor.

3.5 rRVGP Quantification by ELISA (See Note 9)

1. The day before rRVGP analysis, perform ELISA plate sensitization with 200 μL/well IgG D1A primary antibody diluted in carbonate buffer. Cover the plate and incubate at 4 °C for 16 h (*see* **Note 10**).
2. Start the analysis by suspending cell sample in 0.5 mL of suspension buffer; vortex briefly. Add 0.5 mL lysis buffer (*see* **Note 11**), vortex briefly, and incubate at 4 °C for 1 h with periodic gentle mixing every 15 min (*see* **Note 12**). After incubation, centrifuge samples at 8000 × *g* for 10 min at 4 °C.
3. Drain primary antibody solution and block plates with 300 μL/well of blocking solution, for 30 min at 37 °C.
4. Dilute the antigen for quantification in PBS–Tween–BSA to establish the standard curve in seven twofold diluted points from 500 to 7.8 ng/mL.
5. Wash ELISA plate five times with PBS–Tween and transfer 200 μL of samples and standard curve dilutions in duplicates to the wells. Incubate at 37 °C for 1 h.
6. Wash ELISA plate five times with PBS–Tween and add 200 μL/well IgG D1A-PO peroxidase-conjugated antibody diluted in PBS–Tween–BSA. Incubate at 37 °C for 1 h.
7. Wash ELISA plate six times with PBS–Tween, then immediately add 200 μL/well of substrate–chromogen solution. Incubate at room temperature for 30 min in dark.
8. Stop peroxidase reaction by adding 50 μL/well of stopping solution.
9. Measure absorbance in a plate reader at 492 nm.

10. Subtract blank from all measures, plot the absorbance and corresponding antigen concentrations of the standard curve, and generate a linear regression curve and the conversion equation.
11. Calculate the rRVGP content in samples by applying the mean absorbance of the duplicate samples to the conversion equation (*see* **Note 13**).

4 Notes

1. To reach the number of cells required for bioreactor seeding, approximately 10 or 11 shaker flasks with 20 mL of working volume are necessary.
2. For further details, *see* refs. 2, 3.
3. While preparing cell suspension, let the culture medium hydrate the MCs as this allows for a better distribution of the cells in the fixed bed (approximately 10 min.).
4. The working volume of the iCELLis nano is 800 mL: this volume is important to minimize the CO_2 degassing through falling film during the 4 days of the culture. The minimal working volume is 600 mL.
5. Once media is at the required temperature, a recalibration of the DO probe can be performed in one point (end point—100%).
6. The MC sampled has to be treated with 1 mL of lysis solution for 20 min at 37 °C. The citrate lyses the cell membrane and releases the nuclei. Perform the nuclei counting in a hemocytometer, calculating the number of the cells/MC and then multiply by 3544 (number of MS) to obtain the total cell number (*see* ref. 7)
7. To rRVGP synthesis, the S2 cells must be induced with 700 μM $CuSO_4$ at 3 to 5 × 1E6 cells/mL of total medium.
8. As the desired glycoprotein is on the cell membrane, it is necessary to recover the cells from the fixed bed to harvest the product.
9. For further details, *see* ref. 6.
10. The ELISA assay can be performed with lower volumes but at a narrow range of quantification in this specific case. The ELISA could be performed thus with 100 μL volume of antibodies, the same volume for blocking solution and 25 μL of stopping solution.
11. The amount of lysis buffer can be changed depending on sample characteristics. The usual amount of cells used in this

test is 1E6 cells. Higher or lower amounts of cells tend to generate less reproductive results due to incomplete solubilization of RVGP or due to problems of losing cell pellet when manipulating lower amounts of cells. Also, other detergents can be used, as CHAPS 0.6%. However, every detergent used has to be previously tested for assay performance, as in some cases the detergent used can interfere with absorbance or with assay components, especially with the capture antibodies.

12. Avoid bubbles, which are harmful for protein samples.
13. This ELISA assay was designed to quantify the glycoprotein content in rabies vaccines (*see* ref. 8). The intended use was to analyze vaccines by the amount and quality of the main antigen. The use of monoclonal D1 antibodies for both capture (primary antibody) and antigen detection (secondary antibody) is the base of the strategy to quantify only the trimeric form of the molecule, as the antigen site III targeted by D1 is present as a linear epitope in each monomeric RVGP chain. The dimeric form of RVGP is very unstable, so its quantification is not considered in the method.

Acknowledgments

This work was supported in part by Artelis, ATMI LifeSciences, and PALL Life Sciences, and by a grant of CNPq (402439/2013-9). Carlos A Pereira is a recipient of CNPq 1A research fellowship.

References

1. Loeffelholz C, Kaiser SC, Kraume M, Eibl R, Eibl D (2014) Dynamic single-use bioreactors used in modern liter-and m3 – scale biotechnological processes: engineering characteristics and scaling up. Adv Biochem Eng Biotechnol 138:1–44
2. Ventini DC, Astray RM, Lemos MAN, Jorge SAC, Riquelme CC, Suazo CAT, Tonso A, Pereira CA (2010) Recombinant rabies virus glycoprotein synthesis in bioreactor by transfected Drosophila melanogaster S2 cells carrying a constitutive or an inducible promoter. J Biotechnol 146:169–172
3. Ventini-Monteiro D, Dubois S, Astray RM, Castillo J, Pereira CA (2015) Insect cell entrapment, growth and recovering using a single-use fixed-bed bioreactor. Scaling up and recombinant protein production. J Biotechnol 216:110–115
4. Astray RM, Jorge SAC, Pereira CA (2017) Rabies vaccine development by expression of recombinant viral glycoprotein. Arch Virol 162:323–332. (in press)
5. Moraes AM, Jorge SAC, Astray RM, Suazo CAT, Calderón Riquelme CE, Augusto EFP, Tonso A, Pamboukian MM, Piccoli RAM, Barral MF, Pereira CA (2012) Drosophila melanogaster S2 cells for expression of heterologous genes: from gene cloning to bioprocess development. Biotechnol Adv 30:613–628
6. Astray RM, Augusto E, Yokomizo AY, Pereira CA (2008) Analytical approach for extraction of recombinant membrane viral glycoprotein from stably transfected Drosophila melanogaster cells. Biotechnol J 3:98–103
7. van Wezel AL (1967) Growth of cell-strains and primary cells on microcarriers in homogenous culture. Nature 216:64–65
8. Perrin P, Lafon M, Sureau P (1996) Enzyme linked immuno-sorbent assay (ELISA) for the determination of glycoprotein content of rabies vaccines. In: Meslin FX, Kaplan MM, Koprowski H (eds) Laboratory techniques in rabies. WHO, Geneva, pp 383–388

Chapter 8

Cell-Free Production of Protein Biologics Within 24 H

Challise J. Sullivan, Erik D. Pendleton, and John Dresios

Abstract

Protein biologics have emerged as a safe and effective group of drug products that can be used in a variety of medical disorders and clinical settings, including treatment of orphan diseases, personalized medicine, and point-of-care applications. However, the full potential of protein biologics for such applications will not be realized until there are methods available for rapid and cost-effective production of small scale products for individual needs. Here, we describe a modular and scalable method for rapid and adaptable production of protein-based medical products at small doses. The method includes cell-free synthesis of the protein target in a reactor module followed by a fluidic process for protein purification. As a proof of concept, we describe the application of this method for expression and purification of a bioactive pharmaceutically relevant protein biologic, recombinant human erythropoietin, at a single dose within 24 h. This method can be applied toward the development of automated platforms for rapid and adaptive production of protein biologics at the point of care in response to specific medical needs.

Key words Cell-free protein synthesis, Protein expression, Protein biologics, Erythropoietin, Protein purification

1 Introduction

At present, production of protein biologics is primarily realized in cell-based prokaryotic and eukaryotic expression systems, including bacterial, yeast, and mammalian cells. However, these systems require multiple processes, large bioreactors, lengthy production cycles, laborious purification protocols, specialized facilities, and high costs [1–3]. In addition, cell-based production approaches suffer from factors that affect cell viability, therefore are not suitable for expressing toxic proteins or products that can be deleterious to the host when expressed at high levels. Furthermore, these approaches are impractical for employment at the point of need due to their stringent requirements for optimum cell growth, dedicated facilities and long end-to-end processes [4].

Recent advances in synthetic biology have allowed the development of prokaryotic and eukaryotic cell-free protein synthesis (CFPS) systems amenable to rapid, flexible, adaptable, and cost-

Virgínia Picanço-Castro and Kamilla Swiech (eds.), *Recombinant Glycoprotein Production: Methods and Protocols*, Methods in Molecular Biology, vol. 1674, DOI 10.1007/978-1-4939-7312-5_8, © Springer Science+Business Media LLC 2018

effective manufacture of protein products [4–8]. In addition, such systems do not depend on cell viability allowing direction of energy resources toward the expression of specific protein targets as well as high yield expression of proteins that could otherwise be toxic to the host [9–17].

Here, we describe an experimental method for cell-free expression and purification of an active protein biologic (recombinant human erythropoietin, rhEPO) within 24 h. This process is modular, amenable to fluidic automation, and adaptable for rapid production of pharmaceutical proteins [4].

2 Materials

Unless otherwise stated, prepare all solutions using ultrapure water (sensitivity of 18 MΩ cm at 25 °C) and analytical grade reagents.

2.1 DNA Template for Cell Free Protein Synthesis

1. Plasmid: A construct encoding the protein sequence of rhEPO (Epoetin alfa; DrugBank Accession Number DB00016) was generated bearing part of the tobacco mosaic virus Omega sequence [18] and used to generate polymerase chain reaction (PCR) DNA templates for use in coupled transcription/translation in the CFPS system. The construct also contained a T7 promoter and a signal peptide sequence upstream of the mature rhEPO as well as three copies of a FLAG epitope between the signal peptide and the protein coding region [4].
2. Primers: Synthesize a forward primer with the sequence: 5′-GTGATTCATTCTGCTAACCAG-3′, and a reverse primer with the sequence: 5′-T_{50}GTTAGCAGCCGGATCTCAGT-3′.
3. PCR Reagents: High-fidelity DNA polymerase.
4. PCR Purification Kit.
5. 1.2% agarose gel.

2.2 Yeast Cell-Free Protein Synthesis Reaction

1. 100 mM NTP mix: 0.4 M solution of each ATP, GTP, UTP, and CTP in nuclease-free water, adjust pH using 5 N KOH to 7.0–7.2. Combine equal volumes of each 0.4 M solution of ATP, GTP, UTP, and CTP to yield the final 100 mM NTP mix.
2. 12.5 mM 20 amino acid mix: Place 25 mL of ultrapure water in a 50 mL Falcon tube. Weigh and combine components in the following order: 0.234 g of L-valine, 0.408 g of L-tryptophan, 0.330 g of L-phenylalanine, 0.262 g of L-isoleucine, 0.262 g of L-leucine, 0.242 g of L-cysteine, 0.298 g of L-methionine, 0.178 g of L-alanine, 0.348 g of L-arginine, 0.264 g of L-asparagine, 0.266 g of L-aspartic acid, 0.406 g of L-glutamic acid K salt, 0.150 g of L-glycine, 0.292 g of L-glutamine,

0.308 g of L-histidine, 0.365 g of L-lysine, 0.230 g of L-proline, 0.210 g of L-serine, 0.238 g of L-threonine, and 0.362 g of L-tyrosine. Bring volume to 40 mL with ultrapure water, vortex and incubate at 37 °C to get mix into solution (*see* **Note 1**).

3. Creatine phosphokinase: a storage buffer of 50% glycerol, 20 mM HEPES, pH 7.2. Dilute creatine phosphokinase to 5 mg/mL in the storage buffer.
4. T7 polymerase.
5. 80% glycerol.
6. CFPS Master Mix: Combine reagents at the following final concentrations in the 1 mL Reaction Mix as prepared in 3.2: 11% H_2O, 22 mM HEPES, pH 7.4, 120 mM potassium glutamate, 5.5 mM magnesium glutamate, 1.5 mM NTP mix, 0.08 mM 20 amino acid mix, 25 mM creatine phosphate, 1.7 mM DTT, 2 mM putrescine, 0.5 mM spermidine, and 0.4 mM cAMP.
7. 1 M mannitol.
8. Phosphate-buffered saline (PBS) 1× solution.
9. 3.5 kDa molecular weight cutoff regenerated cellulose dialysis membrane.
10. 0.010–0.030 ID polyethylene (PE) tubing fluidic systems.
11. Yeast lysate: prepare yeast extract using *Saccharomyces cerevisiae* source strain S288c [19] as described previously [4, 8] upon growing yeast cells on synthetic complete media (6.7 g/L yeast nitrogen base (YNB), 20 g/L glucose, 50 mM potassium phosphate buffer, pH 5.5, and 2.002 g/L synthetic complete amino acids supplement).
12. Anti-EPO antibody.
13. Goat anti-rabbit HRP secondary antibody.

2.3 Protein Purification

1. Anti-FLAG M2 magnetic beads.
2. TBST: Tris-buffered saline (TBS): 50 mM Tris–HCl and 150 mM NaCl in ultrapure water, pH 7.4. Add 0.1% Tween 20.
3. 0.1 M glycine, pH 3.0.
4. 1 M Tris–HCl, pH 8.0.
5. 5 M NaCl.
6. 1 M $CaCl_2$.
7. Enterokinase.
8. EKapture agarose.
9. SensoLyte Rh110, enterokinase-specific activity assay.
10. SephacrylS200 HR.

2.4 Protein Analyses through Western Blot

1. Sample Buffer, 4× premixed protein sample buffer.
2. 20× MOPS diluted to 1× in deionized water.
3. Reducing Reagent, 20× premixed reducing reagent.
4. Transfer Buffer: 10× Tris–Glycine Buffer diluted to 1× in deionized water + 20% methanol.
5. 10× TBS diluted to 1× in deionized water.
6. TBST: 10× TBS diluted to 1× in deionized water + 0.1% Tween.
7. 4–12% Bis-Tris gels.
8. SimplyBlue SafeStain.

2.5 Protein Activity Assay

1. TF-1 cells.
2. RPMI 1640 media supplemented with 10% Fetal Bovine Serum, 100 U/mL Penicillin/Streptomycin, and 2 ng/mL rhGM-CSF.
3. Cell Titer-Glo reagent (Promega Corporation).
4. Human EPO.

2.6 Protein Synthesis Reactor

The reactor is made of polydimethylsiloxane (PDMS) and is divided into a reaction and feeding chamber (1 mL, each) (Fig. 1a). The two chambers are separated by a 3.5 kDa molecular weight cut-off cellulose dialysis membrane. Both chambers are implemented with 0.010–0.030 ID polyethylene (PE) tubing systems to enable recirculating flow, fluidic connection with the feeding solution and reaction mixture, and reaction material recovery at various time points.

3 Methods

3.1 Generation of PCR Template for Cell-Free Protein Synthesis

1. Set up the PCR reactions in 20 μL reaction volume with final concentrations of 0.2 μM of each primer, 0.1 ng/μL of template DNA, 0.2 mM dNTPs, 0.05 U/μL PhusionHigh-Fidelity DNA Polymerase, and 1× High-Fidelity Buffer.
2. Run the PCR cycle with an initial denaturation at 98 °C for 60 s followed by 35 cycles of 98 °C for 10 s, 60 °C for 22 s, 72 °C for 20 s, and a final extension of 72 °C for 10 min.
3. Purify the PCR product using PCR Purification Kit according to the manufacturer's protocol, elute in 30 μL of Elution Buffer, and pool reactions.
4. Quantify the PCR product using a NanoDrop 1000 or similar equipment, and run on a 1.2% agarose gel to check the size and purity.
5. Aliquot the PCR amplified DNA template and store at −20 °C for use in CFPS reactions if desired (*see* **Note 2**).

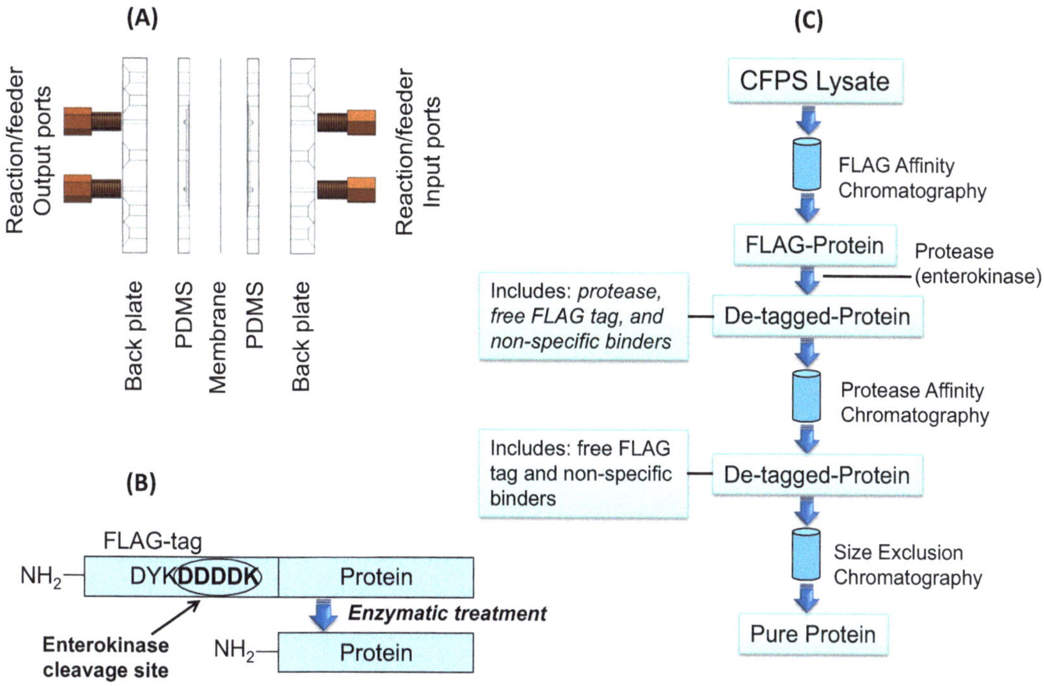

Fig. 1 Cell free protein synthesis (CFPS) and protein purification method. (**a**) Reactor design for CFPS under continuous exchange cell-free format (CECF). This set up involves incubation of the translation mixture in a dialysis chamber for replenishing substrates and removing low molecular-weight products by diffusional exchange across a semipermeable membrane. (**b**) A generalized protein purification method incorporates FLAG-affinity chromatography for protein separation out of a lysate with subsequent removal of the affinity tag through enzymatic (enterokinase) cleavage. (**c**) Protein purification scheme. Protein product in the presence of crude lysate is purified using anti-FLAG monoclonal antibodies immobilized on the surface of magnetic beads. Binding occurs under nondenaturing conditions and protein is eluted by lowering the pH or adding competing FLAG peptides. Protein product is detagged upon treatment with enterokinase which cleaves the FLAG epitope without requiring a specific linker sequence or leaving residual amino acids on the protein product. Enzyme is removed through affinity chromatography and protein target is polished using size exclusion chromatography

3.2 Yeast Cell-Free Protein Synthesis

1. Thaw all translation components on ice.
2. Reaction mixture: Combine the CFPS Master Mix with 0.27 mg/mL creatine phosphokinase, 20 U T7 polymerase, 11% glycerol, 100 ng PCR amplified DNA template, and 2.8 mg/mL S30 yeast extract (*see* **Notes 3–5**) in a 1 mL final volume transcription/translation reaction.
3. Feeder solution for continuous exchange protein synthesis (CECF): Combine the CFPS Master Mix with 0.27 mg/mL creatine phosphokinase, 11% glycerol, and 115 mM mannitol in 10 mL total volume (*see* **Note 6**).
4. Load reaction mixture and feeder solution into the CECF reactor (Fig. 1a) and incubate at room temperature for 4–6 h

under continuous recirculation of the feeder solution at a rate of 1 mL/min (*see* **Notes** 7 and **8**).

5. At the end of the reaction period, assess solubility of the expressed protein by removing 20 μL of reaction mixture, followed by centrifugation at 16,000 × g for 10 min at 4 °C. Remove the supernatant fraction and resuspend the insoluble fraction in PBS. Subject both samples to Western blot analysis (*see* below).
6. Assess the protein by running a Western blot according to standard protocol as follows.
   (a) Prepare 1–5 μL of CFPS final product in Sample Buffer plus Reducing Reagent in a final volume of 20 μL.
   (b) Heat samples for 10 min at 65 °C.
   (c) Load samples on a 4–12% Bis-Tris gel and run in 1× MOPS at 150 V for ~1 h.
   (d) Transfer at 100 V for 45 min in Transfer Buffer.
   (e) Block the membrane for 1 h in 1× TBS + 2% milk, then incubate with the anti-EPO primary antibody at a 1:500 dilution in TBST + 2% milk.
   (f) Wash four times, 5 min each, using TBST, then incubate with the goat anti-rabbit HRP secondary antibody at a 1:20,000 dilution in TBST + 2% milk.
   (g) Wash four times, 5 min each, using TBST and process using ECL Plus Reagents.

3.3 Protein Purification

A schematic description of the protein purification process is depicted on Fig. 1b and c. This approach utilizes FLAG-tag-based affinity chromatography for extraction of the expressed protein out of the cell-free system followed by protease-based detagging, protease enzyme removal through affinity chromatography, and elimination of the affinity tag and other impurities via size exclusion chromatography (*see* **Note 9**).

1. Prepare Anti-FLAG M2 magnetic beads according to the manufacturer's protocol. Briefly, suspend resin through gentle mixing, then remove volume required for the experiment into a new tube and equilibrate suspension with 5 volumes of TBS. Upon thorough mixing, put the tube in the magnetic separator and collect the beads away from the rest of the mixture. Remove the suspension and equilibrate beads by adding 5 volumes of TBS. Mix and place tube in the magnetic separator. Discard the TBS buffer and repeat last step once again.
2. Mix Anti-FLAG M2 magnetic beads with the CFPS mixture at 1:10 v/v. Dilute mixture threefold with TBS in the presence of 0.1% Tween 20, and incubate at room temperature for 4 h, with gentle mixing on a rotating device (*see* **Note 10**).

3. Isolate captured material by placing the tube on a magnetic separator and remove the supernatant.
4. Wash the resin beads with TBS three times using 10× bead volume of TBS Buffer.
5. Elute the FLAG-tag protein using 0.1 M Glycine, pH 3.0 at 5× bead volume. Transfer the eluent immediately to a tube containing 15–30 μL of 1 M Tris, pH 8.0 to reequilibrate to a neutral pH. Test the pH using pH paper to determine where the pH of the mixed solution has neutralized (*see* **Note 11**). At this step, protein recovery may be assessed via Western blot (Fig. 2).
6. Add NaCl to a final concentration of 50 mM and $CaCl_2$ to a final concentration of 2 mM (this is done to adjust the sample to the composition of the enterokinase buffer used in the next step).
7. Add enterokinase enzyme at a ratio of 1 U of recombinant enterokinase to 100 ng of purified FLAG protein sample and incubate at room temperature for 4 h (*see* **Note 12**).
8. Remove enterokinase from the protein sample using Ekapture Agarose, an affinity-based capture method based on an agarose matrix that is embedded with soybean trypsin inhibitor, according to the manufacturer's protocol. According to this protocol, mix EKapture Agarose slurry (use 50 μL of the 50% Ekapture Agarose slurry per one unit of enterokinase) followed by centrifugation at 1000 × *g* for 5 min. Discard the supernatant, resuspend the pellet in 10 vol 1× rEK Cleavage/Capture Buffer, and repeat the previous step. Discard the supernatant and resuspend pellet in 1 vol 1× rEK Cleavage/Capture Buffer.
9. Mix EKapture Agarose and the enterokinase-treated sample and incubate at room temperature for 5 min. Remove the EKapture Agarose through centrifugation at 1000 × *g* for 5 min in a Spin Filter and collect the cleaved target protein into the filtrate tube (*see* **Note 13**).
10. Load protein sample on a 2 mL Sephacryl S200 HR size exclusion chromatography column and purify the detagged protein using a flow rate of 35 μL/min.
11. Test purified products through a 4–12% Bis-Tris gel in 1× MOPS at 150 V for ~1 h. Rinse the gel with deionized water, stain it using SimplyBlue SafeStain for 1 h at room temperature under moderate shaking, and wash it with water for 3 h prior to imaging. In addition, a Western blot analysis can be performed at this step (Fig. 3) if desired as described above (*see* **Note 14**).

3.4 Protein Activity Assay

1. Measure biological activity of purified rhEPO using a TF-1 cells proliferation assay [20]. More specifically, maintain TF-1 cells in T225 flasks using supplemented RPMI 1640 media as

described in Subheading 2.5. Grow the cells at 37 °C in the presence of 5% CO_2.

2. Wash cells three times using 10–20 mL growth medium without GM-CSF and plate cells in a 96-well culture plate with a density of 2×10^4–1×10^5 cells/mL.
3. To each well in the 96-well plate, add serial dilutions of the test sample or control protein (e.g., commercially available human EPO) at a final volume of 100 μL/well. Control sample can be serially diluted over a wide range of concentrations (e.g., tenfold increments ranging from 100 to 0.01 ng/mL).
4. Incubate plates at 37 °C for 48–72 h. Remove plates from 37 °C and allow them to equilibrate to room temperature for approximately 30 min.
5. Prepare Cell Titer-Glo reagent according to the manufacturer's protocol. Add 100 μL of room temperature Cell Titer-Glo reagent per well, and shake the plates at 900 rpm for 2 min. Cover the plate during shaking using a plate cover or adhesive seal.
6. Incubate plates for 10 min at room temperature. Measure luminescence using a microplate reader and analyze data using GraphPad Prism v6 or similar (Fig. 4).

4 Notes

1. When making the 20 amino acid mix, the solution becomes somewhat insoluble after addition of L-tyrosine. This mix may take some time of vigorous vortexing and incubation at 37 °C to bring into solution.
2. When generating the PCR template for use in the CFPS reactions, run 8–16 PCR reactions to pool in order to make enough DNA product to use in CFPS. It is beneficial to make a larger pool to store, rather than to perform the PCR each time that template is needed for CFPS. The templates are stable for long term at −20 °C.
3. When mixing components for the CFPS reactions, it is helpful to prepare a "CFPS Master Mix" containing the HEPES, Potassium Glutamate, Magnesium Glutamate, NTP mix, 20 amino acid mix, creatine phosphate, DTT, putrescine, spermidine, and cAMP. The volumes of these components are quite small to be added accurately per translation reaction. These can be added to the master mix in any order, vortexed and spun down before use.

4. For the remainder of the components of the CFPS reactions not mentioned in the previous note, combine these for each CFPS reaction in the following order: H_2O, CFPS Master Mix, glycerol, PCR template (vortex after adding), creatine phosphokinase, T7 polymerase, and yeast extract (pipette mix really well after adding these final components, do not vortex).
5. All CFPS reagent stock solutions should be aliquoted and stored for long term use at -80 °C with the exception of creatine phosphokinase and T7 polymerase, which should be stored at -20 °C. All CFPS components must be thawed on ice prior to use.
6. Coupled transcription/translation reaction can be performed under various formats, including continuous exchange, semi-continuous exchange or batch systems. The former format can be used as a means to prolong the reaction and generate higher yields of the protein of interest [12]. Similarly, the reaction can take place in a reactor device designed for CECF or a simple tube for the latter format. Removal of byproducts and replenishment of exhausted components in the reaction chamber is realized by continuous dialysis through a semipermeable membrane and a feeding solution (Fig. 1a). The device can be implemented with fluidic modules to allow recirculating flow and fluidic integration with upstream reagents mixing and downstream purification subsystems [4].
7. Incubation time for the CFPS reaction can be optimized by performing time course experiments to determine the optimal incubation for the application.
8. The CFPS reactions can be modified for production of glycosylated proteins by including microsome fractions isolated mammalian cells. These fractions must contain endoplasmic reticulum, Golgi, and Signal Recognition Particle (SRP) allowing for glycosylation of the expressed protein [21, 22].
9. This affinity chromatography method involves incorporation of FLAG, a hydrophilic octapeptide (DYKDDDDK) at the N-terminus of the protein. This sequence contains the inherent enterokinase cleavage site located within the five C-terminal residues of the peptide sequence (DDDDK); cleavage by this enzyme recovers the protein product without carrying extra modifications (e.g., amino acid overhangs) in its N-terminus [4, 23].
10. The ratio of magnetic beads/expressed protein needs to be optimized to maximize target protein binding to the beads. A higher ratio of beads to protein product could increase protein

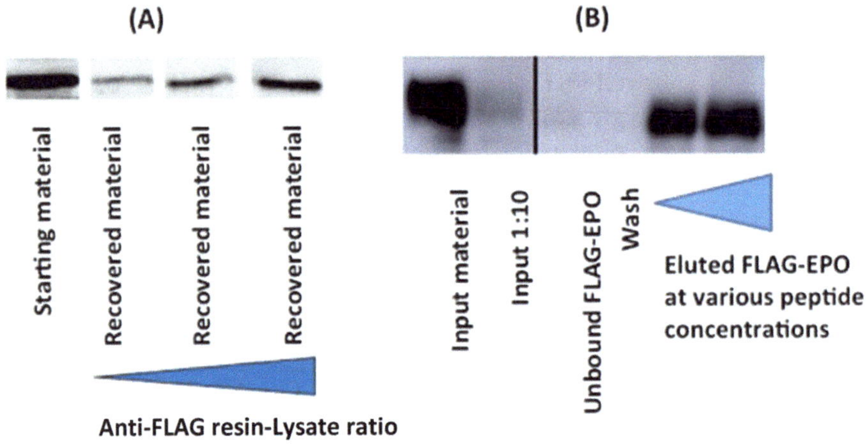

Fig. 2 Protein recovery from FLAG-tag affinity chromatography beads. (**a**) Protein purification yields increases by optimizing the anti-FLAG resin–lysate ratio for maximum protein binding to anti-FLAG beads followed by protein elution of the FLAG-tag protein using 0.1 M Glycine, pH 3.0. (**b**) Protein elution off an anti-FLAG column using various concentrations of competing free FLAG peptide

capture and overall purification yield, however, it may also increase elution volume and necessitate addition of downstream volume reduction steps. Also, incubation time needs to be optimized. Under the conditions described in this article, almost all soluble protein was bound to the beads and about 50% of the bound material was recovered in the eluent (Fig. 2a).

11. An alternative method to elute FLAG-tagged proteins from Anti-FLAG magnetic beads involves use of FLAG peptides. In this case, protein elution is carried out in a neutral buffer using FLAG peptides (final concentration of 100–1000 ng/μL) that compete off the protein for binding to the beads (Fig. 2b).

12. For enterokinase treatment, incubation temperature (4–37 °C), incubation time (up to 16 h) as well as enzyme concentration need to be optimized per application. The optimal conditions for enzymatic treatment will remove the FLAG tag without causing degradation of the protein target through nonspecific cleavage. In addition, before application of this method, the amino acid sequence of the target protein needs to be assessed to ensure the absence of enterokinase recognition sequences within its primary structure. The removal of FLAG-tag is assessed via Western Blot analysis using antibodies against FLAG-tag and antibodies against the protein target. The results in Fig. 3a and b show that enzymatic treatment results in cleavage of the FLAG epitope from the tagged EPO protein with no FLAG-EPO being detected under Western blot analysis.

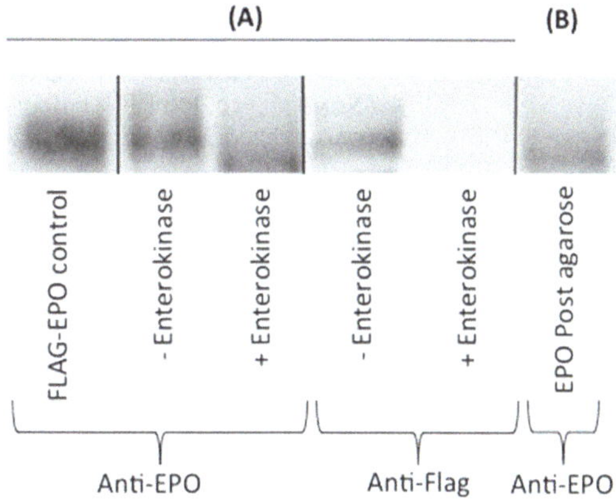

Fig. 3 Removal of FLAG-Tag during purification of cell-free expressed EPO. (**a**) Treatment of FLAG-EPO protein with enterokinase over a period of 4 h followed by Western blot analysis using anti-EPO or anti-FLAG antibodies. (**b**) Enterokinase-treated EPO sample blotted against anti-EPO antibody after enzyme removal through enterokinase agarose affinity chromatography

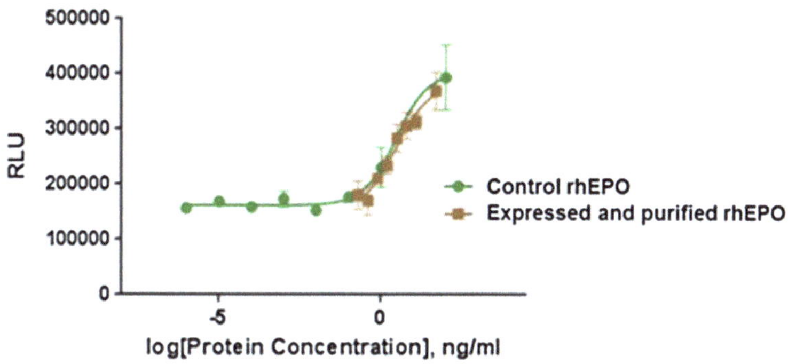

Fig. 4 Biological activity of protein product. rhEPO was synthesized in a CECF reactor hosting a yeast lysate and purified using the method described in this article. The biological activity of rhEPO was evaluated using a standard cell-based proliferation assay monitoring the ability of the target protein to induce proliferation of an EPO-dependent human cell line, TF-1. Purified rhEPO stimulated TF-1 proliferation in vitro at levels that were similar to those of commercial EPO, suggesting that end-to-end produced rhEPO has similar activity to that of the control protein. *RLU* relative light units

13. The efficiency of the enterokinase capture and removal step is assessed using a fluorogenic substrate-based enzymatic method (SensoLyte Rh110 Enterokinase Activity Assay) according to the manufacturer's protocol. Upon enzyme cleavage, this substrate generates a Rhodamine fluorophore (Rhodamine 110) that can be detected at excitation/emission = 490/520 nm.
14. Mass spectrometry analysis can also be used to assess production of a protein with the predicted amino acid sequence as well as confirm the presence of the correct N-terminal protein sequence upon FLAG peptide cleavage by enterokinase during the protein purification process [4]. For this, analyze protein samples in 4–12% Bis-Tris gels, followed by gel staining with SimplyBlue SafeStain (Life Technologies Corporation, CA). Protein bands migrating to the same molecular weight as the comparative control can be visualized using a lightbox and excised. Gel bands are destained and reduced (10 mM DTT) followed by alkylation (55 mM iodoacetamide), and overnight digestion with trypsin or Glu-C before they analyzed by nano-LC-MS/MS. In this study, mass spectrometry analysis was performed at the facilities of The Scripps Research Institute Center for Mass Spectrometry (TSRI, CA).

Acknowledgments

This work was supported by funding from the Defense Advanced Research Projects Agency (N66001-13-C-4024). The views expressed are those of the authors and do not reflect the official policy or position of the Department of Defense or the US Government.

References

1. Spirin AS (2004) High-throughput cell-free systems for synthesis of functionally active proteins. Trends Biotechnol 22:538–545
2. Gilbert M, Albala JS (2002) Accelerating code to function: sizing up the protein production line. Curr Opin Chem Biol 6:102–105
3. Mei Q, Fredrickson CK, Simon A, Khnouf R, Fan ZH (2007) Cell-free protein synthesis in microfluidic array devices. Biotechnol Prog 23:1305–1311
4. Sullivan CJ, Pendleton ED, Sasmor HH, Hicks WL, Farnum JB, Muto M, Amendt EM, Schoborg JA, Martin RW, Clark LG, Anderson MJ, Choudhury A, Fior R, Lo YH, Griffey RH, Chappell SA, Jewett MC, Mauro VP, Dresios J (2016) A cell-free expression and purification process for rapid production of protein biologics. Biotechnol J 11:238–248
5. Goerke AR, Swartz JR (2008) Development of cell-free protein synthesis platforms for disulfide bonded proteins. Biotechnol Bioeng 99:351–367
6. Kanter G, Yang J, Voloshin A, Levy S, Swartz JR, Levy R (2007) Cell-free production of scFv fusion proteins: an efficient approach for personalized lymphoma vaccines. Blood 109:3393–3399
7. Yang J, Kanter G, Voloshin A, Michel-Reydellet N, Velkeen H, Levy R, Swartz JR (2005) Rapid expression of vaccine proteins for B-cell lymphoma in a cell-free system. Biotechnol Bioeng 89:503–511

8. Hodgman CE, Jewett MC (2013) Optimized extract preparation methods and reaction conditions for improved yeast cell-free protein synthesis. Biotechnol Bioeng 110:2643–2654
9. Kim DM, Swartz JR (2001) Regeneration of adenosine triphosphate from glycolytic intermediates for cell-free protein synthesis. Biotechnol Bioeng 74:309–316
10. Katzen F, Chang G, Kudlicki W (2005) The past, present and future of cell-free protein synthesis. Trends Biotechnol 23:150–156
11. Swartz J (2006) Developing cell-free biology for industrial applications. J Ind Microbiol Biotechnol 33:476–485
12. Shirokov VA, Kommer A, Kolb VA, Spirin AS (2007) Continuous-exchange protein-synthesizing systems. Methods Mol Biol 375:19–55
13. Jewett MC, Calhoun KA, Voloshin A, Wuu JJ, Swartz JR (2008) An integrated cell-free metabolic platform for protein production and synthetic biology. Mol Syst Biol 4:220
14. Zawada JF, Yin G, Steiner AR, Yang J, Naresh A, Roy SM, Gold DS, Heinsohn HG, Murray CJ (2011) Microscale to manufacturing scale-up of cell-free cytokine production—a new approach for shortening protein production development timelines. Biotechnol Bioeng 108:1570–1578
15. Carlson ED, Gan R, Hodgman CE, Jewett MC (2012) Cell-free protein synthesis: applications come of age. Biotechnol Adv 30:1185–1194
16. Stech M, Brödel AK, Quast RB, Sachse R, Kubick S (2013) Cell-free systems: functional modules for synthetic and chemical biology. Adv Biochem Eng Biotechnol 137:67–102
17. Schoborg JA, Hodgman CE, Anderson MJ, Jewett MC (2014) Substrate replenishment and byproduct removal improve yeast cell-free protein synthesis. Biotechnol J 9:630–640
18. Gan R, Jewett MC (2014) A combined cell-free transcription-translation system from Saccharomyces cerevisiae for rapid and robust protein synthesis. Biotechnol J 9:641–651
19. Mortimer RK, Johnston JR (1986) Genealogy of principal strains of the yeast genetic stock center. Genetics 113:35–43
20. Kitamura T, Tojo A, Kuwaki T, Chiba S, Miyazono K, Urabe A, Takaku F (1989) Identification and analysis of human erythropoietin receptors on a factor-dependent cell line, TF-1. Blood 73:375–380
21. Dresios J. Sullivan CJ, Pendleton ED, Sasmor HH, Hicks WL, Amendt EM, Griffey RH. Portable fluidic platform for rapid cell-free production of protein biologics. USPTO 2016/0230203A1
22. Brödel AK, Sonnabend A, Kubick S (2014) Cell-free protein expression based on extracts from CHO cells. Biotechnol Bioeng 111:25–36
23. Arnau J, Lauritzen C, Petersen GE, Pedersen J (2006) Current strategies for the use of affinity tags and tag removal for the purification of recombinant proteins. Protein Expr Purif 48:1–13

Chapter 9

Demonstration-Scale High-Cell-Density Fermentation of *Pichia pastoris*

Wan-Cang Liu and Ping Zhu

Abstract

Pichia pastoris has been one of the most successful heterologous overexpression systems in generating proteins for large-scale production through high-cell-density fermentation. However, optimizing conditions of the large-scale high-cell-density fermentation for biochemistry and industrialization is usually a laborious and time-consuming process. Furthermore, it is often difficult to produce authentic proteins in large quantities, which is a major obstacle for functional and structural features analysis and industrial application. For these reasons, we have developed a protocol for efficient demonstration-scale high-cell-density fermentation of *P. pastoris*, which employs a new methanol-feeding strategy—biomass-*stat* strategy and a strategy of increased air pressure instead of pure oxygen supplement. The protocol included three typical stages of glycerol batch fermentation (initial culture phase), glycerol fed-batch fermentation (biomass accumulation phase), and methanol fed-batch fermentation (induction phase), which allows direct online-monitoring of fermentation conditions, including broth pH, temperature, DO, anti-foam generation, and feeding of glycerol and methanol. Using this protocol, production of the recombinant β-xylosidase of *Lentinula edodes* origin in 1000-L scale fermentation can be up to ~900 mg/L or 9.4 mg/g cells (dry cell weight, intracellular expression), with the specific production rate and average specific production of 0.1 mg/g/h and 0.081 mg/g/h, respectively. The methodology described in this protocol can be easily transferred to other systems, and eligible to scale up for a large number of proteins used in either the scientific studies or commercial purposes.

Key words High-cell-density fermentation, Demonstration-scale, *Pichia pastoris*, Recombinant β-xylosidase, Biomass-*stat* strategy, Specific production rate

1 Introduction

High-cell-density fermentation (HCDF) of recombinant *Pichia pastoris* has attracted significant attention in recent years [1]. The P_{AOX1}-regulated fed-batch HCDF of *P. pastoris* is often divided into three stages: glycerol batch stage (S1), glycerol fed-batch stage (S2) and methanol fed-batch stage (S3). S1 and S2 are collectively referred to as the biomass production phase, as producing sufficient biomass is their main objective. S3 is referred to as the induction phase; during S3, methanol is added and the biosynthesis

Virgínia Picanço-Castro and Kamilla Swiech (eds.), *Recombinant Glycoprotein Production: Methods and Protocols*, Methods in Molecular Biology, vol. 1674, DOI 10.1007/978-1-4939-7312-5_9, © Springer Science+Business Media LLC 2018

of the heterologous protein is initiated [2]. Methanol is used as both the P_{AOX1} inducer and sole carbon source. Using optimal methanol feeding strategies, the yields of heterologous proteins could reach as high as 22 g/L [3] for intracellular production of the recombinant hydroxynitrile lyase (42-L fermenter) and 14.8 g/L [4] for extracellular production of the recombinant gelatins (shaking flasks). Furthermore, it is easy for *P. pastoris* to achieve ultra-high cell densities (>100 g/L dry cell weight, DCW or >400 g/L wet cell weight, WCW) in a simple and low-cost base salt medium [5], and therefore, *P. pastoris* is more ideal than other expression systems for the fermentation with fermenters.

P. pastoris has become a versatile expression system for scaling up recombinant protein production [6–8], and a growing number of protein products are hitting the market [9, 10]. Baumgartner et al. investigated the production of recombinant phytohemagglutinin E-form (PHA-E of *Phaseolus vulgaris* origin) in *P. pastoris.* The yield of secreted PHA-E was approximately 100 mg/L from HCDF on the 2-L and 200-L scales [11]. They also produced agglutinin from *Galanthus nivalis* (GNA) on the 200-L scale, and the GNA was secreted at approximately 200 mg/L in the medium [12]. In order to produce an ice-binding protein (rLeIBP) from fed-batch culture of *P. pastoris*, Lee et al. developed a process to scale up HCDF. The recombinant protein yields on the 7-L and 700-L scales were ~272 mg/L and 300 mg/L, respectively [8]. These scaling-up processes were conducted with standard air pressure and oxygen enriched aeration was normally required.

Here, we describe in detail the practical steps for demonstration-scale HCDF of *P. pastoris* which intracellularly accumulates the heterologous bifunctional β-xylosidase/β-glucosidase (LXYL-P1–2) cloned from *L. edodes*. Using a new methanol-feeding strategy—biomass-*stat* strategy and a strategy of increased air pressure instead of pure oxygen supplement in 1000-L scale fermenter, production of the recombinant enzyme can be up to ~900 mg/L or 9.4 mg/g cells (dry cell weight, intracellular expression), with the specific production rate and average specific production of 0.1 mg/g/h and 0.081 mg/g/h, respectively. This information should allow any laboratories skilled in microbial fermentation technology to conduct the HCDF of *P. pastoris*, and provides a proper implementation for the development of heterologous proteins and, a successful paradigm for scaling up HCDF of *P. pastoris* to the demonstration scale.

2 Materials

2.1 Media

1. Stock Solutions: 10 × YNB (13.4% yeast nitrogen base with ammonium sulfate without amino acids), 500 × B (0.02% biotin), 10 × D (20% dextrose), 10 × GY (10% glycerol),

1 M potassium phosphate buffer, pH 6.0 (combine 132 mL of 1 M K_2HPO_4, 868 mL of 1 M KH_2PO_4 and confirm that the pH = 6.0 ± 0.1. If the pH needs to be adjusted, use phosphoric acid or KOH) and 50 mg/mL G418. Filter sterilizes all the stock solutions and store at −4 °C.

2. Yeast extract peptone dextrose (YPD) medium, 10 g yeast extract, 20 g peptone, and 20 g glucose, make up to 1 L. Add 20 g of agar if making YPD slants or plates. For YPD-G418 agar, use 50 mg/mL G418 stock solution to make YPD agar containing G418 at a final concentration of 4.0 mg/mL (*see* **Note 1**).
3. Buffered Glycerol-complex (BMGY) medium, 10 g of yeast extract, 20 g of peptone, 100 mM potassium phosphate buffer pH 6, 13.4 g of yeast nitrogen base with ammonium sulfate w/o amino acids, 400 μg of biotin, and 10 mL of glycerol, make up to 1 L.
4. FM22 medium, 42.9 g of KH_2PO_4, 5 g of $(NH_4)_2SO_4$, 1 g of $CaSO_4{\cdot}2H_2O$, 14.3 g of K_2SO_4, 11.7 g of $MgSO_4{\cdot}7H_2O$, and 40 g of glycerol, make up to 1 L.
5. *Pichia* trace minerals 4 (PTM4) salts solution, 2 g of $CuSO_4{\cdot}5H_2O$, 0.08 g of NaI, 3 g of $MnSO_4{\cdot}H_2O$, 0.2 g of $Na_2MoO_4{\cdot}2H_2O$, 0.02 g of H_3BO_3, 0.5 g of $CaSO_4{\cdot}2H_2O$, 0.5 g of $CoCl_2$, 7 g of $ZnCl_2$, 22 g of $FeSO_4{\cdot}7H_2O$, 0.2 g of biotin, and 1 mL of H_2SO_4, make up to 1 L. Filter sterilized and store in aliquots at −4 °C (*see* **Note 2**).

2.2 Strain

The engineered strain GS115-3.5 K-P1–2 was constructed in our laboratory by transforming the host strain *P. Pastoris* GS115 (Mut^+) with the recombinant plasmid pPIC3.5 K-LXYL-P1–2 (intracellular expression) harboring the sequence encoding LXYL-P1–2 [13]. The strain was stored at −80 °C in 20–30% glycerol solution (*w/v*) before use.

2.3 Fermenters

1. 200-L scale fermenter, which is used for cultivation of the second seed inoculum.
2. 1000-L scale fermenter, which is used for HCDF of *P. pastoris.*

All the fermenters are configured with automatic controllers of dissolved oxygen (DO, %), temperature (°C), pressure (MPa), pH, and the feeding of glycerol and methanol.

2.4 Substrates and Regents

1. 5 mM P-nitrophenyl-*β*-D-xylopyranoside (PNP-Xyl) in 50 mM acetate buffer, pH 5.0.
2. Regents:
   (a) Autoclave srerilized 50% glycerol solution (*w/v*), works with 12 mL/L PTM4.

(b) Absolute methanol, works with 12 mL/L PTM4.
(c) Anti-foam regent, JXPF-1208 (polyether type), sterilized by autoclaving.
(d) Ammonium hydroxide.
(e) Saturated sodium tetraborate ($Na_2B_4O_7$) solution.

3 Methods

3.1 Biomass Measurement

1. Weigh 2 mL centrifuge tubes (m_1).
2. Take 1 mL of aliquots to centrifuge tubes.
3. Centrifuge at 13201 × *g* for 3 min.
4. Wash cell pellets with distilled water.
5. Freeze-dry cells to constant weight and weigh (m_2).
6. Calculate dry cell weight (DCW) (*see* **Note 3**).

3.2 Enzyme Assay

1. Dilute fermentation broth if necessary (*see* **Note 4**).
2. Take 20 μL (V_1) of diluted aliquots to centrifuge tubes.
3. Add 50 μL (V_2) 5 mM PNP-Xyl.
4. Incubate at 50 °C for 20 min (t_1).
5. Add 1 mL (V_3) of saturated $Na_2B_4O_7$ solution to terminate reaction.
6. Centrifuge at 13,201 × *g* for 5 min.
7. Measure absorbance at a wavelength of 405 nm and record (OD_{405}).
8. Calculate enzyme activity (*see* **Note 5**).

3.3 Preparation of the First Seed Inoculum

1. Prepare YPD-G418 agar plates containing 4 mg/mL G418.
2. Streak GS115-3.5 K-LXYL-P1–2 strain from a glycerol stock onto an YPD-G418 agar plate.
3. Cultivate the strain in an incubator at 28–30 °C for 2–3 days.
4. Inoculate a single colony into 500-mL shaking flasks containing 100 mL of the BMGY medium.
5. Incubate the flasks at 28–30 °C for 20 h in a shaking incubator at 220 rpm.

3.4 Preparation of the Second Seed Inoculum

1. Wash 200-L scale fermenter and calibrate DO, temperature, pressure, and pH.
2. Sterilize the fermenter containing 100 L of the FM22 medium at 121 °C for 15 min.
3. Add 4.35 mL/L of PTM4 salts into 100 L of FM22 before inoculation.

4. Inoculate 1 L of the first seed culture into the fermenter.
5. Initiate cultivation at 30 °C, pH 5.0, air pressure 0.10 ± 0.05 MPa (super atmospheric pressure), and DO of 20% saturation until the second seed culture reached a cell density of ~25 g/L DCW (*see* **Note 6**).

3.5 Glycerol Batch Fermentation (S1)

1. Wash 1000-L scale fermenter and calibrate DO, temperature, pressure and pH.
2. Sterilize the fermenter containing 400 L of the FM22 medium at 121 °C for 15 min.
3. Add 4.35 mL/L of PTM4 salts into 400 L of FM22 before inoculation.
4. Inoculate 100 L of the second seed culture into the fermenter.
5. Initiate glycerol batch fermentation at 30 °C, pH 5.0, 0.10 ± 0.05 MPa, and DO of 20% saturation until all the glycerol is completely consumed.

3.6 Glycerol Fed-Batch Fermentation (S2)

1. Feed 50% glycerol, containing 12 mL/L PTM4 salts, at a constant rate of 18 mL/L/h. Fermentation conditions of temperature, pH, pressure, and DO in S2 are the same as those in S1 (*see* **Note 6**).
2. Continue glycerol feeding until the biomass reaches ~400 g/L DCW (*see* **Note 7**).
3. Switch to transitional cultivation without glycerol feeding for ~2 h.

3.7 Methanol Fed-Batch Fermentation (S3)

1. Set induction DO value as 1% (*see* **Note 8**).
2. Feed methanol, containing 12 mL/L PTM4 salts, at a rate of 0.035 mL/g/h (biomass-*stat*) (*see* **Note 9**).
3. Continue fed-batch fermentation until the maximum enzyme activity appears.

3.8 Cell Harvesting

1. Collect fermentation broth.
2. Centrifuge at 10881 g through a disk centrifuge.
3. Wash cells with distilled water.
4. Freeze-dry and store at −20 °C.

4 Notes

1. In most cases, high copy number integration has been shown to be important for high–level expression of foreign protein in *P. pastoris*. Under higher concentration of the antibiotic G418, high gene-copy number transformants following electroporation can be easily isolated. A series of YPD-G418 agars

containing 1 mg/mL, 2 mg/mL and 4 mg/mL G418 are often prepared to screen high copy number transformant. Transformants isolated from 4 mg/mL G418 are mostly preferred; however, this does not mean that the higher the G418 concentration, the higher the number of gene copies, and therefore the higher the expression level of protein [14, 15].

2. For preparation of PTM4, H_2SO_4 is lastly added after all other regents are almost completely dissolved in water. When H_2SO_4 is added, the color of mixture is instantaneously changed from murky to clear solution. After long time storage even at −4 °C, the solution may go to yellow. So,the PTM4 salts should be freshly prepared right before use.

3. The off-line biomass is calculated by the following equation: DCW = $m_2 - m_1$ (unit: g/L). In some cases, DCW is described as wet cell weight (WCW). In our experiment, 4 g of wet cells is nearly equal to 1 g of free-dried cells. On the other hand, biomass can be evaluated by the optical density at 600 nm (OD_{600}).

4. For accurate measurement, the broth is diluted especially at fermentation anaphase, which ensures OD_{405} value within the range of 0.3–0.7.

5. The chromogenic substrate PNP-Xyl is used to measure enzyme activity. This activity is then evaluated by calculating both U/L (volumetric enzyme activity) and U/g (biomass enzyme activity). One unit of enzyme activity is defined as the amount of enzyme required to release 1 nM *p*-nitrophenol from PNP-Xyl per minute at 50 °C and pH 5. Herein, $U/L = \frac{(V1+V2+V3)\times OD405}{17,500\times t1\times V1} \times N \times 10^6$, $U/g = \frac{U/L}{DCW}$, where 17,500 is the molar absorptivity ($L{\cdot}mol^{-1}{\cdot}cm^{-1}$), V_1, V_2, and V_3 are the volume of sample (diluted aliquots), saturated $Na_2B_4O_7$ solution and 5 mM PNP-Xyl solution (mL), respectively, t_1 is the reaction time (min), and *N* is the dilution times of fermentation broth.

6. Tank pressure is set as 0.05 MPa at the early stage. We prefer to maintain DO at certain level by increasing the jar air pressure through aeration and agitation instead of oxygen supplement. Aeration is adjusted within 0.5–2 *vvm*, and agitation is limited to 80% of allowed maximum working value. If DO level can't maintain even conditioned with maximum aeration and agitation, tank pressure would be increasingly adjusted from 0.05 MPa to 0.10 ± 0.05 MPa.

7. The initial induction biomass is one of the main factors that may affect protein expression level in *P. pastoris*. For LXYL-P1–2, we firstly determined the optimum initial induction biomass to be 75 g/L DCW in the pure oxygen supplemented experiment [16]. In the increased air pressure without pure

oxygen supplemented experiment, the optimum initial induction biomass was determined to be 100 g/L DCW [17]. Therefore, only after HCDF process optimization, the optimum initial induction biomass could be determined. At the beginning of the experiment, this value can be set within 75–100 g/L DCW.

8. Similar to initial induction biomass, the optimum induction DO value also should be optimized. Generally, the relative saturated oxygen demand for yeast is not less than 20%; thus, DO of *P. pastoris* HCDF is commonly set as ≥20% (*Pichia* Fermentation Process Guidelines. Invitrogen, USA). However, it does not mean that the higher DO, the better protein can be produced. Sometimes the lower induction DO can even improve the expression of the heterologous proteins. We found that the expression level of LXYL-P1–2 in 1% DO was higher than that in 20% DO [17].
9. At S3, methanol is added under the biomass-*stat* strategy, meaning biomass is maintained at a nearly constant level during the methanol induction phase. Afterward, the methanol feeding rate (v) is determined according to the equation $v = V/m/t$; V is the feeding volume of methanol (mL), m is the biomass of DCW (g/L) and t is the feeding time (h). The main concept of the biomass-*stat* strategy is that biomass is defined and the relationship between biomass and methanol feeding rate is immediately quantified. Meanwhile, the constant methanol-feeding rate is adjusted in order to maximize the production of heterologous proteins. In the current work, v is determined as 0.035 mL/g/h; however, it is not a universal value [17]. For different fermentation processes, the optimum v can be determined through a series of process optimization experiments.

References

1. Schmideder A, Hensler S, Lang M, Stratmann A, Giesecke U, Weuster-Botz D (2016) High-cell-density cultivation and recombinant protein production with *Komagataella pastoris* in stirred-tank bioreactors from milliliter to cubic meter scale. Process Biochem 51(2):177–184
2. Looser V, Bruhlmann B, Bumbak F, Stenger C, Costa M, Camattari A, Fotiadis D, Kovar K (2015) Cultivation strategies to enhance productivity of *Pichia pastoris*: a review. Biotechnol Adv 33(6, part 2):1177–1193
3. Hasslacher M, Schall M, Hayn M, Bona R, Rumbold K, Luckl J, Griengl H, Kohlwein SD, Schwab H (1997) High-level intracellular expression of hydroxynitrile lyase from the tropical rubber tree *Hevea brasiliensis* in microbial hosts. Protein Expr Purif 11(1):61–71
4. Werten MW, van den Bosch TJ, Wind RD, Mooibroek H, de Wolf FA (1999) High-yield secretion of recombinant gelatins by *Pichia pastoris*. Yeast 15(11):1087–1096
5. Athmaram TN, Singh AK, Saraswat S, Srivastava S, Misra P, Kameswara Rao M, Gopalan N, Rao PV (2013) A simple *Pichia pastoris* fermentation and downstream processing strategy for making recombinant pandemic swine origin influenza a virus Hemagglutinin protein. J Ind Microbiol Biotechnol 40(2):245–255
6. Cereghino GP, Cereghino JL, Ilgen C, Cregg JM (2002) Production of recombinant proteins in fermenter cultures of the yeast *Pichia pastoris*. Curr Opin Biotechnol 13(4):329–332
7. Kong N, Mu X, Han H, Yan W (2009) Pilot-scale fermentation, purification, and

characterization of recombinant human Oncostatin M in *Pichia pastoris*. Protein Expr Purif 63(2):134–139

8. Lee JH, Lee SG, Do H, Park JC, Kim E, Choe YH, Han SJ, Kim HJ (2013) Optimization of the pilot-scale production of an ice-binding protein by fed-batch culture of *Pichia pastoris*. Appl Microbiol Biotechnol 97(8):3383–3393
9. Gasser B, Prielhofer R, Marx H, Maurer M, Nocon J, Steiger M, Puxbaum V, Sauer M, Mattanovich D (2013) *Pichia pastoris*: protein production host and model organism for biomedical research. Future Microbiol 8 (2):191–208
10. Bawa Z, Routledge SJ, Jamshad M, Clare M, Sarkar D, Dickerson I, Ganzlin M, Poyner DR, Bill RM (2014) Functional recombinant protein is present in the pre-induction phases of *Pichia pastoris* cultures when grown in bioreactors, but not shake-flasks. Microb Cell Factories 13(1):127
11. Baumgartner P, Raemaekers RJ, Durieux A, Gatehouse A, Davies H, Taylor M (2002) Large-scale production, purification, and characterisation of recombinant Phaseolus vulgaris phytohemagglutinin E-form expressed in the methylotrophic yeast *Pichia pastoris*. Protein Expr Purif 26(3):394–405
12. Baumgartner P, Harper K, Raemaekers RJ, Durieux A, Gatehouse AM, Davies HV, Taylor MA (2003) Large-scale production and purification of recombinant Galanthus nivalis agglutinin (GNA) expressed in the methylotrophic yeast *Pichia pastoris*. Biotechnol Lett 25(15):1281–1285
13. Cheng HL, Zhao RY, Chen TJ, Yu WB, Wang F, Cheng KD, Zhu P (2013) Cloning and characterization of the glycoside hydrolases that remove xylosyl groups from 7-beta-xylosyl-10-deacetyltaxol and its analogues. Mol Cell Proteomics 12(8):2236–2248
14. Pyati P, Fitches E, Gatehouse JA (2014) Optimising expression of the recombinant fusion protein biopesticide ω-hexatoxin-Hv1a/GNA in *Pichia pastoris*: sequence modifications and a simple method for the generation of multi-copy strains. J Ind Microbiol Biotechnol 41(8):1237–1247
15. Wang QH, Liang L, Liu WC, Gong T, Chen JJ, Hou Q, Yang JL, Zhu P (2017) Enhancement of recombinant BmK AngM1 production in Pichia pastoris by regulating gene dosage, co-expression with chaperones and fermenting in fed-batch mode. J Asian Nat Prod Res 19(6):1–14
16. Yu WB, Liang X, Zhu P (2013) High-cell-density fermentation and pilot-scale biocatalytic studies of an engineered yeast expressing the heterologous glycoside hydrolase of 7-beta-xylosyltaxanes. J Ind Microbiol Biotechnol 40(1):133–140
17. Liu WC, Gong T, Wang QH, Liang X, Chen JJ, Zhu P (2016) Scaling-up fermentation of *Pichia pastoris* to demonstration-scale using new methanol-feeding strategy and increased air pressure instead of pure oxygen supplement. Sci Rep 6:18439

Chapter 10

Large-Scale Transient Transfection of Suspension Mammalian Cells for VLP Production

Laura Cervera and Amine A. Kamen

Abstract

Large-scale transient transfection of mammalian cell suspension cultures enables the production of biological products in sufficient quantity and under stringent quality attributes to perform accelerated in vitro evaluations and has the potential to support preclinical or even clinical studies. Here we describe the methodology to produce VLPs in a 3L bioreactor, using suspension HEK 293 cells and PEIPro as a transfection reagent. Cells are grown in the bioreactor to 1×10^6 cells/mL and transfected with a plasmid DNA–PEI complex at a ratio of 1:2. Dissolved oxygen and pH are controlled and are online monitored during the production phase and cell growth and viability can be measured off line taking samples from the bioreactor. If the product is labeled with a fluorescent marker, transfection efficiency can be also assessed using flow cytometry analysis. Typically, the production phase lasts between 48 and 96 h until the product is harvested.

Key words Mammalian cells, Bioreactor, Transient transfection, Suspension culture, VLP production

1 Introduction

Transient transfection is a widely used technique to introduce foreign DNA into mammalian cells. Transient transfection has been typically used at low and medium scale; however, in the past decade, significant efforts have been made toward developing large-scale transient transfection protocols to produce biological products in enough quantity to enable preclinical and clinical studies, as it is the case for viral vectors such as AAV vectors [1].

Several DNA transfection reagents have been used to perform transient transfection, including calcium phosphate, cationic lipids, or cationic polymers. Calcium phosphate (CaPi) has been the most widely used method for transient transfection during the last 30 years [2, 3]. This method is cost-effective and has been used for large-scale transfections. Its only drawback is its incompatibility with suspension culture conditions in serum-free media requiring low-calcium content to prevent cell aggregation [4, 5]. Cationic

Virgínia Picanço-Castro and Kamilla Swiech (eds.), *Recombinant Glycoprotein Production: Methods and Protocols*, Methods in Molecular Biology, vol. 1674, DOI 10.1007/978-1-4939-7312-5_10, © Springer Science+Business Media LLC 2018

lipids, such as Lipofectamine, are very effective in terms of transfection efficiency but very expensive which limits their use in large-scale processes [4]. Following its discovery, the cationic polymer polyethylenimine (PEI) was used as an economic and efficient nonviral gene delivery vehicle into cells in vivo or cultured in vitro [6, 7]. PEI firstly gained attention in the context of gene therapy approaches [8]. There are several types of PEI, with different molecular weights, and the most widely used in transient transfection approaches is the 25kD linear PEI [9–11].

Transient transfection can be used to produce different products such as antibodies, recombinant proteins, virus-like particles and viral vectors. Large-scale production using both PEI and CaPi transfection reagents has been demonstrated at production scales ranging from 1 to 110 L [12]. For instance, successful generation of IgG antibody by transient transfection at 10 L scale in Wave bioreactors and 50 and 100 L scale in stirred tank bioreactors has been reported [13]. Titers exceeding 1 g/L have been attained by transient transfection [14]. The manufacturing of clinical grade complex biopharmaceuticals such as viral gene therapy vectors by transient transfection for human clinical trials has also been well-described (e.g., retroviral, lentiviral, and adeno-associated viral vectors) [15–20].

Virus-like particles are self-assembled particles that mimic the virus structure. However, they do not contain the virus genome and are thus noninfective virus particles. Due to their repetitive organized structure and particulate nature, they are very efficiently taken up by antigen-presenting cells giving rise to a potent immune response with stimulation of both arms of the immune system: humoral and cellular. The latter is often not achieved using conventional vaccines and is essential to combat chronic infections and diseases [21]. These unique VLP properties make them appealing in many aspects as an alternative to the existing vaccines, and are therefore the subject of intensive research representing an advanced vaccine technology platform [22–24].

2 Materials

2.1 For DNA Plasmid Amplification and Purification

1. Plasmid DNA solution pGag-EGFP [25] (*see* **Note 1**).
2. Competent *E.coli* cells.
3. Luria-Broth (LB) medium: 10 g/L peptone 140; 5 g/L yeast extract; and 5 g sodium chloride.
4. Antibiotic (*see* **Note 2**).
5. LB-Agar-antibiotic plates.
6. Megaprep kit (Qiagen).
7. NanoDrop (Thermo Scientific).

2.2 For Bioreactor Preparation and Sterilization

1. Phosphate-buffered saline solution (PBS).
2. Stirred-tank bioreactor (*see* **Note 3**).
3. Tampon solution pH = 7.
4. Tampon solution pH = 4.
5. Pyrex bottles.
6. Silicone tubing.
7. Clamps.
8. 0.22 μm air filters.
9. 0.22 μm liquid filters.
10. Air, nitrogen, and oxygen air lines.
11. Autoclave.
12. 5% w/v $NaHCO_3$ solution.

2.3 For Transient Transfection

1. HEK 293 3F6 cells (*see* **Note 4**).
2. Automated cell counter.
3. SFM4 TransFX-293 (HyClone) media supplemented with 4mM L-Glutamine (HyClone) (*see* **Note 5**).
4. PEIPro polyethylenimine (Polyplus-transfection® SA).
5. Plasmid DNA solution (100–500 ng/μL).
6. 125-mL disposable polycarbonate Erlenmeyer flasks.
7. Orbital shaker.
8. Humidified CO_2 incubator.

3 Methods

3.1 Preparation of Plasmids: Transformation into Competent E. coli Cells

1. Thaw competent *E. coli* cells on ice for 10 min.
2. Add 10% of the cell volume of plasmid DNA.
3. Incubate for 30 min on ice.
4. Induce a thermic shock putting the tubes 30 s at 42°C.
5. Place the tubes on ice for 90 s and add 0.9 mL of LB medium.
6. Incubate tubes at 37 °C for 1 h.
7. Disperse the culture on a LB-Agar-Antibiotic plate.
8. Incubate plates at 37 °C overnight.

3.2 Preparation of Plasmids: Amplification

1. Pick one colony of the plates and transfer it to a 2 mL of LB medium supplemented with the appropriate antibiotic placed in a 50mL falcon tube.
2. Incubate at 37 °C, shaking at 200 rpm for 8 h.

3. Transfer 1 mL of this culture in 1 L of LB medium supplemented with the appropriate antibiotic and distribute in four 1 L Erlenmeyer flasks.
4. Incubate at 37 °C, shaking at 200 rpm for 12–16 h (*see* **Note 6**).
5. Purify the plasmid DNA using the MegaPrep extraction kit (Qiagen). Quantify the purified plasmids by NanoDrop (Thermo Scientific) (*see* **Note 7**).
6. Dilute the plasmid in TE Buffer from the MegaPrep kit or Milli Q water to reach a stock concentration of 100ng/μL (*see* **Note 8**).

3.3 Cell Culture

1. Maintain cell culture in 125-mL disposable polycarbonate Erlenmeyer flasks in 20 mL of culture medium. Flasks need to be shaken at 110 rpm using an orbital shaker placed in an incubator maintained at 37°C in a humidified atmosphere of 5% CO_2 in air.

3.4 Bioreactor Preparation and Sterilization (See Note 9)

1. Calibrate the pump used to add the base in the bioreactor.
2. Calibrate the pH probe and put it in the bioreactor.
3. Fill the bioreactor vessel with PBS. Only the volume needed to soak the probes.
4. Calibrate the oxygen probe (previously connected to the bioreactor for at least 6 h to allow polarization) with the agitation in the bioreactor on.
5. Connect the bottle that will be used for the base ($NaHCO_3$) to one port of the bioreactor.
6. Connect one Pyrex bottle to another port of the bioreactor for inoculation.
7. Connect a sampling device to another port of the bioreactor that has a deep tube to ensure that the sample is taken from the middle of the culture.
8. Add one air filter of 0.22 μm at every air inlet or outlet and protect them from humidity before sterilization.
9. Clamp all the tubing.
10. Close all the ports on the lid of the bioreactor that are not in use.
11. Disconnect the pH and DO probes from the controller cables and protect the caps from humidity.
12. Remove the motor from the bioreactor and protect the cap from humidity.
13. Slightly open the bottles and open one of the ports on the lid of the bioreactor.

14. Autoclave the bioreactor vessel and all the bottles attached during 30 min at 121 °C.
15. Once the sterilization cycle is finished, close the bottles and the opened one port on the lid of the bioreactor.
16. Unclamp all the tubing.
17. Remove all the protection from humidity.
18. Place the motor on the bioreactor.
19. Connect the DO and pH probes to their respective controllers in the bioreactor stand.
20. Inside the biosafety hood filter the base ($NaHCO_3$) using a 0.22 μm filter and add it to the addition bottle. Pass the tube that goes from the base bottle to the bioreactor through the base pump and prime the tubing.

3.5 Inoculation of the Bioreactor

Prepare all the solutions inside the biological safety hood.

1. Empty the PBS that is inside the bioreactor using the bottle that will be used for inoculation.
2. Exchange the bottle filled with PBS with an empty bottle inside the biosafety hood.
3. In the biological safety hood; tare the 2-L glass bottle prepared for the inoculation and pour 1350 mL of SFM4 TransFX-293 (HyClone) media supplemented with 4 mM L-Glutamine (HyClone). Transfer the content of the bottle into the bioreactor by gravity or by overpressuring the bottle eventually. During the addition of the media clamp the inoculation tube and unclamp the inoculation tube only when the transfer into the bioreactor is carried out. When the addition of the media is completed clump again the tube.
4. Switch on the agitation (80–120 rpm) (*see* **Note 10**), temperature (37 °C), pH (7.2) and oxygen/air (40% of dissolved oxygen) and CO_2 controllers and wait until all the set points are reached. It is especially important that the temperature is at the optimal value to not add the cells in a medium too cold or too hot, as this could affect in their growth.
5. In the same inoculation bottle add the volume of inoculum necessary to reach the desired final concentration inside the bioreactor. This volume will depend on the growth in the amplification culture. As example, if the desired final concentration is 0.16×10^6 cells/mL a volume of 350 mL of inoculum needs to be added in the bioreactor following the same procedure as Subheading 3.3, **step 1**. Before transferring the cells inside the bioreactor switch off the agitation (*see* **Note 11**).
6. Resume the agitation and in the same inoculation bottle and using the same procedure as in Subheading 3.3, **step 1**, add

100 mL of SFM4 TransFX-293 (HyClone) media supplemented with 4mM L-Glutamine (HyClone) to clean the remaining of cells that might be stuck in any part of the tubing.

7. Follow the cell growth in the bioreactor to reach the desired cell density for transfection. In this process, the cell growth takes 48–72 h to reach 1.0×10^6 cells/mL (*see* **Note 12**).

3.6 Transient Transfection

See Fig. 1 for an overview of transfection process in bioreactor. Open all the bottles inside the biosafety hood.

1. Check cell viability. For cell transfection, the optimal viability should be higher than 80%.
2. For transient transfection, the reaction volume must be 10% of the final volume. In this case for a bioreactor with a final working volume of 2 L then a 200-mL reaction must be prepared.
3. Preparation of the complexes: In a 50-mL falcon tube mix the DNA for a final 1 μg plasmid/mL culture (4 mL) with 42 mL of SFM4 TransFX-293 (HyClone) media supplemented with 4mM L-Glutamine (HyClone).
4. Vortex for 10 s.
5. Add the PEIPro (Polyplus-transfection® SA) (4 mL) for a final mass ratio of DNA–PEI 1:2 (*see* **Note 13**).
6. Vortex for 10 s.
7. Incubate for 10–15 min to generate the complexes. Do not incubate for more than 15 min (*see* **Note 14**).
8. Transfer the complexes into the inoculation bottle. Use gravity or overpressure to transfer the complexes into the bioreactor.
9. In the same inoculation bottle add 100 mL of remaining volume of SFM4 TransFX-293 (HyClone) media supplemented with

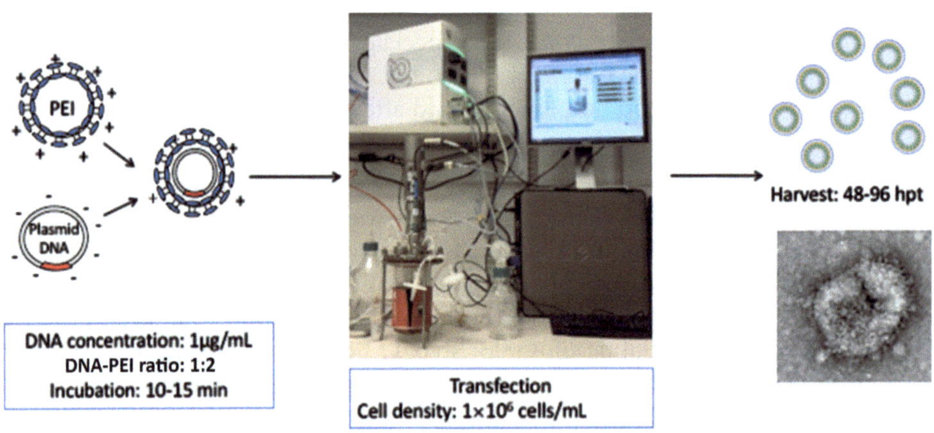

Fig. 1 Schematic overview of the transient transfection protocol at bioreactor scale

4 mM L-Glutamine (HyClone) to complete the 200 mL. This step ensures a complete transfer of the complexes into the bioreactor.

3.7 Production Monitoring

Online data acquisition is considered optional, but it can be used to document the process and demonstrate run reproducibility. Online monitoring parameters include:

- DO.
- pH.
- Temperature.
- Gas flow rates.
- Capacitance.

Off-line monitoring parameters might include:

- Flow cytometry analysis to determine the transfection efficiency (*see* **Note 15**).
- Cell density by automated cell counter that uses the established Trypan Blue exclusion method. In addition to cell concentration, some automated cell counter records viability and average cell diameter.

3.8 Harvest

The optimal time for harvest is between 48 and 96 h post transfection depending on the protein expressed before significant cell lysis occurs.

1. Harvest the culture using the bottle for inoculation and creating overpressure by clamping all the tubes except the one that goes to the inoculation bottle.
2. Separate the cells from the supernatant by centrifuging 15 min at 300 × *g* and 4 °C.
3. Decant supernatant and recover the VLPs (*see* **Note 16**).
4. Proceed with purification and quantification of the VLPs.

4 Notes

1. VLPs for different diseases have been developed: Hepatitis B [26], papillomavirus [27], influenza virus A [28], human immunodeficiency virus [29–31], human parvovirus [32], Norwalk virus [33], and rotavirus [34], among others. In this protocol the Gag protein of the HIV virus is produced. Gag protein self-assembles into particles that bud from the cell taking part of its lipid membrane generating enveloped VLPs [35]. The expression of the EGFP protein fused with the Gag protein enables for easy follow-up of the process and quantification of VLPs [36].

2. The selection of the antibiotic will depend on the plasmid resistance.
3. There are several stirred tank bioreactor brands in the market, for example, Applikon, Sartorius, EMD Millipore, and Eppendorf. The main features of the bioreactor are agitation and the controllers of temperature, pH, and dissolved oxygen. Online monitoring and data acquisition systems are considered optional. A computer with data acquisition hardware/software that can record several signals simultaneously online should be installed.
4. This protocol is optimized for HEK293-3F6 cell line, however it can be used for other types of cells (CHO, CAP) requiring some adaptation and optimization for better results.
5. Make sure that the medium you are using is compatible with PEI transient transfection. There are some serum-free media that contain anticlumping agents such as heparin or dextran sulfate polyanions that interfere with transfection. For example, the concentrations of dextran sulfate in Excel medium completely inhibit transfection (*see* refs. 30, 37). This limitation can be overcome by high concentrations of cells or polyplexes (*see* ref. 38).
6. If the culture is grown for less than 12 h contamination with RNA can occur and if it lasts for more than 16 h there is a possibility of contamination with proteins.
7. Check the 260/280 ratio. The optimal is around 1.8. Less means that there is contamination with proteins and a higher number means that there is a contamination with RNA.
8. A good practice is to perform a DNA gel with the plasmid digested with restriction enzymes to ensure its identity. Add one sample of undigested plasmid to observe that it is mainly in its supercoiled form.
9. This section of the protocol is applicable for the classical glass vessels. Nowadays, there are single-use bioreactors available that do not require previous sterilization [39]. Examples of this type of bioreactors are Wave (BD biosciences), Moebius (EMD Millipore), CelliGen® BLU series (Eppendorf), and single-use bioreactors from Sartorius.
10. Agitation set point will vary depending on the vessel shape.
11. Make sure the tube is not clumped before initiating the transfer.
12. To attain a good transfection efficiency cells have to be in the early-mid phase of the growth curve in order to assure a good competent state (*see* refs. 11, 30).

13. This ratio can be optimized for every product. Also consider that when more than one plasmid is used for transfection the total mass of plasmid should be 1 μg per mL of culture.
14. Make sure that the incubation time of the polyplexes is the same between batches. It is a critical point to have reproducible results between experiments.
15. The transfection efficiency can be followed by flow cytometry if the plasmid DNA expresses a fluorescent reporter protein or by mixing the working plasmid with 5% of a plasmid coding for a fluorescent protein.
16. Some VLPs are enveloped VLPs (e.g., Influenza and HIV 1), meaning that they are budding from the cells and released in the cell culture supernatant. In the case of nonenveloped VLPs, a cell lysis procedure will be needed to recover the product from the cell pellet.

Acknowledgments

The authors would like to acknowledge the funding from NSERC (RGPIN-2015-05132 and STPGP-462995-14) and Canada Research Chair (CRC-2403940).

References

1. Chahal PS, Schulze E, Tran R et al (2014) Production of adeno-associated virus (AAV) serotypes by transient transfection of HEK293 cell suspension cultures for gene delivery. J Virol Methods 196:163–173
2. Meissner P, Pick H, Kulangara A et al (2001) Transient gene expression: recombinant protein production with suspension-adapted HEK293-EBNA cells. Biotechnol Bioeng 75:197–203
3. Jordan M, Schallhorn A, Wurm FM (1996) Transfecting mammalian cells: optimization of critical parameters affecting calcium-phosphate precipitate formation. Nucleic Acids Res 24:596–601
4. Geisse S (2009) Reflections on more than 10 years of TGE approaches. Protein Expr Purif 64:99–107
5. Jordan M, Wurm F (2004) Transfection of adherent and suspended cells by calcium phosphate. Methods 33:136–143
6. Tait AS, Brown CJ, Galbraith DJ et al (2004) Transient production of recombinant proteins by Chinese hamster ovary cells using polyethyleneimine/DNA complexes in combination with microtubule disrupting anti-mitotic agents. Biotechnol Bioeng 88:707–721
7. Boussif O, Lezoualc'h F, Zanta MA et al (1995) A versatile vector for gene and oligonucleotide transfer into cells in culture and in vivo: polyethylenimine. Proc Natl Acad Sci U S A 92:7297–7301
8. Lungwitz U, Breunig M, Blunk T et al (2005) Polyethylenimine-based non-viral gene delivery systems. Eur J Pharm Biopharm 60:247–266
9. Durocher Y, Perret S, Kamen A (2002) High-level and high-throughput recombinant protein production by transient transfection of suspension-growing human 293-EBNA1 cells. Nucleic Acids Res 30:E9
10. Derouazi M, Girard P, Van Tilborgh F et al (2004) Serum-free large-scale transient transfection of CHO cells. Biotechnolo Bioeng 87:537–545
11. Carpentier E, Paris S, Kamen A et al (2007) Limiting factors governing protein expression following polyethylenimine-mediated gene transfer in HEK293-EBNA1 cells. J Biotechnol 128:268–280

12. Baldi L, Muller N, Picasso S et al (2005) Transient gene expression in suspension HEK-293 cells: application to large-scale protein production. Biotechnol Prog 21:148–153
13. Tuvesson O, Uhe C, Rozkov A et al (2008) Development of a generic transient transfection process at 100 L scale. Cytotechnology 56:123–136
14. Backliwal G, Hildinger M, Chenuet S et al (2008) Rational vector design and multi-pathway modulation of HEK 293E cells yield recombinant antibody titers exceeding 1 g/l by transient transfection under serum-free conditions. Nucleic Acids Res 36:e96
15. Ausubel LJ, Hall C, Sharma A et al (2012) Production of CGMP-grade lentiviral vectors. BioProcess Int 10:32–43
16. Cornetta K, Reeves L, Cross S (2008) Production of retroviral vectors for clinical use. Methods Mol Biol 433:17–32
17. Merten OW, Charrier S, Laroudie N et al (2011) Large-scale manufacture and characterization of a lentiviral vector produced for clinical ex vivo gene therapy application. Hum Gene Ther 22:343–356
18. Wright JF (2008) Manufacturing and characterizing AAV-based vectors for use in clinical studies. Gene Ther 15:840–848
19. Wright JF (2009) Transient transfection methods for clinical adeno-associated viral vector production. Hum Gene Ther 20:698–706
20. Ayuso E, Mingozzi F, Bosch F (2010) Production, purification and characterization of adeno-associated vectors. Curr Gene Ther 10:1–14
21. Berzofsky JA, Ahlers JD et al (2004) Progress on new vaccine strategies against chronic viral infections. J Clin Invest 114:450–462
22. Grgacic EVL, Anderson DA (2006) Virus-like particles: passport to immune recognition. Methods 40:60–65
23. Ludwig C, Wagner R (2007) Virus-like particles-universal molecular toolboxes. Curr Opin Biotechnol 18:537–545
24. Noad R, Roy P (2003) Virus-like particles as immunogens. Trends Microbiol 11:438–444
25. Hermida-Matsumoto L, Resh MD (2000) Localization of human immunodeficiency virus type 1 Gag and Env at the plasma membrane by confocal imaging. J Virol 74:8670–8679
26. Zhou W, Bi J, Janson JC et al (2006) Molecular characterization of recombinant Hepatitis B surface antigen from Chinese hamster ovary and Hansenula polymorpha cells by high-performance size exclusion chromatography and multi-angle laser light scattering. J Chromatogr B Analyt Technol Biomed Life Sci 838:71–77
27. Harper DM, Franco EL, Wheeler C et al (2004) Efficacy of a bivalent L1 virus-like particle vaccine in prevention of infection with human papillomavirus types 16 and 18 in young women: a randomised controlled trial. Lancet 364:1757–1765
28. Bright RA, Carter DM, Daniluk S et al (2007) Influenza virus-like particles elicit broader immune responses than whole virion inactivated influenza virus or recombinant hemagglutinin. Vaccine 25:3871–3878
29. Zhang X, Wang X, Zhao D et al (2011) Design and immunogenicity assessment of HIV-1 virus-like particles as a candidate vaccine. Sci China Life Sci 54:1042–1047
30. Cervera L, Gutiérrez-Granados S, Martínez M et al (2013) Generation of HIV-1 Gag VLPs by transient transfection of HEK 293 suspension cell cultures using an optimized animal-derived component free medium. J Biotechnol 166:152–165
31. Buonaguro L, Tornesello ML, Tagliamonte M et al (2006) Baculovirus-derived human immunodeficiency virus type 1 virus-like particles activate dendritic cells and induce ex vivo T-cell responses. J Virol 80:9134–9143
32. Roldão A, Mellado MCM, Castilho LR et al (2010) Virus-like particles in vaccine development. Expert Rev Vaccines 9:1149–1176
33. Herbst-Kralovetz M, Mason HS, Chen Q (2010) Norwalk virus-like particles as vaccines. Expert Rev Vaccines 9:299–307
34. Vieira HLA, Estêvão C, Roldão A et al (2005) Triple layered rotavirus VLP production: kinetics of vector replication, mRNA stability and recombinant protein production. J Biotechnol 120:72–82
35. Buonaguro L, Buonaguro FM, Tornesello ML et al (2001) High efficient production of Pr55gag virus-like particles expressing multiple HIV-1 epitopes, including a gp120 protein derived from an Ugandan HIV-1 isolate of subtype A. Antiviral Res 49:35–47
36. Gutiérrez-Granados S, Cervera L, Gòdia F et al (2013) Development and validation of a quantitation assay for fluorescently tagged HIV-1 virus-like particles. J Virol Methods 193:85–95
37. Geng ZH, Nudson W, Davis L et al (2007) Optimizing medium for transient transfection. In: Smith R (ed) Cell technology for cell products. Springer Netherlands, Dordrecht

38. Backliwal G, Hildinger M, Hasija V et al (2008) High-density transfection with HEK-293 cells allows doubling of transient titers and removes need for a priori DNA complex formation with PEI. Biotechnol Bioeng 99:721–727
39. Eibl R, Kaiser S, Lombriser R et al (2010) Disposable bioreactors: the current state-of-the-art and recommended applications in biotechnology. Appl Microbiol Biotechnol 86:41–49

Chapter 11

Bioreactor-Based Production of Glycoproteins in Plant Cell Suspension Cultures

Tanja Holland and Johannes Felix Buyel

Abstract

Recombinant glycoproteins such as monoclonal antibodies have a major impact on modern healthcare systems, e.g., as the active pharmaceutical ingredients in anticancer drugs. A specific glycan profile is often necessary to achieve certain desirable activities, such as the effector functions of an antibody, receptor binding or a sufficient serum half-life. However, many expression systems produce glycan profiles that differ substantially from the preferred form (usually the form found in humans) or produce a diverse array of glycans with a range of in vivo activities, thus necessitating laborious and costly separation and purification processes. In contrast, protein glycosylation in plant cells is much more homogeneous than other systems, with only one or two dominant forms. Additionally, these glycan profiles tend to remain stable when the process and cultivation conditions are changed, making plant cells an ideal expression system to produce recombinant glycoproteins with uniform glycan profiles in a consistent manner. This chapter describes a protocol that uses fermentations using plant cell cultures to produce glycosylated proteins using two different types of bioreactors, a classical autoclavable STR 3-L and a wave reactor.

Key words Plant cell culture, Protein glycosylation, Stirred tank reactor, Cultivation conditions, Process monitoring

Abbreviations

cdO_2	Controller output for the oxygen control loop
dO_2	Dissolved oxygen
DW	Dry weight
FW	Fresh weight
OUR	Oxygen uptake rate
PCV	Packed cell volume
RT	Room temperature
STR	Stirred tank reactor
VVM	Volume per volume and minutes

Virgínia Picanço-Castro and Kamilla Swiech (eds.), *Recombinant Glycoprotein Production: Methods and Protocols*, Methods in Molecular Biology, vol. 1674, DOI 10.1007/978-1-4939-7312-5_11, © Springer Science+Business Media LLC 2018

1 Introduction

1.1 Background

Recombinant proteins have many day-to-day applications, e.g., as enzymes in the food and detergent industries, and as special ingredients in cosmetics [1–3]. However, the most important application of recombinant proteins is their role as active pharmaceutical ingredients (APIs) in drug formulations, e.g., for cancer therapy [4, 5]. The complex structure of recombinant proteins allows them to act specifically on a particular drug target, which can reduce the number and/or severity of side effects compared to simple, chemically synthesized small molecules such as acetylsalicylic acid [6, 7]. Unlike small molecules, recombinant proteins may also possess enzymatic activity, allowing the treatment of metabolic disorders such as Gaucher's disease [8]. The activity of recombinant proteins often depends on the type of glycans added during and after translation. Glycosylation can have a direct impact on activity, e.g., by affecting the binding affinity of an antibody to its receptor and thus its effector functions [9, 10], or controlling whether a protein can interact with its receptor at all [8]. The impact may also be indirect, e.g., by increasing protease resistance [11] and hence the serum half-life [12], thus prolonging the opportunity for proteins to exert their specific functions. However, the glycosylation of a protein can differ according to the expression host. Prokaryotic expression systems typically lack the ability to produce glycans, whereas yeasts and other simple eukaryotes often produce simple, repetitive structures such as high-mannose glycans. In contrast, higher eukaryotes such as insects, mammals and plants can decorate proteins with complex and hybrid glycans that have specific functions in vivo [13, 14]. Plant cell cultures are advantageous as an expression system because they combine the safety profile of plants with the contained cultivation typical of traditional fermenter-based processes [15]. Plant cells also tend to add homogenous and reproducible glycan structures to recombinant proteins [14, 16]. These advantages underlie the commercial success of ELELYSO (taliglucerase alfa), a recombinant form of the human enzyme glucocerebrosidase used to treat Gaucher's disease, which was the first recombinant pharmaceutical protein produced in plant cells to be approved for human use [17, 18].

1.2 Characteristics of Plant Cell Cultures

Plant cells tend not to grow as single cells but rather as clusters [19, 20]. These clusters can reach several millimeters in diameter and can grow either by cell division or cell elongation. The latter results in a wet biomass and volume increase, especially due to the expansion of the vacuole. Depending on the growth phase, plant cells have a spherical to cylindrical shape. Cells in the exponential growth phase undergo rapid division and adopt spherical or elliptical forms with a length of 50–100 μm, whereas cells at the end of the exponential growth phase grow mainly by elongation and tend to be more cylindrical, with a length of up to 200 μm [19, 21].

The tendency of plant cells to form aggregates can differ widely according to the cell type, age and growth conditions. Large cell clusters should be avoided when adapting cells to grow in suspension cultures because cells in the center of the clusters may suffer oxygen and nutrient limitation decreasing the product yield and cell viability. Identifying a suitable cell line, i.e., one exhibiting homogeneous growth with small cell cluster, can require the screening of a large number of suspension cultures derived from different calli. Furthermore, subcultures should only be prepared from cells in the exponential growth phase, where cell division reduces the cluster size.

When screening the expression of a recombinant protein, it is important to note that the productivity and growth characteristics of cell lines at the callus stage do not necessarily correspond to those observed in suspension culture. Fluorescent marker proteins can be used for the rapid identification of positive transgenic clones at the callus stage, but it is good practice to continue screening the suspension cell culture for recombinant protein expression under realistic production conditions, e.g., in stirred tank reactors (STRs), to ensure a compromise between protein production and cell growth rates [22, 23]. In any case, a major benefit of plant cell cultures over mammalian cells is the dominance of specific glycosylation patterns for secreted proteins or those retained in the endoplasmic reticulum. Holland et al. reported a highly homogeneous glycan profile comprising two major glycoforms (GnMXFi and GnGnXF) that together comprised 97% of the total for a monoclonal antibody [24]. The remaining 3% was vacuolar glycan profiles. For the same antibody an ER-targeting resulted in a 77% yield of the desired glycan prolife (Man9, Man8). The remaining glycan forms were attributed to modifications in the Golgi apparatus [22] (*see* Table 1).

1.3 Bioreactor Systems

Plant cells have been cultivated in various bioreactor formats, including STRs [25–27], wave reactors [28], wave and undertow reactors [29], bubble column reactors [29], single-use bubble column reactors [30], air-lift reactors [31], membrane reactors [32], and rotating drum reactors [33]. In many reports, STRs with large impellers and ring spargers have been used to reduce shear stress, delivering maximum biomass values of 60–70% packed cell volume (PCV). The different reactor designs have been discussed and compared elsewhere [34, 35].

As in other cell culture-based processes, and regardless of the bioreactor design, three main fermentation strategies are available: batch, fed-batch, and continuous. Batch fermentation is the most common process [34]. The reactor is filled with medium and then inoculated, and thereafter becomes a closed system except for the supply of oxygen and perhaps base/acid to control the pH. The cell

Table 1
Characteristics of plant cell cultures

Parameter	Property
Shape	Spherical or cylindrical
Size	20–200 μm
Cell growth	Cell division or cell elongation
Cell wall	Yes
Doubling time	16–100 h
Cell aggregation or clumping	Yes, cluster size in mm range
Oxygen uptake rate	5–10 mmol/L/h
K_la required in bioreactor operations	10–50·h^{-1}
Protein localization	Intracellular or secreted

culture undergoes a series of defined growth phases: lag, exponential, and stationary. Cell growth thus occurs under changing and sometimes unfavorable conditions, e.g., nutrient inhibition during the early stages of cultivation and nutrient limitation during the later stages. A fed-batch fermentation can overcome these problems by starting like a classical batch fermentation but with a lower nutrient concentration before switching to an additional feed mode once certain conditions are met, e.g., the levels of specific nutrients fall below a predefined threshold. A common drawback of batch and fed-batch fermentations is the unfavorable ratio of setup time to actual run time, i.e., system preparation, sterilization before cultivation, and final cleaning can account for 20–30% of the duration of a fermentation run. Although this can be compensated by investments in personnel and infrastructure, continuous fermentation strategies such as perfusion and chemostat processes achieve a higher space-time yield. In both these strategies, the cells in the bioreactor are continuously supplied with fresh medium while an equivalent amount of spent medium is withdrawn, resulting in a constant culture volume. The difference between the perfusion and chemostat processes is that the cell-free fermentation broth is harvested from the perfusion culture whereas the cell-containing bleed is collected during a chemostat process [36, 37]. The cell density therefore increases during perfusion and dead cells may accumulate, whereas the cell density remains constant in a chemostat allowing the process to run for longer with the cells maintained in the exponential growth phase [38], making this strategy an excellent choice for large-scale production [39, 40]. Despite the benefits of continuous fermentation, such processes are sensitive to influences from factors such as microbial contamination, genetic instability of the plant cell line, and changes in the biotic phase of the bioreactor [41]. To avoid these drawbacks, semicontinuous fermentation strategies have been developed in which a proportion of the fermentation broth is removed once and replaced with fresh medium [42, 43]. The merits of these different cultivation strategies have recently been reviewed [34].

Here, we discuss batch fermentations using plant cell cultures and provide protocols for two different reactor types: a classical autoclavable 3-L STR and a wave reactor. We also describe the steps needed to prepare each bioreactor and the preparation of a preculture which is used as the initial inoculum.

2 Materials

2.1 General Equipment

- Sterile inoculation bottle with a lower outlet equipped with transfer tubing and a sterile connector.
- Microscope, 10×/25×/40× objective lenses on a rotating turret and a 10× ocular lens.
- Centrifuge.
- Vacuum pump.
- 10-mL centrifuge tubes.
- Filter paper, 4–12 μm pore size.
- Aluminum foil.
- Sterile sample vessels.
- Bunsen burner.
- Filter capsule (0.45–0.20 μm) for the sterile filtration of cultivation medium.

2.2 Solutions and Media

Prepare all solutions with purified water at room temperature (~22 °C). Sterilize all media by autoclaving (121 °C, 20 min, and 1.0 bar).

- Trypan Blue solution, 0.4% (m/v) (ready to use).
- Preculture in the exponential growth phase to inoculate the bioreactor (Subheading 3.1).
- Antifoam stock: prepare a 10% (v/v) Pluronic L61 solution in purified water and store at room temperature.
- Thiamine stock: prepare a 4.0 mg/mL thiamine–HCl stock solution in purified water, filter-sterilize and store at 4 °C.
- 2,4-dichlorophenoxyacetic acid (2,4-D) stock: prepare a 2.0 mg/mL 2,4-D stock solution in dimethylsulfoxide (DMSO) and store at 4 °C.
- Murashige and Skoog (MS) medium: 4.3 g/L MS salt mix, 30 g/L sucrose, 250 μL/L thiamine stock, 100 μL/L 2,4-D

stock, 200 mg/L KH_2PO_4, 100 mg/L myo-inositol, and 0.01% (v/v) antifoam stock; pH range 5.2–5.8, adjusted with 1.0 M sodium hydroxide or potassium hydroxide (*see* **Note 1**).

2.3 Bioreactor

- 3-L autoclavable STR (Applikon Biotechnology, Schiedam, Netherlands) with a pitched blade impeller; a sintered metal L-shaped sparger; a pH sensor; a dO_2 sensor; a Pt-100; and a thermometer pocket for temperature measurement; a sampling sensor with sterile replaceable sampling vessels; and a septum holder.

  Optional: biomass sensor [21].
  Control unit: ez-control (Applikon).
  Supervisory control and data acquisition (SCADA) software: BioXpert XP (Applikon).
- Flexsafe RM optical bag wave bioreactor with optical sensors.
  Control unit: BIOSTAT control tower (Sartorius, Göttingen, Germany).
  Rocker: BIOSTAT RM (Sartorius).
  SCADA software: MFCS (Sartorius).

3 Methods

3.1 Preculture Preparation

The preculture used for inoculation should be in the exponential growth phase. To determine the PCV or fresh weight (FW) of exponentially growing cells, a growth curve should be recorded from shake-flask cultures based on samples each consisting of at least three biological replicates.

1. Inoculate at least three shake flasks with 5–10% (v/v) of a routine culture.
2. Take a 0 h sample and after 72 h of cultivation samples should be taken every 24 h.
3. Measure the %PCV by transferring a 10-mL sample of the fermentation broth to a graduated tube and centrifuging (3 min, 4000 × *g*, room temperature (RT, ~22 °C). The % PCV is the ratio of the volume of the cell pellet to the volume of the cells + medium × 100%. Determine the FW by vacuum filtering 10 mL of the fermentation broth through filter paper (pore size 4–12 μm) and weighing the cells retained on the filter paper. Then, dry the cells at 60 °C until the weight stays constant (at least 24 h) to determine the dry weight (DW).
4. Plot the %PCV, FW, and DW against the cultivation time to obtain the growth curve (Fig. 1).
5. Either fit an exponential function to the data and use the derivative of that fit to plot the growth rate, or visually

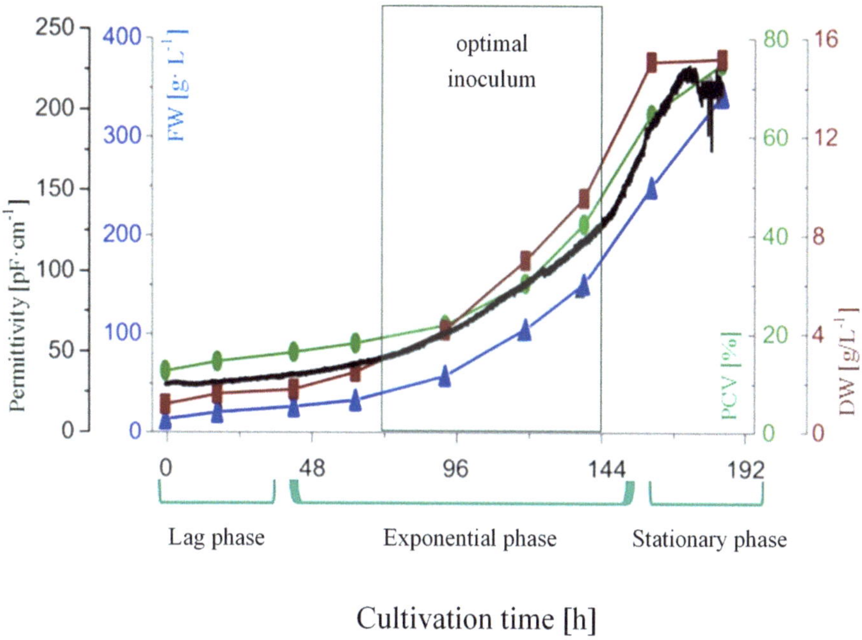

Fig. 1 Typical growth curve of tobacco BY-2 cells in suspension culture. The biomass signal (permittivity [21]), fresh weight (FW), dry weight (DW), and packed cell volume (%PCV) are plotted against the cultivation time [h]. The optimal inoculum for subsequent fermentation, e.g., in an STR, forms in the time window highlighted by the *green box* (the mid-exponential phase)

determine the optimal harvest time for the preculture, i.e., the mid-exponential phase.

3.2 Preparation of a Batch Fermentation: 3-L Autoclavable STR

1. Set points and control loop:

Temperature: the optimal set point is 26 °C and the range is 15–28 °C. Use a control loop for cooling and heating.
Stirrer speed: the range is 100–350 rpm, which can be maintained either at a fixed speed or in a cascade control loop, e.g., taking the dO_2 into account.

The dO_2 range is 20–30% saturation. The typical oxygen uptake rate (OUR) of plant cell cultures is 2–10 mmol O_2/L/h [35, 44]. The dO_2 can be maintained at 20% saturation by automatically pulsing pressurized air into the fermenter or by using a constant aeration rate and a controlled stirrer speed. Alternatively, a cascade can be established by pulsing pressurized air followed by increasing the stirrer speed.
Aeration rate: the range is 0.01–0.5 volume per volume and minute (vvm).

The pH range is 5.2–5.8. Monitoring the pH is necessary to check if characteristic regime is being followed, however controlling the pH is neither necessary nor favorable (*see* **Note 2**).

2. To ensure that all detergent and ethanol residues from earlier cleaning cycles are removed, the reactor vessel and all tubes should be flushed with a sufficient amount of purified water, e.g., 4–6 L for a 2-L STR (*see* **Note 3**).
3. Calibrate the pH sensor with two reference solutions covering the pH range.
4. Configure the bioreactor with all sensors, an inoculation line with a sterile connector, the aeration line with a sintered metal L-sparger, and a sampling line (*see* **Note 4**), making sure that all unused ports are closed with blind stoppers.
5. Install the air filter in the aeration line and check if the filter and aeration line is free.
6. Prepare the culture medium and transfer it to the bioreactor.
7. Before sterilization of the medium and fermenter, close all tubes in contact with the medium using a hose clamp, close the open end of all air filters with aluminum foil, and check for pressure compensation to ensure that the air outlet is open.
8. Autoclave the bioreactor together with all glass bottles (sampling vessels and inoculation bottle).
9. After autoclaving, connect the reactor and all sensors to the controller unit. Initiate the control loops for temperature and stirrer speed. Initiate aeration overnight before fermentation commences to saturate the medium with O_2.
10. Calibrate the dO_2 sensor by two-point calibration (0% and 100%). The 0% calibration point can either be achieved by using N_2 gas to purge all the oxygen from the fermentation medium or simply by disconnecting the sensor and using mock signal as the 0% value. Even though the second method is rather fast, it reduces the accuracy of the measurement to about ±5% dO_2 which can be tolerated if no oxygen limitation is expected during the fermentation. The 100% set point is obtained after overnight aeration of the medium before inoculation (*see* **Note 5**).
11. Begin data recording immediately prior to inoculation.

3.3 Preparation of a Batch Fermentation: Wave Bioreactor

1. Set points and control loop:

Temperature: the optimal set point is 26 °C and the range is 15–28 °C.

The shaking frequency should be 20–30 Hz.

The shaking angle should be 8–10°.

The dO_2 range should be 20–30% saturation and should be measured with the single-use optical sensor installed in the bag. The aeration rate should be 0.15–0.50 L/min using pressurized air.

The pH range should be 5.2–5.8. Check whether the wave reactor requires additional external pH monitoring because the range of the single-use optical sensor installed in the bag is often limited to 6.5–8.5.

2. Insert a Flexsafe RM optical bag in the rocker and connect the optical sensor and air tubes to the bag and to the controller unit.
3. Enter the calibration data from the optical sensors in the bag into the controller unit and configure the set point and control loops.
4. Prepare the cultivation medium and transfer the medium to the wave reactor bag by sterile filtration (0.45–0.20 μm) using a peristaltic pump.
5. Initiate the control loops for temperature and rocker frequency. Initiate aeration overnight before the fermentation commences to saturate the medium with O_2.
6. Begin data recording immediately prior to inoculation.

3.4 The Fermentation Process (STR and Wave Bioreactors)

1. Transfer the preculture into the inoculation bottle under a laminar flow hood and couple the sterile connector either under laminar flow or if possible using a Bunsen burner. When using a Bunsen burner, ensure that the connector has cooled sufficiently before coupling to avoid damaging the plant cells. The connector should be warm but not hot when touched, i.e., <40 °C (*see* **Note 6**). Start the inoculation by gravity flow (*see* **Note 6**).
2. Take a 0 h sample (~25–30 mL) after inoculation and stain with Trypan Blue by adding 0.1 mL 0.4% (m/v) Trypan Blue solution to 0.9 mL fermentation broth and incubating for 5 min. Afterward, determine the number of dead (stained) cells under a microscope and check for osmotic stress (plasmolysis) in the living cells (Fig. 2; *see* **Note 7**). The proportion of dead cells should not exceed 5%. Plasmolysis can prolong the lag phase because the cells require more time to adapt to the new medium.
3. Measure the %PCV, FW, and DW of a 10-mL sample as explained in Subheading 3.1, **step 3**.
4. Use the growth rate determined for the preculture (Subheading 3.1) to define the sampling interval. Take one or two samples during each doubling cycle of the cells. For example, if a culture has a doubling time of 24 h, take one or two samples during this period. Note the fermentation time (process time) for every sample point based on the SCADA software.
5. The termination of each fermentation process should be carefully planned. The late stationary phase is not suitable for

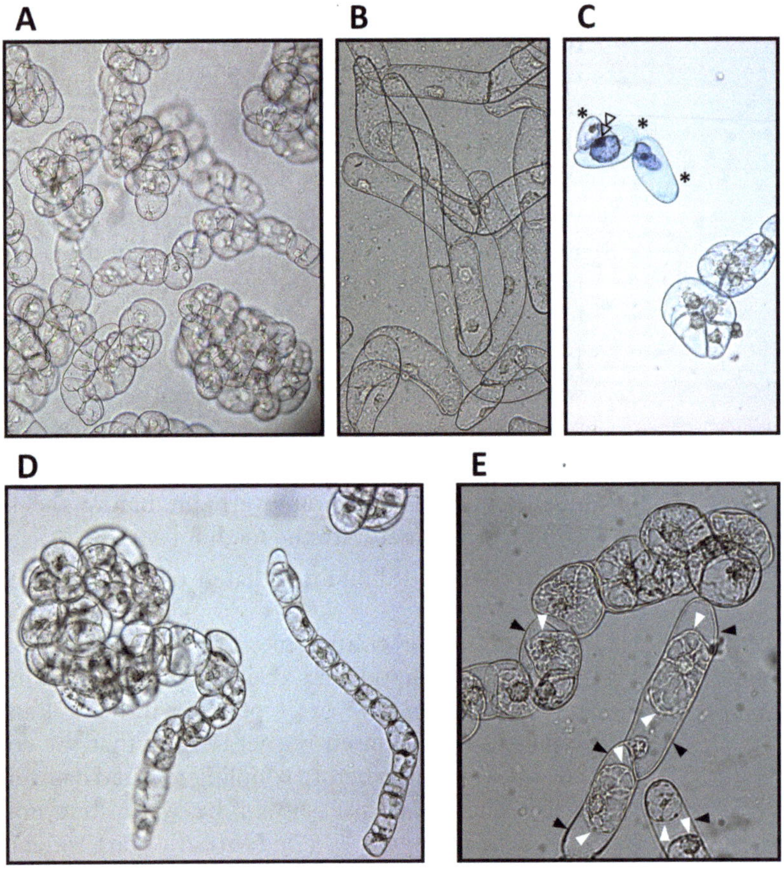

Fig. 2 Morphological characteristics of tobacco BY-2 cells in suspension culture. (**a**) Spherical cells in the exponential growth phase (5 days post-inoculation). (**b**) Cylindrical cells in the stationary phase (10 days post-inoculation). (**c**) Vital staining of a 7-day-old culture with Trypan Blue; the dead cells are stained in blue (*asterisk*) with more intensely stained nuclei (*double arrow*). (**d**) Different forms of aggregates, i.e., large clusters and long chains. (**e**) Plasmolysis of BY-2 cells caused by an increase in the osmolarity of the medium. Cell walls are marked with *black arrows* and cell membranes with *white arrows*

harvesting because intracellular proteolytic activity increases toward the end of cultivation (reducing the yield of intracellular proteins) and the risk of cell lysis increases (which can release incompletely processed proteins with intracellular glycoforms into the medium, and these are difficult to separate from the true product). The best harvest time is therefore the late exponential/early stationary phase. The dO_2 signal is an accessible indicator of the fermentation status. If a control loop with automatic pulses of pressurized air or a cascade is used to control the dO_2, the controller output provides an immediate readout. The controller output falls rapidly once the cells enter stationary phase due to the decline in metabolic activity, resulting in a

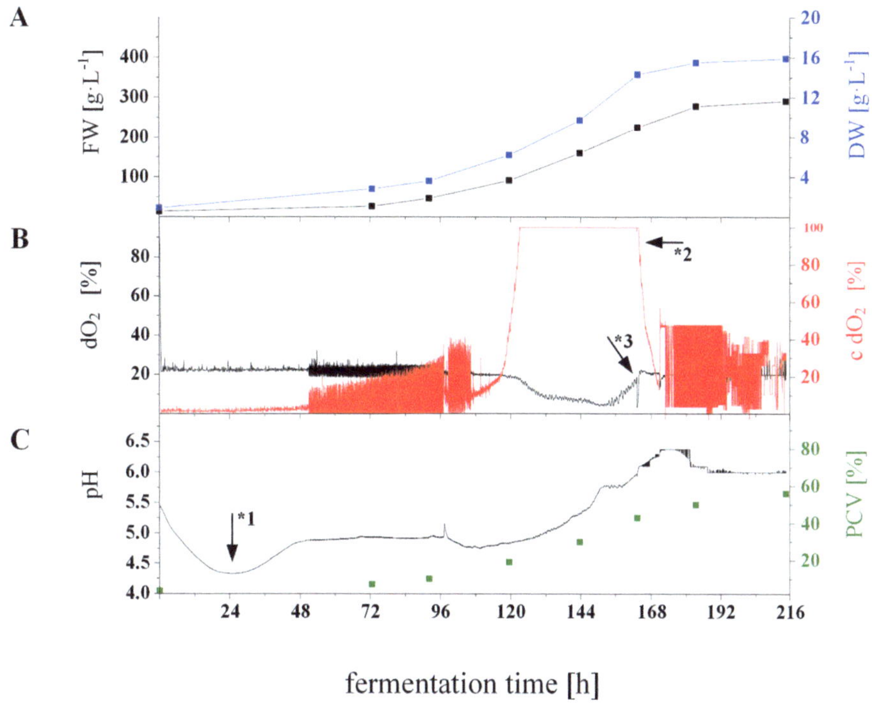

Fig. 3 Characteristics of tobacco BY-2 cell batch fermentation in a stirred tank reactor. (**a**) Fresh weight (FW, *black*) and dry weight (DW, *blue*) plotted against the fermentation time. (**b**) Dissolved oxygen concentration (dO_2, *black*) and the dO_2 controller output value (cdO_2) plotted against the fermentation time. (**c**) The pH (*black*) and percentage packed cell volume (%PCV, *green*) plotted against the fermentation time. Key events: **1*, typical pH drop; **2*, sharp decrease in the cdO_2 signal at the beginning of the stationary phase; **3*, increase in the dO_2 signal at the beginning of the stationary phase

reduced OUR (Fig. 3, *2). If constant air flow with a controlled stirrer speed is used instead, then the stirrer speed, controller output or the dO_2 signal itself can be used to determine the fermentation status and thus the optimal harvest time. The stirrer speed and controller output will decrease at the end of the fermentation whereas the dO_2 signal will increase once the cells enter the stationary phase (Figs. 3,*3; 4,*1). Notably, the dO_2 signal can oscillate during the transition from exponential growth phase to the stationary phase and readings should be accepted only when the signal is stable or increasing steadily (*see* **Note 8**).

6. Separate the biomass from the culture supernatant by vacuum filtration. Harvest the culture supernatant if the target is a secreted protein, or the filtered cells if the target is an intracellular protein. Intracellular and secreted proteins usually carry different glycan structures (Fig. 5).

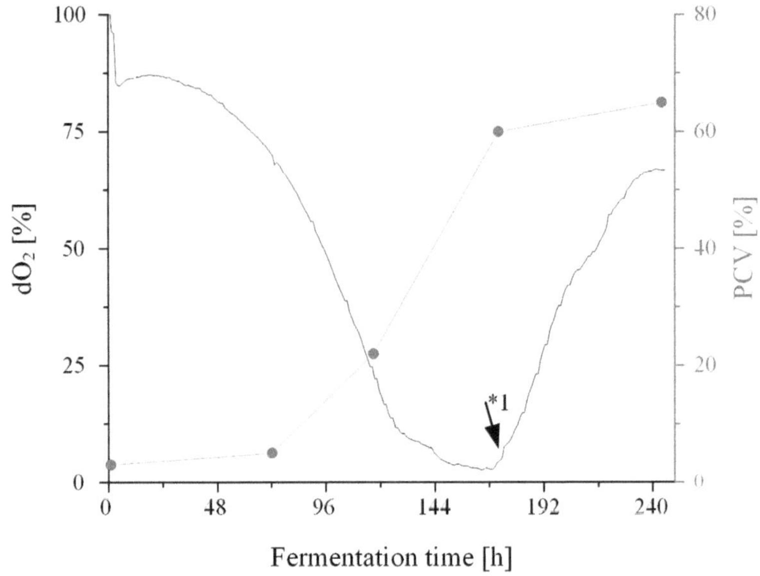

Fig. 4 Characteristics of tobacco BY-2 cell batch fermentation in a wave bioreactor. The dissolved oxygen concentration (dO_2, *black*) and percentage packed cell volume (%PCV, *green*) are plotted against the fermentation time. Key events: **1*, increase in the dO_2 signal at the beginning of the stationary phase

4 Notes

1. If the pH of the cultivation medium is in the range 5.2–5.8, once all components have been combined, then further adjustment is unnecessary. Otherwise, a small volume of sodium hydroxide or potassium hydroxide solution can be used to adjust the pH with minimal impact on the osmolarity of the medium.
2. In many cases, controlling the pH of a plant cell culture is unnecessary or even counterproductive because this can inhibit the growth of the plant cells (our unpublished data). However, monitoring the pH can be useful because many plant cell cultures show a characteristic pH profile during early growth stages (Fig. 3, *1) which can be used for quality control [45]. For example, the pH initially can drop to 4.0 and then increases to 5.0–5.5 within the first 24–48 h. The initial pH drop, which is caused by an uptake of nutrients such as phosphate (our unpublished data), can be used to confirm the vitality of the cells in the inoculum. If the initial pH drop is absent, the culture is likely to grow more slowly than expected if at all. Indeed, in the case of tobacco BY-2 cells, we found that in the absence of a pH drop there was no growth in at least 50% of the cultures (our unpublished data).

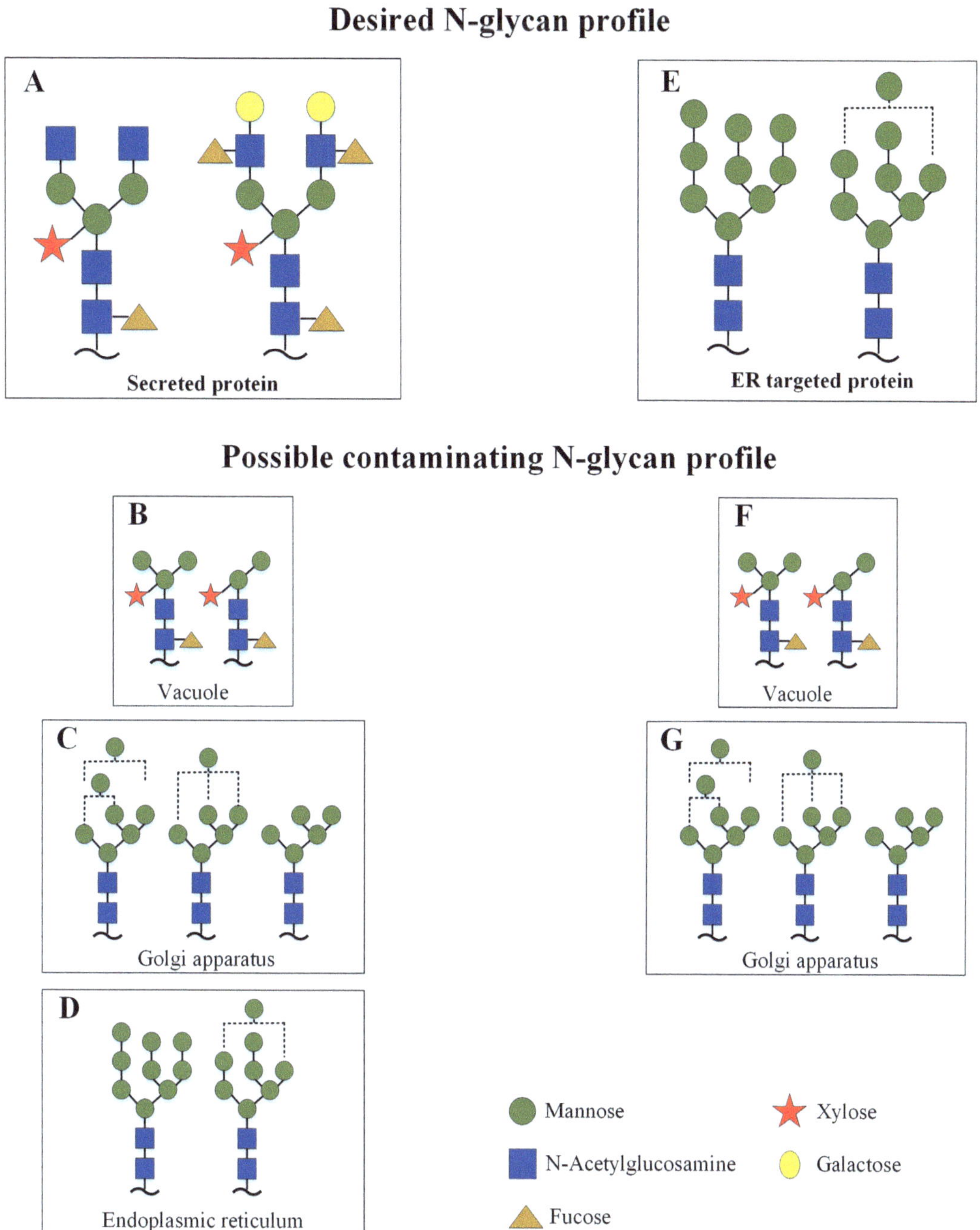

Fig. 5 The N-glycan profile of a recombinant protein depends on protein targeting in the plant cell. For a secreted protein, the desired complex N-glycan profiles are shown in (**a**). Depending on the cell line and the enzymatic repertoire of the cell culture, other variants may be synthesized. The desired N-glycan profile for proteins retained in the endoplasmic reticulum (ER) is shown in (**e**). Depending on the harvest time point, various contaminating intracellular N-glycans may also be present. In the case of secreted proteins, cell lysis can trigger the release of intracellular proteins with incompletely processed N-glycans (**b**–**d**). Potential aberrant N-glycan profiles found on proteins retained in the ER are shown in (**f**) and (**g**) and may be caused by a failure to recognize ER-retention/retrieval signals such as the KDEL tag and the consequential trafficking of the protein to other cellular compartments

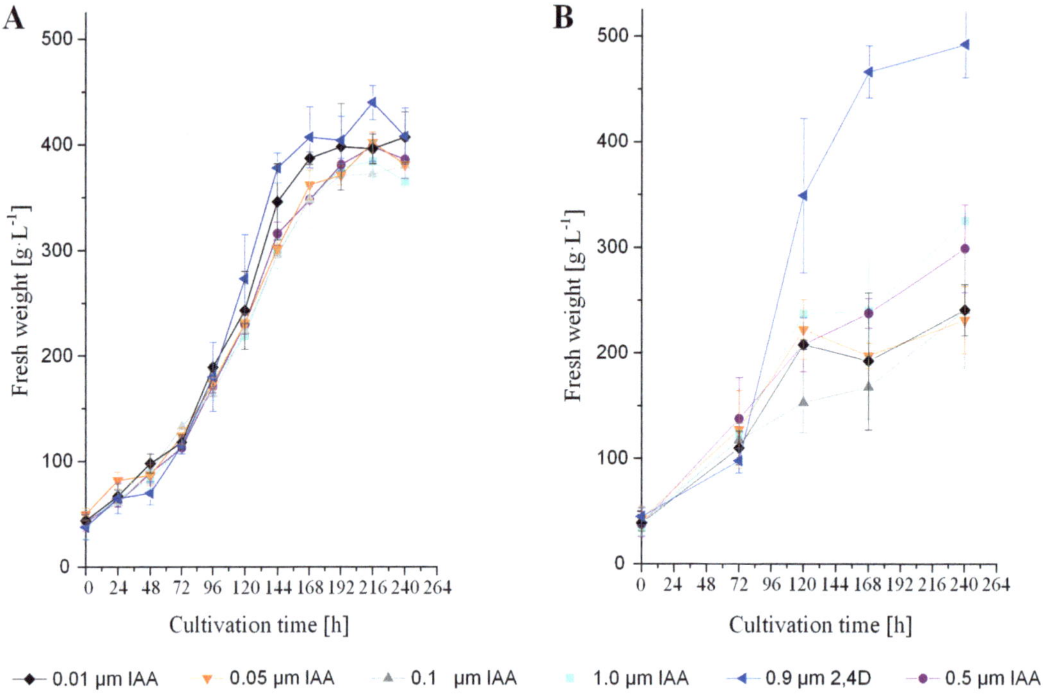

Fig. 6 Influence of minimal residual ethanol concentrations on the growth of tobacco BY-2 cells in suspension culture. BY-2 cells were cultivated in MS medium containing different concentrations of indole-3-acetic acid (IAA) as a substitute for 2,4-dichlorophenoxyacetic acid (2,4D, control) dissolved in either 1 M sodium hydroxide (**a**) or 70% ethanol (**b**). The 2,4D was always dissolved in dimethylsulfoxide (DMSO). IAA dissolved in ethanol reduced the final biomass fresh weight by 54 ± 9% ($n = 5$) compared to IAA dissolved in sodium hydroxide. *Error bars* indicate the standard deviation of three biological replicates

3. Because plant cells tend to grow on surfaces, cleaning the equipment with water is often insufficient to remove all residual cells and debris before the next run. If detergents or alcohols are used to increase the efficiency of cleaning, the reactor and all tubes must be flushed with water to ensure residual detergents and alcohols are removed because plant cells are highly sensitive to these chemicals (Fig. 6).

4. Plant cells typically grow as clusters that can reach several millimeters in diameter, thus increasing the risk of blockages in the tubes and pipes used to interconnect the fermenter and peripheral devices such as the inoculum flask. Furthermore, plant cells tend to adhere to and then grow upon surfaces such as the reactor wall, parts of the impeller and the inner surface of the pipes and tubes, which can negatively affect the hydrodynamic regime in the reactor and further increase the risk of blockages (Fig. 7). This phenomenon should be taken into account when considering which gauge of tubing is most appropriate, i.e., tubes/pipes with larger internal diameters should be selected when culturing plant cells that tend to

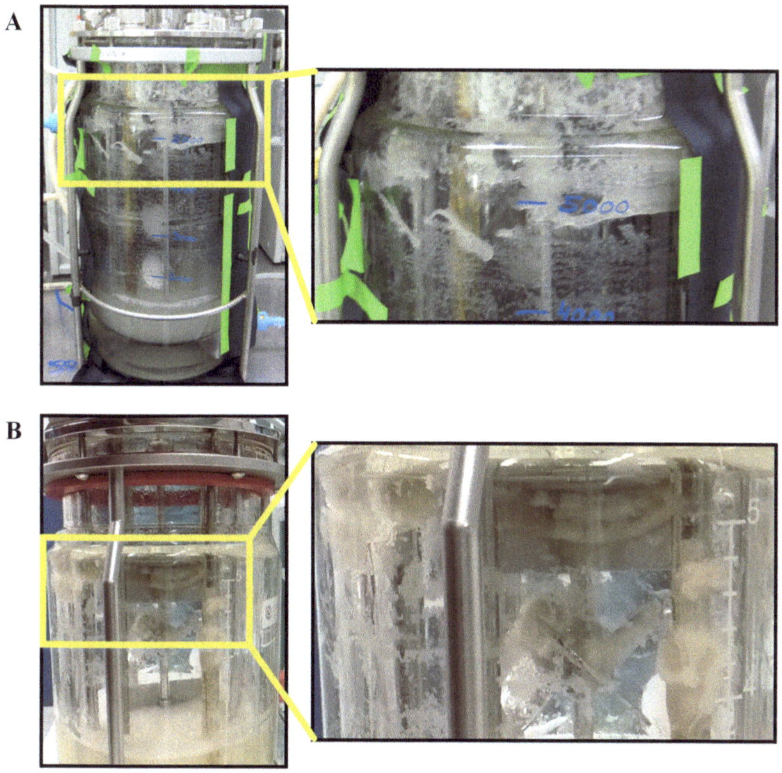

Fig. 7 The adhesive growth of plant cells. (**a**) Adhesive growth of BY-2 cells on the vessel wall during semicontinuous fermentation after 2 weeks of cultivation. (**b**) Adhesive growth of BY-2 cells during semicontinuous fermentation after 3 months of cultivation. Cells have grown on the reactor wall; the stirrer blades; and between the reactor wall and the sensors

form larger aggregates. However, the can be constraints regarding the number of ports with a sufficient diameter preventing the use of wide-bore tubing as a default solution for the problem. The growth of plant cells on surfaces can also make the fermentation less reproducible because adherent cells are not completely surrounded by the nutrient medium and may not grow or synthesize proteins as efficiently as cells in suspension (Fig. 7). The cells in adherent clusters may ultimately undergo apoptosis, and this can negatively affect the performance of suspended cells if the apoptotic clusters detach, fall back into the medium and release pro-apoptotic effector molecules.

5. To calibrate the dO_2-Sensor at the 0% point two methods can be used, either by using Nitrogen to completely remove the oxygen or by disconnecting the sensor and using this signal as the 0%. The second method is rather fast but leads to lower accuracy, meaning dO_2-Levels <5% cannot be determined accurately. If the process is running at a strict O_2-limiting

condition and an oxygen monitoring and control at <5% dO_2 is necessary, the calibration needs to be carried out with a nitrogen purge. In contrast, the less accurate but faster method can be used if the set point of the process is >15% dO_2 and remains above this threshold over the entire process.

6. Sterile connectors are typically joined to the reactor in the heat of a Bunsen burner flame. To avoid subjecting plant cells to a heat shock, make sure that the connector has cooled down to 40 °C or below (warm but not too hot to touch) before inoculation. If the connectors are too hot, the plant cells can be damaged or killed, reducing the effective inoculation density and prolonging the lag phase. Heat shock can also affect plant cell metabolism, which will reduce the reproducibility of the fermentation process.
7. Inoculation by gravity flow is preferable over using a peristaltic pump because the latter can cause cell damage, reducing the effective inoculation density and causing issues with reproducibility. To inoculate a bioreactor by gravity flow, lift the inoculation bottle until the lower outlet is higher than the liquid level in the bioreactor. If the %PCV of the preculture is greater than 40%, then the culture should be diluted with fresh medium before inoculation to reduce the risk of blocking narrow (6–8 mm) connecting tubes.
8. A reference set of typical plant cell morphologies should be established for each cultivated cell type, e.g., by the microscopic analysis of cells from typical fermentation stages (lag, exponential, and stationary). Such a set may also include cells that have been purposely exposed to specific types of stress. This will facilitate process monitoring and the timely identification of suboptimal fermentation conditions that may arise due to faulty sensors or defective equipment. For example, cells will typically change their morphology from a spherical to a cylindrical form during the transition from cell division to cell elongation [21]. Furthermore, changes in osmolarity can affect cell morphology up to the point of plasmolysis, e.g., if there are differences between the fermentation medium and the medium used in the preculture. Such shifts can also be observed by microscopic analysis (Fig. 4).

Acknowledgments

The authors acknowledge Dr. Richard M Twyman for editorial assistance. This work was funded in part the Fraunhofer-Gesellschaft Internal Programs under Grant No. Attract 125-600164. The authors have no conflict of interest to declare.

References

1. Nasto B (2007) Biotech at the beauty counter. Nat Biotechnol 25(6):617–619
2. Vojcic L, Pitzler C, Korfer G, Jakob F, Ronny M, Maurer KH, Schwaneberg U (2015) Advances in protease engineering for laundry detergents. New Biotechnol 32(6):629–634. doi:10.1016/j.nbt.2014.12.010
3. Li Q, Yi L, Marek P, Iverson BL (2013) Commercial proteases: present and future. FEBS Lett 587(8):1155–1163. doi:10.1016/j.febslet.2012.12.019
4. Schuster AC, Burghardt M, Alfarhan A, Bueno A, Hedrich R, Leide J, Thomas J, Riederer M (2016) Effectiveness of cuticular transpiration barriers in a desert plant at controlling water loss at high temperatures. AoB Plants 8. doi:10.1093/aobpla/plw027
5. Spiegel H, Stöger E, Twyman RM, Buyel JF (2016) Current status and perspectives of the molecular farming landscape. In: Kermode AR (ed) Molecular pharming: applications, challenges and emerging areas. Wiley-VCH, Weinheim
6. Buyel JF How plants can contribute to the supply of anti-cancer compounds. In: Malik S (ed) Biotechnoloy and production of anti-cancer compounds. Springer, Berlin
7. Caspi RR (2008) Immunotherapy of autoimmunity and cancer: the penalty for success. Nat Rev Immunol 8(12):970–976. doi:10.1038/nri2438
8. Shaaltiel Y, Gingis-Velitski S, Tzaban S, Fiks N, Tekoah Y, Aviezer D (2015) Plant-based oral delivery of beta-glucocerebrosidase as an enzyme replacement therapy for Gaucher's disease. Plant Biotechnol J 13(8):1033–1040. doi:10.1111/pbi.12366
9. Chen LQ, Drake MR, Resch MG, Greene ER, Himmel ME, Chaffey PK, Beckham GT, Tan ZP (2014) Specificity of O-glycosylation in enhancing the stability and cellulose binding affinity of Family 1 carbohydrate-binding modules. Proc Natl Acad Sci U S A 111(21):7612–7617. doi:10.1073/pnas.1402518111
10. Hayes JM, Cosgrave EF, Struwe WB, Wormald M, Davey GP, Jefferis R, Rudd PM (2014) Glycosylation and Fc receptors. Curr Top Microbiol Immunol 382:165–199. doi:10.1007/978-3-319-07911-0_8
11. Sareneva T, Pirhonen J, Cantell K, Julkunen I (1995) N-glycosylation of human interferon-gamma: glycans at Asn-25 are critical for protease resistance. Biochem J 308(Pt 1):9–14
12. Sinclair AM, Elliott S (2005) Glycoengineering: the effect of glycosylation on the properties of therapeutic proteins. J Pharm Sci 94(8):1626–1635. doi:10.1002/jps.20319
13. Strasser R (2016) Plant protein glycosylation. Glycobiology. doi:10.1093/glycob/cww023
14. Fischer R, Buyel JF, Holland T, Sack M, Schillberg S, Twyman RM Glyco-engineering of plant-based expression systems. In: Reichl U (ed) Glycobiotechnology, vol 10. Advances in biochemical engineering/biotechnology. Springer, Berlin
15. Tuse D (2011) Safety of plant-made pharmaceuticals product development and regulatory considerations based on case studies of two autologous human cancer vaccines. Hum Vaccines 7(3):322–330. doi:10.4161/Hv.7.3.14213
16. Gomord V, Fitchette AC, Menu-Bouaouiche L, Saint-Jore-Dupas C, Plasson C, Michaud D, Faye L (2010) Plant-specific glycosylation patterns in the context of therapeutic protein production. Plant Biotechnol J 8(5):564–587. doi:10.1111/j.1467-7652.2009.00497.x
17. Mor TS (2015) Molecular pharming's foot in the FDA's door: protalix's trailblazing story. Biotechnol Lett 37(11):2147–2150. doi:10.1007/s10529-015-1908-z
18. Bethencourt V (2009) Virus stalls Genzyme plant. Nat Biotechnol 27(8):681–681. doi:10.1038/nbt0809-681a
19. Mavituna F, Park JM (1987) Size distribution of plant cell aggregates in batch culture. Chem Eng J 35(1):B9–B14. doi:10.1016/0300-9467(87)80045-X
20. Nagata T, Hasezawa S, Depicker A (2013) Tobacco BY-2 Cells. Springer, Berlin
21. Holland T, Blessing D, Hellwig S, Sack M (2013) The in-line measurement of plant cell biomass using radio frequency impedance spectroscopy as a component of process analytical technology. Biotechnol J 8(10):1231–1240. doi:10.1002/biot.201300125
22. Holland T (2013) Development of plant suspension cultures with regard to industrial production of biopharmaceuticals, Dissertation/Ph.D. thesis. Aachen, RWTH Aachen University. http://publications.rwth-aachen.de/record/229173/files/4829.pdf
23. Santos RB, Abranches R, Fischer R, Sack M, Holland T (2016) Putting the spotlight back on plant suspension cultures. Front Plant Sci 7:297. doi:10.3389/fpls.2016.00297
24. Holland T, Sack M, Rademacher T, Schmale K, Altmann F, Stadlmann J, Fischer R, Hellwig S (2010) Optimal nitrogen supply as a key to increased and sustained production of a

monoclonal full-size antibody in BY-2-suspension culture. Biotechnol Bioeng 107 (2):278–289. doi:10.1002/bit.22800
25. Trexler MM, McDonald KA, Jackman AP (2005) A cyclical semicontinuous process for production of human alpha 1-antitrypsin using metabolically induced plant cell suspension cultures. Biotechnol Prog 21(2):321–328. doi:10.1021/bp0498692
26. Hooker BS, Lee JM, An G (1990) Cultivation of plant cells in a stirred vessel: effect of impeller design. Biotechnol Bioeng 35(3):296–304. doi:10.1002/bit.260350311
27. Doran PM (1999) Design of mixing systems for plant cell suspensions in stirred reactors. Biotechnol Prog 15(3):319–335. doi:10.1021/bp990042v
28. Eibl R, Werner S, Eibl D (2009) Bag bioreactor based on wave-induced motion: characteristics and applications. Adv Biochem Eng Biotechnol 115:55–87. doi:10.1007/10_2008_15
29. Terrier B, Courtois D, Henault N, Cuvier A, Bastin M, Aknin A, Dubreuil J, Petiard V (2007) Two new disposable bioreactors for plant cell culture: the wave and undertow bioreactor and the slug bubble bioreactor. Biotechnol Bioeng 96(5):914–923. doi:10.1002/bit.21187
30. Shaaltiel Y, Bartfeld D, Hashmueli S, Baum G, Brill-Almon E, Galili G, Dym O, Boldin-Adamsky SA, Silman I, Sussman JL, Futerman AH, Aviezer D (2007) Production of glucocerebrosidase with terminal mannose glycans for enzyme replacement therapy of Gaucher's disease using a plant cell system. Plant Biotechnol J 5(5):579–590. doi:10.1111/j.1467-7652.2007.00263.x
31. Wen SW, Jun He B, Liang H, Sun S (1996) A perfusion air-lift bioreactor for high density plant cell cultivation and secreted protein production. J Biotechnol 50(2):225–233. doi:10.1016/0168-1656(96)01568-4
32. McDonald KA, Hong LM, Trombly DM, Xie Q, Jackman AP (2005) Production of human alpha-1-antitrypsin from transgenic rice cell culture in a membrane bioreactor. Biotechnol Prog 21(3):728–734. doi:10.1021/bp0496676
33. Tanaka H, Nishijima F, Suwa M, Iwamoto T (1983) Rotating drum fermentor for plant cell suspension cultures. Biotechnol Bioeng 25 (10):2359–2370. doi:10.1002/bit.260251007
34. Xu J, Ge X, Dolan MC (2011) Towards high-yield production of pharmaceutical proteins with plant cell suspension cultures. Biotechnol Adv 29(3):278–299. doi:10.1016/j.biotechadv.2011.01.002
35. Huang TK, McDonald KA (2012) Bioreactor systems for in vitro production of foreign proteins using plant cell cultures. Biotechnol Adv 30(2):398–409. doi:10.1016/j.biotechadv.2011.07.016
36. WW S, Arias R (2003) Continuous plant cell perfusion culture: bioreactor characterization and secreted enzyme production. J Biosci Bioeng 95(1):13–20. doi:10.1016/S1389-1723(03)80142-1
37. De Dobbeleer C, Cloutier M, Fouilland M, Legros R, Jolicoeur M (2006) A high-rate perfusion bioreactor for plant cells. Biotechnol Bioeng 95(6):1126–1137. doi:10.1002/bit.21077
38. Chmiel H (2011) Bioprozesstechnik, vol 3. Springer, Spektrum. doi:10.1007/978-3-8274-2477-8
39. van Gulik WM, ten Hoopen HJ, Heijnen JJ (2001) The application of continuous culture for plant cell suspensions. Enzym Microb Technol 28(9–10):796–805
40. Miller RA, Shyluk JP, Gamborg OL, Kirkpatrick JW (1968) Phytostat for continuous culture and automatic sampling of plant-cell suspensions. Science 159(3814):540–542
41. Winkler M (1990) Chemical engineering problems in biotechnology, vol 1. Springer, Amsterdam
42. Huang TK, Plesha MA, McDonald KA (2010) Semicontinuous bioreactor production of a recombinant human therapeutic protein using a chemically inducible viral amplicon expression system in transgenic plant cell suspension cultures. Biotechnol Bioeng 106(3):408–421. doi:10.1002/bit.22713
43. Hogue RS, Lee JM, An G (1990) Production of a foreign protein product with genetically modified plant cells. Enzym Microb Technol 12(7):533–538
44. Georgiev MI, Weber J, Maciuk A (2009) Bioprocessing of plant cell cultures for mass production of targeted compounds. Appl Microbiol Biotechnol 83(5):809–823. doi:10.1007/s00253-009-2049-x
45. Schmale K, Rademacher T, Fischer R, Hellwig S (2006) Towards industrial usefulness - cryo-cell-banking of transgenic BY-2 cell cultures. J Biotechnol 124(1):302–311. doi:10.1016/j.jbiotec.2006.01.012

Chapter 12

Fed-Batch CHO Cell Culture for Lab-Scale Antibody Production

Yuzhou Fan, Daniel Ley, and Mikael Rørdam Andersen

Abstract

Fed-batch culture is the most commonly used upstream process in industry today for recombinant monoclonal antibody production using Chinese hamster ovary (CHO) cells. Developing and optimizing this process in the lab is crucial for establishing process knowledge, which enables rapid and predictable tech-transfer to manufacturing scale. In this chapter, we describe stepwise how to carry out fed-batch CHO cell culture for lab-scale antibody production.

Key words Antibody production, CHO cells, Fed-batch culture

1 Introduction

Chinese hamster ovary (CHO) cell was originally isolated in 1957 [1] and first adapted to growth in serum-free media in 1977 [2]. To compromise a general downside of low productivity in most mammalian systems, DHFR or glutamine synthethase (GS)-mediated gene amplification system was developed for industrial application. Currently, CHO-K1, CHO-S, CHO-DXB11 [3], and CHO-DG44 [4], for example, are the most commonly used parental cell lines for generating many recombinant protein producing CHO cell lines. The history of the different lineages of CHO is well summarized in Wurm [5].

CHO cells remain the preferred mammalian expression system for therapeutic protein production in the pharmaceutical industry [6], due to the ability to perform human-like posttranslational protein modifications, particularly glycosylation, which affect efficacy, stability, immunogenicity, and clearance rate of the therapeutic protein [7–9]. Monoclonal antibodies (mAbs) constitute the bulk of products manufactured using CHO cells. Six of the top-10 best-selling biopharmaceuticals on the market in 2013 were mAbs, including Humira (adalimumab), a mAb against rheumatoid

Virgínia Picanço-Castro and Kamilla Swiech (eds.), *Recombinant Glycoprotein Production: Methods and Protocols*, Methods in Molecular Biology, vol. 1674, DOI 10.1007/978-1-4939-7312-5_12, © Springer Science+Business Media LLC 2018

arthritis; Remicade (infliximab), a mAb against Crohn's disease; and Mabthera/Rituxan (rituximab), a mAb against lymphoma [6]. Production processes have steadily improved through advances in cell line engineering, medium development and bioprocess control [10]. As a result, product titers of 5–7 g/L are routinely achieved within 2 weeks of process duration [11, 12]. Until now, CHO cells derived mAbs have been widely approval by FDA (Food and Drug Administration) or EMA (European Medicines Agency) [13]. Regulatory considerations for production of mAbs include workflow documentation according to GMP standards, in order to ensure consistent product supply with reproducible quality. Furthermore, the mAb must be characterized with respect to structural integrity, specificity, and potency using immunochemical and biological assays [14]. Production cell lines should be tested for infections (e.g., virus and mycoplasma) and contamination of animal-derived raw materials to avoid transfer to patients [15].

From serum-based to serum-free and lately to chemically defined media, medium development has led to progress in commercially available media over the last decades [16–18]. Chemically defined media usually contains amino acids, vitamins, trace elements, inorganic salts, lipids, and insulin or insulin-like growth factors. Medium development has been one of the key factors in optimizing growth behavior and productivity for CHO cells and improving product quality [19]. Today, commercially available media serve as the basis for media development, which choose the optimal blend of media components by taking cell line, process, media components, cell metabolism, and by-product accumulation into a holistic consideration, in order to achieve a best possible balance between improving cell growth and increasing productivity [20].

Fed-batch cell culture processes support high volumetric productivity at low operational complexity [21]. In a fed-batch process, a basal medium supports initial cell growth, and a feed medium prevent nutrient depletion, while extending the productive growth phase. Process parameters such as temperature, pH, pO_2, and agitation rate define the physical environment, while the choice of growth medium, feeds, and feeding-strategy define the chemical environment suitable for protein production. Nutrient control is typically performed manually, using offline nutrient analysis and bolus feeds introduced at 24 h intervals [22, 23]. In an attempt to increase process control and reduce batch-to-batch variation, manufacturers apply sensor technology such as mass and Raman spectroscopy, which enable online measurement of cell characteristics as well as nutrient and by-product concentrations [24]. Implementation of such quality-by-design (QbD) methods enhances development of sophisticated control systems and improves process knowledge. Scale-down models are commonly used for screening of raw materials and parameters in process characterization, enabling process transfer to pilot and manufacturing scale. Frequently, design-of-experiment (DoE) approaches (e.g., Plackett–Burman designs and response surface methodology) are applied to identify optimal nutrient concentrations or process parameters that maximize product titers and quality [25]. Manufacturers invest extensive resources on medium and process optimization, as it needs to be performed on an individual basis for each cell line [26].

Antibody production process in CHO cell culture consists of upstream production, downstream processing, and product analytics. In this chapter, we focus on the part of upstream production of mAbs and present some representative methods involved in lab-scale fed-batch CHO cell culture.

2 Materials

2.1 Equipments

1. A mammalian cell counting system, e.g., nucleocounter NC-200 (Chemometec, Denmark).
2. Bioreactors (0.5–1 L working volume) coupled with pH and pO_2 sensors; multi-pump module; mass-flow controlled gassing system; and temperature and agitation control unit.
3. Process control computer equipped with bioreactor control software.
4. Vacuum aspiration system.
5. Automated chemistry analyzer for cell culture.
6. Octet QK384 equipped with Protein A biosensors (ForteBio, Menlo Park, CA).
7. Humidified CO_2 incubator.
8. Laminar flow hood.

2.2 Materials for Preculture, Bioreactor Preparation, Inoculation, and Sampling

1. Cryopreserved production cell line (1×10^7cells/mL) from culture medium with 10% DMSO stored at −150 °C or lower.
2. Commercially available chemically defined serum-free medium (*see* **Note 1**) and feeds (*see* **Note 2**) for CHO cell culture.
3. Base: sterile NaOH, $NaHCO_3$, or Na_2CO_3 solution in the concentration range of 0.5–2 M.
4. 45% D-(+)-glucose solution.
5. 15 and 50 mL centrifuge tubes.
6. Shake flasks.
7. Phosphate buffered saline (PBS) solution (10×).
8. Single use pipettes (1–25 mL) and automatic pipette.
9. 250 mL conical centrifuge tubes.
10. 50 mL Syringe (luer-lok).

11. 100 mL Beakers.
12. Buffers for pH calibration (pH 7 and 4).
13. 3 M KCl solution.
14. 70% ethanol.
15. 2 mL Eppendorf tubes.
16. 0.22 μm low protein binding filter.
17. MilliQ water.
18. Tin foil.

3 Methods

3.1 Cell Thawing and Preculture Maintenance

1. Thaw a vial of cells from working cell bank in a water bath at 37 °C.
2. Resuspend the cells in 10 mL preheated culture medium at 37 °C in 15 mL tube. Spin down cells at 200 × *g* and discard the supernatant to remove DMSO from the medium.
3. Resuspend cells in 5 mL preheated culture medium and transfer to 125 mL shake flask. Add additional 15 mL of preheated culture medium to final volume of 20 mL.
4. Maintain cell culture in a humidified incubator operated at 37 °C, 5% CO_2, and 120 rpm.
5. Perform cell count and viability measurement after 1 h.
6. Maintain cells in shake flasks for one week before inoculation in bioreactors (2 or 3-day passage). Perform cell count and viability assay measurement before each passage. Seed cells at 0.3×10^6 viable cells/mL after each passage in incrementally larger shake flasks to obtain enough cells to inoculate bioreactors, depending on desired scale of experiment.

3.2 Bioreactor Preparation

The bioreactor system (Fig. 1) should be assembled and sterilized according to the manufacturer's instructions. For cell culture in the DASGIP system (Eppendorf), the following workflow is recommended:

1. Start calibration of the pH probe using the DASGIP control software.
2. Connect the pH probe to the control box using the color coded cable.
3. Prepare a beaker with pH buffer (pH = 7), insert pH probe and control-box temperature probe in the buffer and allow the signal to stabilize for 15 min or until reading is stable. The probes should be submerged 4 cm into the buffer.

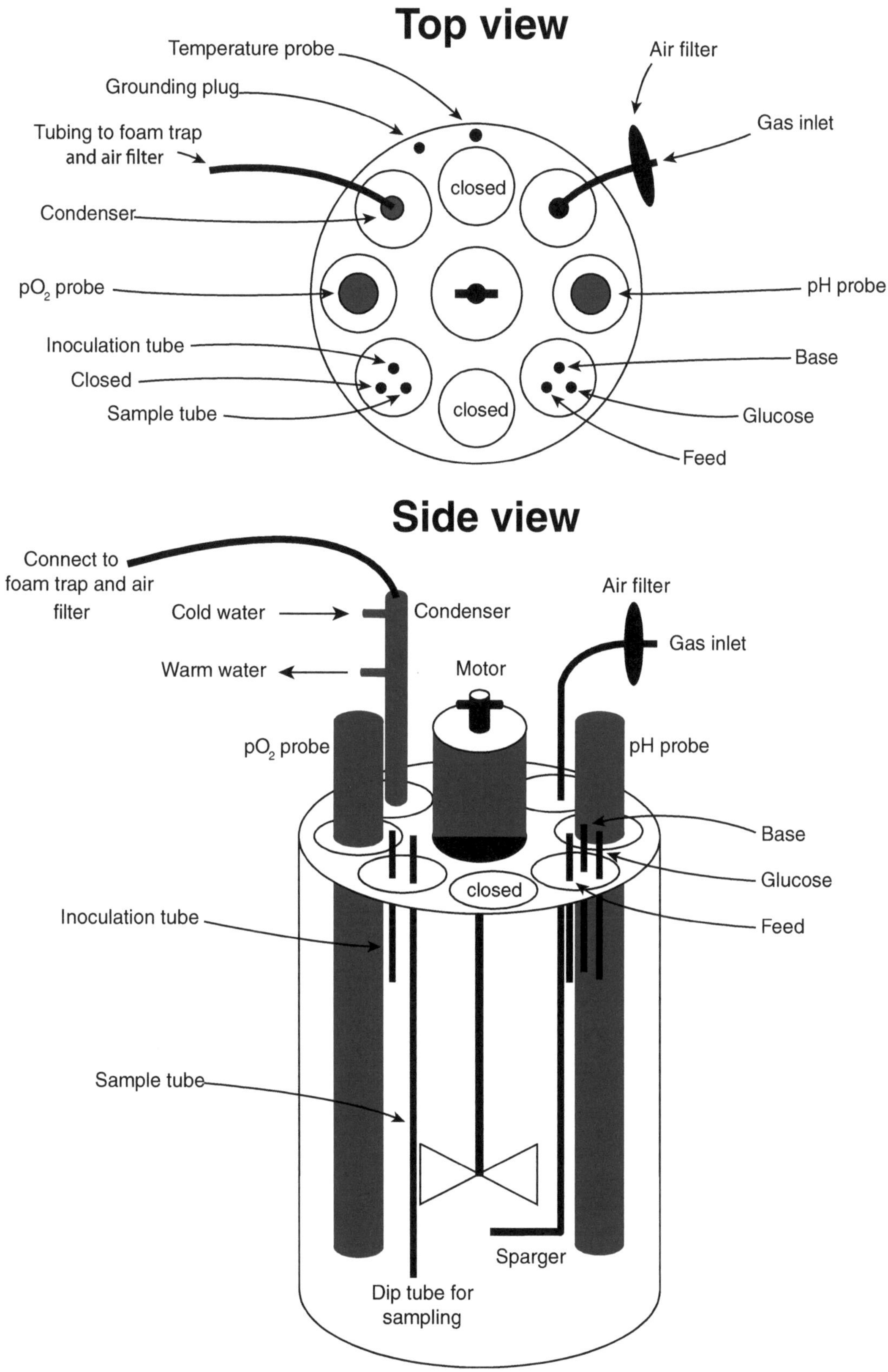

Fig. 1 Overview of bioreactor assembly

4. Once the signal is stable, click "calibrate offset" on the control panel on the computer screen. Wait for the yellow lights to turn green before proceeding.
5. Rinse the probes with water.
6. Prepare a beaker with pH buffer (pH = 4), insert pH probe and control-box temperature probe in the buffer and allow the signal to stabilize.
7. Once the reading is stable, click "calibrate slope" on the control panel on the computer screen. Wait for the yellow lights to turn green before proceeding.
8. Click stop. The control software will prompt for pO_2 calibration. Do *NOT* click yes, wait for **step 23** before proceeding with this.
9. Assemble the bioreactor: Insert pH probe, pO_2 probe (*see* **Note 3**), and condenser in the available slots and slide them gently into position.
10. Make sure tubes are inserted in the correct positions and gaskets are tightly closed. Attach clamps to dip tubes and check if air filters are tightly attached.
11. Add 5 mL water in the base, feed, and glucose bottles. Close lids, but not tightly. Connect the base, feed, and glucose bottles to the bioreactor using the base, feed and glucose tubes.
12. Unscrew the lid of the bioreactor and add 200 mL of PBS (1×). Close lid again, but not tightly.
13. Wrap the following extruding parts in double-layered tin foil: air filters, electrodes, and sampling valves.
14. Autoclave the bioreactor.
15. After autoclaving, quickly close all lids tightly. Check if all the tubes are intact and remove tin foil.
16. In a sterile environment (laminar flow hood), carefully remove the lids from the base, feed and glucose bottles and remove the liquid using a 1 mL pipette attached to a vacuum suction unit. Add 100 mL base solution to the base bottle and close the lid tightly. Add sufficient amount of feed (*see* **Note 4**) to the feed bottle(s) and close the lid tightly. Add 50 mL of glucose solution to the glucose bottle and close the lid tightly.
17. In a sterile environment, carefully remove the lid of the bioreactor and remove liquid from the sterilized bioreactor with a 1 mL pipette attached to a vacuum suction unit. Add 400 mL sterile growth medium to the bioreactor and close the lid tightly.
18. Insert the bioreactor into the bioblock module and gently slide it into the corresponding position.

19. Connect the cables to the pH and pO_2 probes.
20. Connect air supply to the air filter.
21. Insert temperature probe into the corresponding tube.
22. Attach and connect the overhead drive and connect the cable for grounding.
23. Start pO_2 calibration on the control computer.
24. First, start stirring at 200 rpm, then heating at 37 °C and air supply with 100% atmospheric air (21% O_2). If cell culture is performed with alternative agitation rate or temperature, apply those settings instead.
25. Attach the red and blue cooling tubes from the bioblock to the condenser unit and turn on the condenser.
26. Wait for the oxygen pressure and the signal to stabilize on the control computer. This step may take several hours and can be performed overnight.
27. Once the oxygen saturation reading has stabilized, press start on the control computer to calibrate the probe to 100% dissolved oxygen (DO).
28. For the second calibration point at 0%, disconnect the cable from the pO_2 probe and wait for the oxygen saturation reading to be stable at zero. This may only take several few seconds (*see* **Note 5**).
29. Once the reading has stabilized, press "start" on the control computer to calibrate the probe to 0% O_2. Wait for the computer to show "ready" again.
30. Reconnect the cable to the pO_2 probe and check if the measured value stabilizes at 100%. If not, redo **steps 26** and **27**.
31. Click "finish" to end the calibration procedure.
32. Optionally, calibrate pumps (*see* **Note 6**).
33. Set the process parameters as follows (*see* **Note** 7):
    (a) Set agitation rate at 200 rpm.
    (b) Set temperature at 37 °C.
    (c) Start air supply and set air flow at 0.6sL/h.
    (d) Start DO control with set point at DO = 50%.
    (e) Start pH control on the bioreactor control system (set point = 7.1 with 0.2 dead band).
34. Pump base, feed, and glucose through the tubing manually until the tubes are filled with liquid.
35. Wait for the set points of agitation rate, temperature, air flow, DO, and pH to be reached and stabilized. This typically takes 1–2 h.

3.3 Bioreactor Inoculation

1. Determine viable cell count and viability in the preculture carried out in shake flasks.
2. Calculate the amount of cells needed to inoculate the bioreactor at the desired seeding density (3–5 $\times$ 10^5 cells/mL) (*see* **Note 8**).
3. Extract the culture volume corresponding to the amount of cells calculated in **step 2**. Spin down cells at 200 $\times$ *g* for 5 min in a 250 mL conical centrifuge tube.
4. Remove all but 10 mL supernatant and resuspend cells.
5. In a sterile environment, aspirate the cell suspension with a 50 mL syringe. Suck in approximately 10 mL of air.
6. Wash gloves in 70% ethanol to sterilize surface and block the opening of the syringe with the thumb. While blocking the opening of the syringe with the thumb, walk to the bioreactor and sterilize the inoculation valve with 70% ethanol.
7. In a quick controlled movement, remove the thumb from the syringe opening and connect it to the inoculation valve. Turn the syringe upright, with the piston pointing upward, and slowly inject the cell suspension down into the bioreactor. Proceed by blowing the air through to empty the tube of cells.
8. Press "inoculate" on the control computer software.

3.4 Bioreactor Sampling and Analysis

Daily sampling is usually needed along the fed-batch process in order to monitor cell growth and metabolism. Other samples for monitoring, for example, mAb titer and mAb quality (e.g., N-glycosylation) can be taken either at the end of fed-batch process or along with the process with desired frequency. The following steps are recommended:

1. Prepare enough 2 mL Eppendorf tubes for sampling.
2. Open sampling valve on the bioreactor and cleanse both cap and valve with 70% ethanol.
3. Withdraw 2 mL of culture from the bioreactor using a 6 mL syringe. Discard the syringe with the culture to the waste.
4. Take a new 6 mL syringe and withdraw 2 mL of culture again from the bioreactor. Add the 2 mL culture into the prepared Eppendorf tube. This sample will be used for further analysis, including for example viable cell count and viability measurement, extracellular metabolites in the culture and product titer analysis. If more analyses are needed, increase the sample volume and the syringe size accordingly. Usually the following volumes are required for each analysis:
   (a) Viable cell count and viability measurement: 0.5 mL culture sample is needed for cell count and viability

measurement using automated cell counter (nucleocounter NC-200, for example).

(b) Metabolism analysis: 1.5 mL cell culture supernatant is needed for measuring extracellular metabolites (e.g., glucose, glutamine, glutamate, lactate, and NH_4^+) in the cell culture using an automated chemistry analyzer (Bioprofile 100plus, for example).

(c) One-hundred to five-hundred microliters of cell culture supernatant is needed for measuring IgG titer using Octet QK384 equipped with Protein A biosensors (*see* **Note 9**).

(d) mAb quality analysis: The sampling volume is often large (10–15 mL). Frequent sampling along the cultivation process is not recommended and may lead to oversampling (*see* **Note 10**).

3.5 Feeding the Cell Culture

Base, glucose, and feed media are usually fed along the cell culture. Base is typically fed automatically and continuously when the bioreactor system regulates pH in the cell culture. Glucose and feed media are often fed manually once a day or once every second day, after performing cell count and metabolism analysis.

1. Feeding with glucose: calculate the volume to be added, according to the volume of the cell culture and the remaining glucose concentration in the cell culture from the metabolism analysis. Spike glucose into the bioreactor to obtain 8 g/L when glucose concentration in the cell culture decreases below 4 g/L in the culture (*see* **Note 11**).
2. Feeding with feed media: the feeding strategy can be quite different, when different cell line, basal media, and feed are used. Three types of feed media and their corresponding feeding strategies are recommended in Table 1 (*see* **Note 12**).

Table 1
Feed media and feeding strategies

Basal medium	Feed media and feeding strategies	
ActiCHO P	ActiCHO A; Daily feeding from day 3; feed volume in 2% of the initial culture volume every time	ActiCHO B; Daily feeding from day 3; feed volume in 0.2% of the initial culture volume every time
CD CHO	Efficient Feed A; fed on day 4, 6, 8, 10, 12; feed volume in 10% of the initial culture volume every time	FunctionMax; fed on day 4, 6, 8, 10, 12; feed volume in 3.3% of the initial culture volume every time
PowerCHO-2 CD	Cell Boost 6; fed on day 4, 6, 8, 10, 12; feed volume in 10% of the initial culture volume every time	

3. To add glucose and feed media into bioreactor, turn on the corresponding pump manually. If the pump has been calibrated, calculated the time for pumping the amount needed into the bioreactor. To control the volume of glucose or feed medium pumping into the bioreactor, weigh the glucose or feed bottle on a scale before and after feeding. 1 g of feed medium is usually equivalent to 1 mL of feed medium. Please note that 1.187 g of 45% D-(+)-glucose solution is equivalent to 1 mL of 45% D-(+)-glucose.

3.6 Bioreactor Harvesting, Shutdown and Cleaning

The harvest process is recommended before viability <60%. Depending on cell line, media, feed, and process parameters chosen, fed-batch culture duration is usually about 12–16 days. The following steps could be followed when harvesting:

1. End data recording, stirring, pumping, and air supply by clicking "Finish" on the top right of the control screen for the bioreactor.
2. Turn off the condenser on the bioblock and remove the blue and red tube.
3. Disconnect pH and pO_2 electrodes, temperature probe, air supply, base tube, feed tube, ground wire, and motor from the bioreactor.
4. Pour 40 mL of the cell suspension from the bioreactor to a 50 mL centrifuge tube. If larger volumes of culture are needed for downstream processing, use one or many larger centrifuge tubes instead.
5. Centrifuge the tube at 1000 × *g* for 10 min. Collect the supernatant.
6. Filtrate the supernatant through a 0.22 μm low protein binding filter.
7. Store the filtrated supernatant at −20 °C until further processing (*see* **Note 13**).
8. Disconnect the base and feed tubes from the base and feed bottles. Place the one end of these tubes in a beaker containing about 300 mL distilled water. Place the other end in an empty beaker and start pumping water through the tube by starting the control unit manually. Let the pump flush the tube for an hour.
9. Empty the remaining cell culture from the bioreactor into a biowaste bottle.
10. Clean the pH electrode with 70% ethanol. Collect the ethanol into the biowaste bottle. Gently wipe it dry with tissue paper.
11. Repeat **step 8** one time with distilled water.

12. Put the cap back on the pH electrode with 3 M KCl buffer completely filling the cap.
13. Clean the pO_2 electrode with 70% ethanol. Collect the ethanol into the biowaste bottle. Gently wipe it dry with tissue paper. Do NOT touch the bottom of the electrode.
14. Repeat **step 11** one time with distilled water.
15. Put the cap on the pO_2 electrode.
16. Rinse the inside of the bioreactor twice with 70% ethanol and collect the ethanol into the biowaste bottle.
17. Repeat **step 14** one time with distilled water.
18. Rinse the base bottle and attached tubes with distilled water. Place on paper tissue to dry.
19. Close the base injection valve on the bioreactor by inserting a plug.
20. When the base tube has been rinsed for an hour, remove the end submerged in the clean water, and transfer it to a beaker containing 70% ethanol. Turn off the pump and remove the tube from the pump. Leave the tubes on paper tissue to dry.

4 Notes

1. This protocol is based on chemically defined serum-free medium. Choice of medium can be varied depending on adaptation of production cell line and desired culture process. Common commercially available basal media for CHO cell culture include: CD CHO (Life Technologies), PowerCHO-2 CD (Lonza), and ActiCHO™ P (GE Healthcare). Supplementation of L-glutamine (4–8 mM) may be required if the medium is supplied without L-glutamine, unless a non-glutamine synthetase engineered cell line is employed. Optionally, an anticlumping agent can be added (0.1–0.5%) to the medium if cell aggregation is observed.
2. Commercially available feeds for CHO cell culture include: ActiCHO Feed A and Feed B (GE Healthcare, usually used in combination with basal medium ActiCHO™ P); Efficient-Feed™ A and FunctionMAX™ (Life Technologies, usually used in combination with basal medium CD CHO); and HyClone™ Cell Boost™ 6 Supplement (GE Healthcare, usually used in combination with PowerCHO-2 CD).
3. Electrolyte solution inside the pO_2 probe needs to be changed every month according to the manufacturer's instructions. To achieve best possible performance, electrolyte solution could also be change before every bioreactor run. It is also recommended to perform offline calibration on pO_2 probe in milli-Q

water every 1–2 months to ensure its correct function. Apply 100% air to saturate oxygen in milli-Q water for 100% DO calibration. Apply N_2 to remove all dissolved oxygen in milli-Q water for 0% DO calibration.

4. Feed alternatives: 150 mL ActiCHO Feed A and 25 mL ActiCHO Feed B; 250 mL EfficientFeed A and 85 mL FunctionMax; and 250 mL HyClone™ Cell Boost™ 6 Supplement with a concentration of 36 g/L.
5. Calibration of pO_2 probe at 0% can also be done by constantly applying N_2 to the medium. However, this may cause precipitation of some medium components, and, thus, it is practically not often used.
6. Pumps calibration can be performed determining pump flow rates across multiple pump-speed settings, based on time-constant measurement of the transferred mass. Flow rate settings used for pump calibration depend on diameter of pump tubing. Flow rates can be modified to suit larger or smaller scale experiments by adjusting flow rate set points. For low flow rate set points, beware of evaporation.
7. Process parameters are usually optimized in a cell line-dependent manner to achieve high product titer and acceptable product quality. Physical parameters (e.g., temperature, gas flow rate and agitation speed), chemical parameters (e.g., DO and carbon dioxide, pH, osmolality, by-product concentration, and metabolite levels), and biological parameters (e.g., seeding density, viable cell count, and viability) are all essential process parameters to be considered when optimizing a fed-batch culture process. The key of process optimization is to understand relationships between process parameters and reproducibility, as well as culture performance and product quality attributes. DoE approaches are often used to minimize the number of experiment to be performed, and to identify the best possible operation windows for various process parameters. The following operation windows of process parameters are recommended when running process optimization experiments:
   (a) Agitation rate from 80 to 200 rpm is a typical range for CHO cell culture.
   (b) Temperature shift from 37 to 30–35 °C at 2–7 days post inoculation can be considered as a way to retain cells in G1 phase and delay the onset of apoptosis.
   (c) pH is usually controlled with a combination of CO_2 sparging and base addition. Typically pH set point between 7 and 7.3 is preferred. The dead band around the set point would determine how often to drift culture pH by triggering the addition of base or CO_2.

(d) DO is typically set between 20 and 50% of air saturation. Limitation of dissolved oxygen may increase lactate accumulation in the cell culture, while high dissolved oxygen concentrations may lead to cytotoxicity.

8. Seeding densities are usually between 3 and 5 $\times$ 10^5 cells/mL. However, depending on the purpose, seeding density outside this operation window may also be preceded. Higher seeding density may reduce the exponential growth phase, increase the maximum cell density and shorten the culture duration. On the other hand, lower seeding density may improve the robustness of the process.
9. mAb titer can be analyzed using a Octet QK384, according to manufacturer instructions. It is a biolayer interferometry-based system that can directly measure the binding rate of mAb from cell culture supernatant to protein A biosensor in order to quantify the mAb in a 96-well plate format.
10. mAb quality analysis may include intact mass, ion exchange, size exclusion, and glycosylation analysis (*see* refs. 27, 28). However, they are independent methods themselves, which will not be described in detail in this chapter.
11. The amount and frequency of feeding glucose may various in different cell lines and at different growth phases. DoE approach is often necessary to define an optimal feeding criteria and strategy.
12. Changing culture medium, feed supplement medium and feeding strategy (e.g., feeding amount and frequency) will considerably modulate cell culture components. Several methods can be used in combination with a DoE setup for medium optimization, for example component titration, spent medium analysis, and medium blending (*see* refs. 29, 30).
13. Further processing of the produced mAb consists of downstream processing, e.g., protein A purification, buffer exchange, and formulation [27, 28] as well as the mAb quality analysis described in **Note 10**.

References

1. Puck TT, Cieciura SJ, Robinson A (1958) Genetics of somatic mammalian cells. III. Long-term cultivation of euploid cells from human and animal subjects. J Exp Med 108 (6):945–956
2. Hamilton WG, Ham RG (1977) Clonal growth of chinese hamster cell lines in protein-free media. In Vitro 13(9):537–547
3. Urlaub G, Chasin LA (1980) Isolation of Chinese hamster cell mutants deficient in dihydrofolate reductase activity. Proc Natl Acad Sci U S A 77(7):4216–4220
4. Urlaub G, Kas E, Carothers AM, Chasin LA (1983) Deletion of the diploid dihydrofolate reductase locus from cultured mammalian cells. Cell 33(2):405–412
5. Wurm F (2013) CHO quasispecies—implications for manufacturing processes. Processes 1(3):296–311. doi:10.3390/pr1030296

6. Walsh G (2014) Biopharmaceutical benchmarks 2014. Nat Biotechnol 32(10):992–1000. doi:10.1038/nbt.3040
7. Butler M (2006) Optimisation of the cellular metabolism of glycosylation for recombinant proteins produced by mammalian cell systems. Cytotechnology 50(1–3):57–76. doi:10.1007/s10616-005-4537-x
8. Costa AR, Rodrigues ME, Henriques M, Oliveira R, Azeredo J (2014) Glycosylation: impact, control and improvement during therapeutic protein production. Crit Rev Biotechnol 34(4):281–299. doi:10.3109/07388551.2013.793649
9. Hossler P, Khattak SF, Li ZJ (2009) Optimal and consistent protein glycosylation in mammalian cell culture. Glycobiology 19 (9):936–949. doi:10.1093/glycob/cwp079
10. Reinhart D, Damjanovic L, Kaisermayer C, Kunert R (2015) Benchmarking of commercially available CHO cell culture media for antibody production. Appl Microbiol Biotechnol 99(11):4645–4657. doi:10.1007/s00253-015-6514-4
11. Jayapal K, Wlaschin K, Yap M, Hu W (2007) Recombinant protein therapeutics from CHO cells – 20 years and counting. Chem Eng Prog 103:40–47
12. Yu M, Hu Z, Pacis E, Vijayasankaran N, Shen A, Li F (2011) Understanding the intracellular effect of enhanced nutrient feeding toward high titer antibody production process. Biotechnol Bioeng 108(5):1078–1088. doi:10.1002/bit.23031
13. Kim JY, Kim YG, Lee GM (2012) CHO cells in biotechnology for production of recombinant proteins: current state and further potential. Appl Microbiol Biotechnol 93(3):917–930. doi:10.1007/s00253-011-3758-5
14. Kurz M (2007) Upstream processing: regulatory considerations regarding quality aspects of monoclonal antibodies. BioPharm Int 20(7)
15. Marcus-Sekura C, Kozak R (1999) Continuous cell lines and contaminant testing in novel therapeutics. In: Oxender DL, Post LE (eds) Modern biotechnology: from laboratory to human testing. Springer, London
16. Birch JR, Racher AJ (2006) Antibody production. Adv Drug Deliv Rev 58(5–6):671–685. doi:10.1016/j.addr.2005.12.006
17. Zhang H, Wang H, Liu M, Zhang T, Zhang J, Wang X, Xiang W (2013) Rational development of a serum-free medium and fed-batch process for a GS-CHO cell line expressing recombinant antibody. Cytotechnology 65(3):363–378. doi:10.1007/s10616-012-9488-4
18. Kim SH, Lee GM (2009) Development of serum-free medium supplemented with hydrolysates for the production of therapeutic antibodies in CHO cell cultures using design of experiments. Appl Microbiol Biotechnol 83 (4):639–648. doi:10.1007/s00253-009-1903-1
19. Jordan M, Voisard D, Berthoud A, Tercier L, Kleuser B, Baer G, Broly H (2013) Cell culture medium improvement by rigorous shuffling of components using media blending. Cytotechnology 65(1):31–40. doi:10.1007/s10616-012-9462-1
20. Gronemeyer P, Ditz R, Strube J (2014) Trends in upstream and downstream process development for antibody manufacturing. Bioengineering 1(4):188–212. doi:10.3390/bioengineering1040188
21. Fike R (2009) Nutrient supplementation strategies for biopharmaceutical production, part 1: Identifying a formulation. Bioprocess Int 7 (10):44–51
22. Xie L, Wang DI (1994) Fed-batch cultivation of animal cells using different medium design concepts and feeding strategies. Biotechnol Bioeng 43(11):1175–1189. doi:10.1002/bit.260431123
23. Xie L, Wang DI (1994) Applications of improved stoichiometric model in medium design and fed-batch cultivation of animal cells in bioreactor. Cytotechnology 15 (1–3):17–29
24. Lopes JA, Costa PF, Alves TP, Menezes JC (2004) Chemometrics in bioprocess engineering: process analytical technology (PAT) applications. Chemometr Intell Lab 74(2):269–275
25. Gonzalez-Leal IJ, Carrillo-Cocom LM, Ramirez-Medrano A, Lopez-Pacheco F, Bulnes-Abundis D, Webb-Vargas Y, Alvarez MM (2011) Use of a Plackett-Burman statistical design to determine the effect of selected amino acids on monoclonal antibody production in CHO cells. Biotechnol Prog 27 (6):1709–1717. doi:10.1002/btpr.674
26. Wurm FM (2004) Production of recombinant protein therapeutics in cultivated mammalian cells. Nat Biotechnol 22(11):1393–1398. doi:10.1038/nbt1026
27. Fan Y, Jimenez Del Val I, Muller C, Wagtberg Sen J, Rasmussen SK, Kontoravdi C, Weilguny D, Andersen MR (2015) Amino acid and glucose metabolism in fed-batch CHO cell culture affects antibody production and glycosylation. Biotechnol Bioeng 112(3):521–535. doi:10.1002/bit.25450
28. Kildegaard HF, Fan Y, Sen JW, Larsen B, Andersen MR (2016) Glycoprofiling effects of

media additives on IgG produced by CHO cells in fed-batch bioreactors. Biotechnol Bioeng 113(2):359–366. doi:10.1002/bit.25715

29. Vijayasankaran N, Li J, Shawley R, Chen A, Shiratori M, Gawlitzek M (2010) Animal cell culture media. In: Flickinger MC (ed) Encyclopedia of industrial biotechnology, bioprocess, bioseparation, and cell technology. John Wiley & Sons, Inc, Hoboken, NJ
30. Fletcher T (2005) Designing culture media for recombinant protein production: a rational approach. BioProcess Int 3(1):30–36

Chapter 13

Strategies to Develop Therapeutic N- and O-Hyperglycosylated Proteins

Agustina Gugliotta, Natalia Ceaglio, Marina Etcheverrigaray, Ricardo Kratje, and Marcos Oggero

Abstract

Glycoengineering by N- and/or O-hyperglycosylation represents a procedure to introduce potential sites for adding N- and/or O-glycosyl structures to proteins with the aim of producing biotherapeutics with improved pharmacodynamic and pharmacokinetic properties. In this chapter, a detailed description of the steps routinely performed to generate new proteins having high content of N- and/or O-glycosyl moieties is carried out. The rational strategy involves the initial stage of designing N- and/or O-hyperglycosylated muteins to be expressed by mammalian cells and includes the upstream and downstream processing stages necessary to develop hyperglycosylated versions of the proteins of interest with the purpose of beginning the long road toward producing biobetters.

Key words Glycoengineering, Hyperglycosylation, Biotherapeutics, Site-directed mutagenesis, Peptide fusion, Pharmacokinetics

1 Introduction

Many therapeutic proteins suffer from low in vivo activity mainly produced by their rapid clearance or their short half-life in blood. Different strategies have arisen in order to solve this problem. The engineering of proteins through adding new glycosyl moieties has provided novel proteins with improved biological properties. Also, this approach has minimized the structural modifications, the loss of bioactivity, the immunogenicity or the difficulties during their manufacturing that were commonly observed when using other methods such as chemical conjugation with polyethylene glycol or fusion with other proteins such as albumin [1].

Elliott and Sinclair [2] introduced the concept that glycoengineering could be a procedure not only to optimize the glycosylation or to remodel the glycan structures attached to a protein, but also a set of methods to incorporate N- and/or O-glycosyl residues to

Virgínia Picanço-Castro and Kamilla Swiech (eds.), *Recombinant Glycoprotein Production: Methods and Protocols*, Methods in Molecular Biology, vol. 1674, DOI 10.1007/978-1-4939-7312-5_13, © Springer Science+Business Media LLC 2018

produce glycoproteins with enhanced biological properties, particularly, those related to their pharmacodynamics and pharmacokinetics.

The addition of N- or O-glycosylation sites has been extensively performed into diverse proteins from human origin which were produced in mammalian cell cultures, including erythropoietin (EPO) [3, 4], follicle stimulating hormone (FSH) [5, 6], interferon-α2b (IFN-α2b) [7–9], IFN-β [10–12], among others. In this regard, two second-generation glycosylated biopharmaceuticals have successfully arrived on the market of mammalian cell-produced biobetters: darbepoetin alfa or Novel Erythropoiesis Stimulating Protein (NESP™) derived from hEPO [3] and corifollitropin alfa or Elonva™ derived from hFSH [13]. Both proteins incorporated several glycosylation sites through the application of procedures that can be gathered with the name of glycoengineering by hyperglycosylation.

N-Glycosylation is a cotranslational event that proceeds in the endoplasmic reticulum (ER), being initiated by the transfer of a glycosidic precursor to the consensus sequence Asn-Xxx-Ser/Thr of the nascent polypeptide (where Xxx represents any amino acid, except for Pro). The oligosaccharide is further extended and modified in the ER and Golgi apparatus in order to yield different oligosaccharide structures [14, 15].

O-Glycosylation is a posttranslational event that takes place in the Golgi through a covalent modification of the amino acids Ser or Thr but, differently from N-glycosylation, the consensus sequence is not evident. Nevertheless, it is well known for its preference for Ser/Thr/Pro-rich domains, where Thr is the residue with the highest probability of being glycosylated [16, 17].

Novel glycoproteins displaying new N-glycosyl moieties can be generated by introducing or modifying some amino acids to create a consensus sequence for N-glycosylation or by fusing natural or synthetic peptides containing more than one of the mentioned sequences. Regarding the generation of glycoproteins that incorporate O-glycosyl moieties, they should be obtained by fusing peptide domains from proteins that naturally contain a dense concentration of O-glycans, since the consensus sequence for O-glycosylation is not obvious [9, 18, 19].

In this chapter, a thorough description of the materials and methods routinely used to develop N- and O-hyperglycosylated therapeutic proteins is carried out, employing human IFN-α2b as a model to exemplify them.

2 Materials

2.1 Mutein Generation by PCR

1. High-fidelity DNA polymerase.
2. dNTPs.
3. Thermocycler.

4. Equipment and buffers for casting 1% (w/v) agarose gel electrophoresis.
5. QIAquick Gel Extration Kit (QIAGEN).
6. Transfer lentiviral vector pLenti CMV GFP Puro (Addgene; Plasmid number: #17448).
7. Ligase.
8. TOP10 *E. coli*.

2.2 Cell Line Generation and Clone Selection

2.2.1 Lentivirus (LV) Production

1. Human embryonic kidney 293 cell line (HEK293T/17, ATCC® CRL-11268).
2. Dulbecco's modified Eagle's medium (DMEM) (Gibco®), supplemented with 10% (v/v) fetal bovine serum (FBS, Gibco®), heat-inactivated at 56 °C for 30 min.
3. Third generation lentiviral packaging construct (pMDLg/pRRE), VSV-G expressing construct (pMD2.G), and the Rev. expressing construct (pRSV-Rev) (Addgene; Plasmid numbers: #12251, #12259, and #12253, respectively).
4. Transfer lentiviral vector pLV-GOI (gene of interest).
5. Lipofectamine® 2000 reagent.
6. Syringes, filter units (0.45 μm).

2.2.2 Cell Line Generation

1. Chinese hamster ovary cell line (CHO-K1, ATCC® CCL-61).
2. 1:1 mixture of DMEM medium/Ham's F12 medium supplemented with 5% (v/v) FBS.
3. Lentiviral particles carrying the GOI.
4. 10 cm diameter plates.

2.2.3 Clone Selection

1. Cloning medium: 1:1 mixture of DMEM medium/Ham's F12 medium supplemented with 20% (v/v) FBS.
2. Equipment, buffers, and antibodies necessary for the protein quantification through ELISA.
3. Cell culture plates and flasks.

2.3 Production Under Adherent Conditions

1. Mixture of 1:1 DMEM medium/Ham's F12 medium supplemented with 0.5% (v/v) FBS.
2. Multilayer flasks, Cell Factory systems, or roller bottles.

2.4 Purification

All the buffers for purification must be prepared using ultra pure water. All reagents must be of analytical grade.

1. Filter units (0.45 μm).
2. Triton X-100 for sample preparation.

3. Equilibration buffer: 0.3% (v/v) Triton X-100 in 25 mM Tris–HCl pH 7.5.
4. Washing buffer 1: 0.5 M NaCl, 0.2% (v/v) Triton X-100 in 25 mM Tris–HCl pH 7.5.
5. Washing buffer 2: 0.15 M NaCl.
6. Elution buffer: 0.1 M glycine pH 2.
7. Neutralization solution: 1 M Tris–HCl pH 8.
8. Conservation solution: ethanol 20% (v/v) in water or 0.02% (w/v) sodium azide in 0.1 M Tris–HCl pH 7.
9. Amicon Ultra-4 centrifugal filter units with a cut-off membrane depending on the molecular mass of the protein.
10. XK 16/20 Column (16 × 2.0 cm, GE Healthcare).
11. Equipment for liquid chromatography ÄKTA purifier (GE Healthcare).

2.5 Physicochemical Characterization

2.5.1 SDS-PAGE Followed by Coomassie Blue Staining

1. Equipment and buffers for casting SDS-polyacrylamide gel electrophoresis.
2. Separation gel: 15% (w/v) acrylamide plus N, N′-methylene-bisacrylamide mix (97:3).
3. Stacking gel: 5% (w/v) acrylamide plus N, N′-methylenebisacrylamide mix (97:3).
4. Loading buffer: 50 mM Tris–HCl, 2% SDS (w/v), 10% glycerol (v/v), 5% β-mercaptoethanol (v/v), Bromophenol Blue 0.05% (w/v), pH 6.8.
5. Running buffer: 25 mM Tris, 0.2 M glycine; 0.1% (w/v) SDS.
6. Coomassie Blue solution: 0.1% (w/v) Coomassie Brilliant Blue R-250, 40% (v/v) methanol, 10% (v/v) acetic acid.
7. Distaining solution: 15% (v/v) methanol and 10% (v/v) acetic acid.

2.5.2 Isoelectric Focusing (IEF) Followed by Coomassie Blue Staining

1. Equipment and buffers for casting IEF polyacrylamide/urea gels.
2. IEF gel: 8% (w/v) acrylamide plus N, N′-methylenebisacrylamide mix (97:3) containing urea 7 M.
3. Ampholytes.
4. Coomassie Blue colloidal solution: 0.1% (w/v) Coomassie Brilliant Blue R-250, 20% (v/v) methanol, and 5% (v/v) acetic acid.

2.5.3 Sialic Acids Quantification

1. Mobile phase A: NaOH 100 mM.
2. Mobile phase B: AcONa 500 mM/NaOH 100 mM.
3. Vacuum Concentrator Plus (Eppendorf) or other equivalent.

4. Water bath at 80 °C.
5. High-pH anion-exchange chromatography with pulsed amperometric detection (HPAEC-PAD) using a DIONEX ICS-5000 system (Thermo Scientific Dionex).
6. Sialic acid standards for quantification: acetylneuraminic acid (Neu5Ac) and glycolylneuraminic acid (Neu5Gc) (Calbiochem).
7. CarboPac™ PA20 column (Thermo Scientific Dionex).

2.6 Pharmacokinetic Analysis

1. Female Wistar rats of 200 g.
2. Syringes and needles for animal injection and blood withdrawn.
3. Modified therapeutic protein.
4. Equipment and buffers for the quantification of the protein.

3 Methods

3.1 Design of N- and O-Hyperglycosylated Muteins

1. An N-hyperglycosylated protein can be produced by modifying certain amino acids of its sequence in order to generate the consensus sequence for N-glycosylation: Asn-Xxx-Ser/Thr (where Xxx could be any amino acid except Pro). The approach consists in introducing the consensus sequence into the protein sequence by site-directed mutagenesis of its cDNA. For that, make a theoretical study of the protein sequence and structure in order to determine which amino acids can be modified to obtain the consensus sequence (*see* **Note 1**). Although many amino acid residues can be mutated, the most conservative approach is to modify the least number of residues as possible. For this, three different strategies can be followed:
   (a) Localize every Ser and Thr residues in the protein sequence and exchange the amino acid residue in position −2 for Asn.
   (b) Localize all Asn residues in the protein sequence and mutate the residue in position +2 by Thr (*see* **Notes 2** and **3**).
   (c) An alternative pathway is to design small peptides encompassing such consensus sequence and to fuse them to the terminal ends of the protein.
2. For O-hyperglycosylation, as the consensus sequence is not known, seek for peptide domains from natural proteins that are known to contain a dense concentration of O-glycans or make a theoretical arrangement of different well-identified O-glycosylation domains into one peptide and fused it to one or both ends of the protein sequence. The introduction of the peptide can be performed using mutagenesis by simple

insertion. Evaluate which extreme of the sequence is more suitable for the fusion of the peptide with potential O-glycosylation sites.

3. For both types of glycoengineering procedures make the following analyses:
   (a) Predict N- and O-glycosylation efficiency of the theoretically designed proteins using the NetNGlyc 1.0 and the NetOGlyc 4.0 Servers, respectively (*see* **Note 4**).
   (b) Evaluate the tertiary structure of the protein to avoid making mutations that might cause a significant distortion of it, which might lead to the loss of protein's bioactivity (*see* **Notes 5** and **6**).
   (c) Also evaluate the interaction/recognition/binding sites of the protein of interest (POI) to its biological counterpart by means of bibliographic analysis and by using a protein 3D-structure viewer (such as JMOL, PyMOL, RASMOL, or Swiss PDB Viewer, among others). This should be done in order to prevent introducing modifications in such areas.

3.2 Mutein Generation by PCR

1. Obtain the cDNA of the POI and, if peptide fusions will be carried out, also obtain the cDNA of the corresponding peptide sequences by using a natural source or by synthesis.
2. Design forward and reverse internal primers for site-directed mutagenesis in order to create potential N-glycosylation sites. These must contain the nucleotides necessary for the amino acid replacement (Fig. 1a) (*see* **Note 7**).
3. Design internal primers for mutagenesis by simple insertion. For N-terminal fusion, the sequences of the selected signal peptide and the peptide itself must be amplified together considering the incorporation of some nucleotides from the N-terminal end of the mature POI sequence. These nucleotides should be present in the internal primer (Fig. 1b). For C-terminal fusion, the peptide must be amplified using a primer that incorporates some nucleotides from the C-terminus of the protein (Fig. 1c).
4. Use the same set of external primers for the complete sequence amplification of the new muteins (*see* **Note 8**).
5. Prepare two different tubes for PCR 1. Each of them must contain 1 μl of the internal and external primers prepared at 50 μM concentration and 1 μl of the template prepared at 100 μg/ml. A high fidelity Pfx DNA Polymerase must be used.
6. Analyze fragments from PCR 1 using 1% (w/v) agarose gel electrophoresis. Dissolve 1 mg of agarose in 100 ml 1× TAE buffer. Add 10 μl of SYBR® safe stain. Pour this preparation

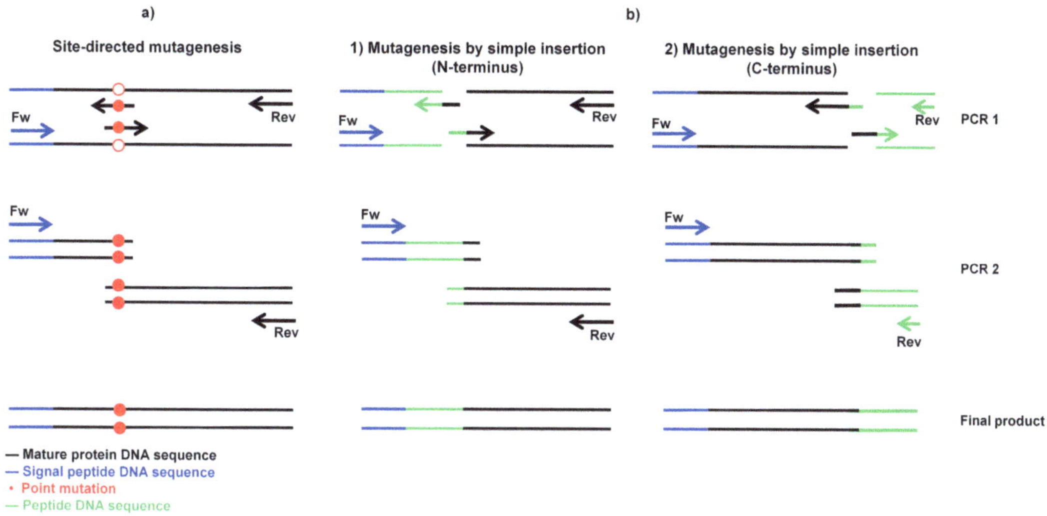

Fig. 1 Strategies for cDNA modification in order to add potential sites for N- and/or O-glycosylation. Two different strategies can be used in order to add potential glycosylation sites: (**a**) Site-directed mutagenesis to introduce a point mutation (*red*) in the cDNA. (**b**) Mutagenesis by simple insertion. Depending on the insertion site, the procedure can be carried out by: (**b1**) The introduction of a peptide (*green*) in the N-terminus of the protein, between the signal peptide (*blue*) and the mature protein sequences (*black*). (**b2**) The fusion of a peptide (*green*) in the C-terminus of the protein sequence. For all the examples the internal and external, forward (Fw) and reverse (Rev) primers, are illustrated. PCR 1 and PCR 2 steps as well as the final product are depicted

into the agarose gel mold and insert an appropriate lane comb. Once it is gelled seed the samples and run the gel at 80 V for approximately 30 min. Visualize using UV transilluminator.

7. Purify the products of the PCR 1 from the gel using the QIAquick® gel extraction kit.
8. Perform PCR 2 in order to join the two fragments already obtained in PCR 1. The protocol has two steps. In the first step the same amount in moles of both fragments are heated and hybridized using an appropriate temperature. The missing nucleotides are added in order to complete the entire sequence. The first step normally takes approximately 10 cycles. In the second step 1 μl of the external primers prepared at 50 μM should be added in order to amplify the whole sequence and incorporate the desire restriction sites for cloning.
9. Analyze the product from PCR 2 using agarose gel electrophoresis as described in **step 6** of this section.
10. Purify the product of PCR 2 from the gel as described in **step 7** of this section.
11. Digest the purified fragment with the corresponding restriction enzymes and ligate it to the plasmid pLenti CMV GFP

puro already digested with proper enzymes. Use standard protocols in both procedures.

12. Transform competent TOP10 *E. coli* with the ligation product (pLenti-GOI). Seed the bacteria over LB agar containing an appropriate antibiotic for the selection of the transformants.
13. Choose five bacteria colonies, to extract the plasmid DNA and to verify the presence of the mutated fragment by restriction with different enzymes.
14. Verify the correct introduction of the mutations by DNA sequencing.

3.3 Cell Line Generation and Clone Selection

3.3.1 LVs Production

1. The first day of the assay, seed HEK293T cells at a concentration of 4×10^5 cell/ml in 10 cm diameter plates and incubate ON at 37 °C and 5% CO_2.
2. The second day, prepare the reagents for transfection. In one sterile tube add 2.5 ml of DMEM medium and 50 μl of Lipofectamine® 2000 reagent. In a second sterile tube add the following plasmids: 2.5 μg of pRSV-Rev., 3.6 μg of pMD2.G, 6.5 μg of pMDLg/pRRE and 10 μg of the transference vector containing the mutated fragment (pLenti-GOI).
3. Mix the content of both tubes, vortex and incubate the mixture at RT during 20 min, in order to allow the formation of the DNA-lipid complexes.
4. During complexes formation, wash the cells smoothly using 5 ml of DMEM medium. Be careful to avoid cell detachment.
5. Remove the supernatant from the cells and add the DNA–lipid complexes over them by dripping.
6. After 5 h, add 5 ml of fresh DMEM medium containing 20% (v/v) FBS. Incubate the cells at 37 °C and 5%CO_2 during 48 h.
7. Harvest the supernatant containing the LVs and clarify it by centrifugation at $100 \times g$ for 10 min. Filter them using a 0.45 μm filter unit.
8. LVs must be kept at −80 °C. Considering that the lentiviral titer falls down with the time, immediately use them for transduction if it is possible.

3.3.2 Cell Line Generation

1. Seed CHO-K1 cells at a concentration of 3×10^4 cell/ml in a 6-well plate and incubate them over night (ON) at 37 °C and 5% CO_2 for 24 h.
2. Discard the supernatant from the cell culture and add 1 ml of the LVs to the monolayer. A high MOI is desirable to generate a cell line with high productivity. Incubate ON at 37 °C and 5% CO_2.
3. Remove the supernatant containing the LVs and add fresh medium to the cells. Incubate them under the same conditions until the monolayer reaches confluence.

4. Start the selection of the cell line using a low concentration of the corresponding agent, whose resistance is conferred by a gene encoded in the transference vector. Increase gradually the concentration of the antibiotic until death of the entire culture is observed. The cell line that resists to the maximum concentration of the selection agent should correspond to the one with the maximum productivity (*see* **Note 9**). Use a well containing nontransduced cells as a negative control.
5. Collect the supernatant from the selected cell line and determine the POI concentration by ELISA to calculate the specific productivity (*see* **Note 10**).
6. To evaluate the success of the glycosylation strategy, analyze the molecular mass of the protein by SDS-PAGE/western blot using specific antibodies against the POI.

3.3.3 Clone Selection

1. Seed the selected cell line in 96-well plates using two different cell proportions: 1 and 0.2 cells/well. Use a special medium for cloning. Prepare five plates.
2. After 1 week, observe each well of the plates in order to determine those where a single clone can be observed.
3. After 15 days, harvest the supernatant of each clone and analyze the POI concentration by ELISA. Choose those wells having the corresponding clone, which is able to produce the maximum POI concentration.
4. Amplify the selected clone/s and generate the master and working cell banks. The size of them would depend on the frequency of use and on the production stability, which might determine the length of the production process.

3.4 Production Under Adherent Conditions

1. Seed the cells in a proper device for adherent growth in a concentration of 5×10^5 cell/ml. Multilayer flasks, Cell Factory systems, or roller bottles can be chosen. Use DMEM medium containing 5% (v/v) FBS to initiate the cell culture.
2. After 24 h, discard the supernatant and add fresh medium containing only 0.5% (v/v) FBS (production medium) (*see* **Note 11**). For cell lines with low adherent capacity, substrates can be pretreated 1 h at 37 °C and ON at 4 °C with poly-lysine 5 μg/ml (prepared in sterile water) in order to improve their attachment.
3. Harvest the supernatant each 48 or 72 h, centrifuge them at $100 \times g$ for 10 min and preserve them under −20 °C. After harvesting, replace the supernatant by the production medium. At the end of the production step, pool all the harvests (*see* **Note 12**).

3.5 Purification

1. Purify the hyperglycosylated POI using a chromatographic protocol. To our experience, immunoaffinity chromatography represents a suitable method to purify the whole set of glycoforms derived from the glycosylated muteins (*see* **Note 13**). Consequently, the following procedure corresponds to an immunoaffinity chromatography.
2. Prepare the harvest for loading to the immunoaffinity matrix (*see* **Note 14**). Adjust pH to 7.5 and add Triton X-100 to a final concentration of 0.3% (v/v). Filter the sample through a 0.45 μm cellulose acetate membrane filter.
3. Equilibrate the column with 5 bed volumes (BV) of equilibration buffer at a flow rate of 0.5 cm/min.
4. Load the previously prepared sample to the immunoadsorbent column maintaining a flow rate of 0.5 cm/min.
5. Wash the column with 5 BV of washing buffer 1 and 8 BV of washing buffer 2, at the same flow rate.
6. Elute the matrix-retained mutein by pH decrease using elution buffer (*see* **Note 15**). Collect fractions corresponding to a half of BV at the same flow rate. The pH in the eluted fractions might be immediately neutralized with neutralization solution. Elution may proceed until the absorbance signal is nearly zero.
7. Apply a conservation solution to the immunoaffinity matrix in order to preserve it from contamination.
8. Quantify the POI in the elution fractions by the same ELISA previously used for supernatant quantification.
9. Pool the fractions containing the highest POI concentrations.
10. Diafilter and concentrate the sample against an appropriate vehicle (*see* **Note 16**). Use Amicon Ultra-4 centrifugal filter units with a membrane having a molecular mass (MM) cut-off depending on the POI MM.

    Determine the protein concentration using a spectrophotometric method like absorbance measurement at $\lambda = 280$ nm.
11. Analyze the purity of the protein in SDS-PAGE gels followed by Coomassie Blue staining (*see* **step 1 from 3.3.6**). The purity can be determined by band densitometry.

3.6 Physicochemical Characterization

To evaluate the extent of physicochemical changes in the POI after glycoengineering procedures, many different simple techniques that allow quick analyses of the new proteins can be performed, e.g., apparent molecular mass determination, isoform distribution and sialic acid content (*see* **Note 17**).

3.6.1 SDS-PAGE Followed by Coomassie Blue Staining

1. Assemble the system for vertical gel electrophoresis and fill it with water in order to check that there is no leak.

2. Prepare a denaturing SDS–polyacrylamide gel with a thickness of 1.5 mm. First, pour the separation gel preparation inside the vertical support. Once it is gelled, add the stacking gel mixture.
3. Place the gel in the running module and immerse this support into the buffer tank containing the running buffer. It is recommended to set the buffer tank inside a bowl with ice water in order to preserve the run at low temperature.
4. Prepare the samples by diluting 5 μg of the protein in loading buffer. Boil the samples for 10 min. The molecular mass marker should be prepared following the manufacturer's instructions. Dispense samples and marker into different lanes.
5. Close the system with the lid and connect it to a power supply. Perform the electrophoretic run at 200 V constantly during 60 min.
6. After the run is finished, remove the gel from its support and immerse it into Coomassie Blue solution for 1 min.
7. Distain the gel with distaining solution in order to remove the background. Clear bands for each sample should be detected (Fig. 2a).

3.6.2 IEF Followed by Coomassie Blue Staining

1. Assemble the gel support and fill it with water in order to check that there is no leak.
2. Prepare a 1-mm thick gel for IEF procedure. A pH range has to be established using ampholytes that must be added during the gel preparation. For hyperglycosylated proteins a low pH range would be suitable (*see* **Note 18**).
3. Prefocus the gel at 10 W, 2000 V and 100 mA for 1 h. Use anolyte and catholyte solutions to wet the IEF electrode strips.
4. Prepare samples by diafiltering them using a low ionic strength buffer (*see* **Note 19**).
5. Dilute the samples in water and apply them on the sample application strip placed at 1 cm from cathode above the gel. Approximately 20 μg of each protein should be loaded in order to be able to detect the whole set of isoforms.
6. Focus the samples using the same conditions as the prefocusing step but during 45 min.
7. Remove the gel from the support and stain the samples using a Coomassie Blue colloidal solution for 30 min.
8. Discard the staining solution and wash carefully with water. Many different bands might be observed for each hyperglycosylated protein. The number of isoforms must be higher for the new molecules compared to the native one. Also, the migration distance toward the anode will be higher for the

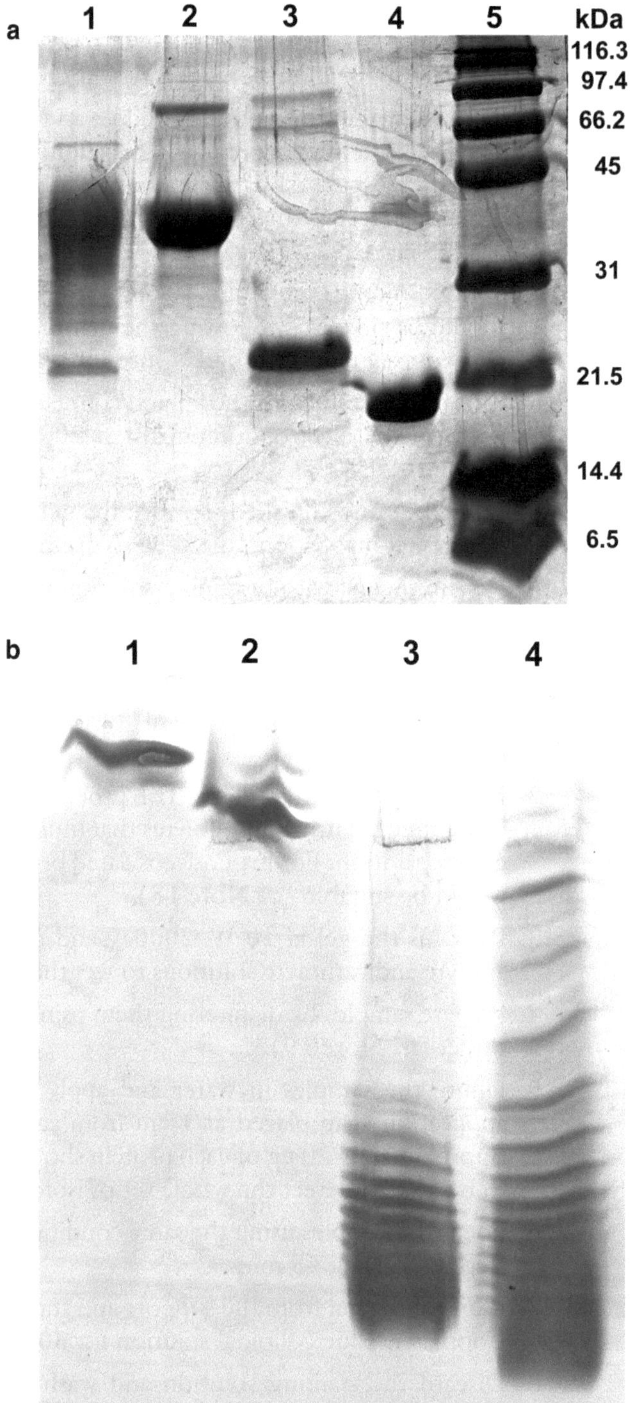

Fig. 2 Apparent molecular mass (**a**) and isoform distribution (**b**) analysis of N- and O-hyperglycosylated human IFN-α2b variants through SDS-PAGE followed by Coomassie Blue staining. (**a**) shows an example of an N-hyperglycosylated (IFN4N) and an O-hyperglycosylated (CTP-IFN-CTP) human IFN-α2b. IFN4N (*lane 1*) was obtained by introducing four potential N-glycosylation sites by site-directed mutagenesis and CTP-IFN-CTP (*lane 2*) was obtained by fusing the carboxyl-terminal peptide (CTP), derived from the β-subunit of the human

hyperglycosylated version, reaching an area that has a lower pH compared to that reached by the nonmodified protein (Fig. 2b).

3.6.3 Quantification of Sialic Acids

1. Dry 20 μg of protein using a vacuum concentration device as listed in Subheading 2. Remove the sialic acids from the protein by acid hydrolysis with 0.1 M HCl (200 μl) during 1 h at 80 °C.
2. Dry the product of hydrolysis and resuspend the pellet with 20 μl of milliQ water.
3. Prepare the Neu5Ac and Neu5Gc standard curves (Calbiochem) from 0.0625 to 5 pmol/μl and from 0.002 to 1 pmol/μl, respectively.
4. Analyze samples in triplicates and curves in duplicates by HPAEC-PAD using a DIONEX ICS-5000 system equipped with a CarboPac™ PA20 column. The column must be previously equilibrated using mobile phases A and B in a 90:10 proportion, respectively. The volume of injection for samples and standards is 100 μl. Elution of both species of sialic acids is performed using the gradient of A and B solutions described in the following Table 1.
5. Use the standard curves for quantification of sialic acids derived from the samples. Build graphs of standard curves representing area under the curve versus mass of sialic acid (pmol).

Table 1
Gradient of A and B solutions

	Solution	
Time (min)	**A (%)**	**B (%)**
0	90	10
4	90	10
14	50	50
18	90	10
24	90	10

Fig. 2 (continued) chorionic gonadotropin, to both extremes of the IFN. Their molecular masses were higher than those observed for the wild type molecule (*lane 3*) and the nonglycosylated one (*lane 4*). (**b**) exhibits the isoform distribution of nonglycosylated IFN (*lane 1*), wild type IFN (*lane 2*), CTP-IFN-CTP (*lane 3*), and IFN4N (*lane 4*). The hyperglycosylated IFN variants showed a higher isoform content with a lower isoelectric point

3.7 Pharmacokinetic Analysis

1. Wistar rats must be housed in a temperature-controlled room at 23 °C, with a 12 h light–dark cycle and free access to food and water. At least eight rats of 200 g must be used for the analysis of each protein.
2. Randomly divide the rats in two groups of four animals and assign different time schedule for blood withdrawn after injection. This design should be considered for minimizing the impact of blood withdrawn on each animal.
3. Inject a unique dose of the POI subcutaneously. Mass or biological activity units can be used for this experiment. The amount of protein injected would depend on the sensitivity and detection limit of the method that will be used for its quantification.
4. Take blood samples at the times post-injection previously scheduled. Collect the samples on heparin, centrifuge at 2400 × *g* during 5 min and store at −20 °C until analysis.
5. Analyze the samples according to the established method.
6. Graph the amount of quantitated protein versus time (Fig. 3).
7. Determine if the molecule follows a monocompartmental, bicompartmental, or noncompartmental model (*see* **Note 20**). Calculate parameters such as maximum plasma

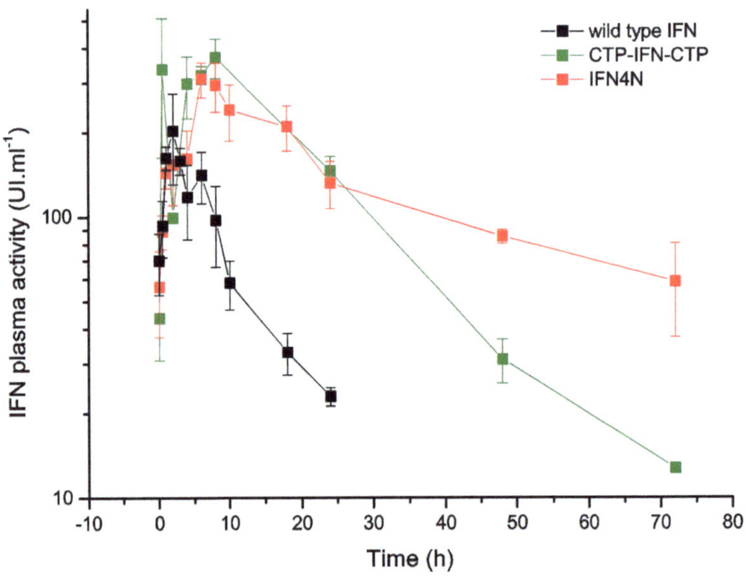

Fig. 3 Pharmacokinetic analysis of N- and O-hyperglycosylated IFN variants. Wild type IFN (*black*), CTP-IFN-CTP (*green*) and IFN4N (*red*) pharmacokinetic plasma profiles were evaluated in Wistar rats after subcutaneous administration. Blood samples were collected at different time points and the in vitro antiviral activity was analyzed. Results are expressed as the average ± SD. The pharmacokinetic properties of the protein improved as its apparent molecular mass increased

concentration (C_{max}), time to reach the C_{max} (T_{max}), terminal half-life ($t_{1/2}$) and apparent plasma clearance (CL_{app}).

8. Statistical analyses must be performed to determine the existence of significant differences between the modified and unmodified molecule.

4 Notes

1. A theoretical study to carefully select the amino acids to be mutated should be performed in order to: (a) generate consensus sites with a high probability of glycosylation, and (b) avoid altering the 3D structure and stability of the protein, with the aim of preserving its biological activity.
2. Thr is chosen because it has been previously described that there is a preference for N-X-T sites over N-X-S, resulting in an occupation degree between two and three times higher [20].
3. Take into account other factors that affect N-glycosylation efficiency: (a) the presence of Pro in the second position or immediately after the consensus site completely blocks N-glycosylation; (b) due to their size and hydrophobicity, amino acid residues such as Leu, Glu, Asp, and Trp may not be suitable to occupy the second position of the consensus tripeptide; (c) consensus sites near the C-terminal end of the protein are less efficiently glycosylated, mainly because of the reduced amount of time available to add carbohydrate to an almost completely synthesized protein [21].
4. Other online N-glycosylation prediction servers can be employed, such as GlycoEP Server [22] and NGlycPred [23]. Also, O-glycosylation can be predicted by other servers, e.g., CKSAAP_OGlySite [24] and ISOGlyP [25].
5. The first residue of the consensus sequence (Asn or the amino acid to be mutated to Asn) should exhibit a high solvent accessibility. This parameter can be evaluated using online algorithms, such as ASAview (http://www.abren.net/asaview/) or NetSurfP v1.1 (http://www.cbs.dtu.dk/services/NetSurfP/), which render information about the exposure degree of the residue to which the glycan chain would be attached.
6. In the case of N-glycosylation, which constitutes a cotranslational process, if the residue to which the glycan is attached is buried in the 3D structure of the protein, its mutation or the addition of an oligosaccharide chain may alter the correct folding of the protein, leading to a potential loss of biological activity.

7. Internal primers should be designed of an appropriate length considering the GC content and the number of mismatches, so that a T_m higher than 50 °C is obtained.
8. External primers should contain the restriction sites necessary for cloning in an appropriate plasmid for lentivirus (LVs) assembly.
9. For pLenti CMV GFP Puro vector, puromycin from *Streptomyces alboniger* is used as selective agent. Resistance to puromycin is conferred by the Pac gene encoding a puromycin N-acetyl-transferase that was found in a Streptomyces producer strain and that is encoded in pLenti vector. The appropriate concentration of antibiotic for selecting stable cell lines is different for each cell type. If the concentration for the desired cell type is unknown, a titration experiment must be performed. Typically, 1–10 μg/ml is sufficient to kill most untransduced mammalian cell types. Puromycin concentration can be gradually increased up to 200–300 μg/ml for highly resistant transduced cell lines, depending on the promoter strength and the copy number of the transgene integrated in the genome
10. Different ELISA formats can be used, including competition and sandwich assays. In general, sandwich assays using a mAb for capture and a pAb for detection exhibit a lower detection limit. Take into account that if the protein has been mutated in amino acids relevant for the binding to the antibodies used for the ELISA assay, the signal might be reduced or the assay might not be used for quantification. Also consider that a highly glycosidic content could lead to a lower antibody recognition when a sandwich system is carried out [7]. In this case a competition design using a mAb would be appropriated because the binding is ruled by the recognition of only one epitope.
11. Considering that most of the adherent cell lines used for production can duplicate in 24 h, by that time the culture would contain 1×10^6 cell/ml. That means monolayer confluence and, in general, is the best phase for production.
12. The length of the production process depends mainly on the protein stability and the production characteristics of the clone.
13. The physicochemical characteristics of the POI generally change after the introduction of new glycan moieties. If a purification protocol based in different ion exchange/hydrophobic/size exclusion chromatographic steps is routinely used for the purification of the POI, consider that the binding/elution conditions and the final purity can change. Also, not all glycoforms might be recovered from these procedures. Besides, immunoaffinity purification can be a suitable option for the purification of different muteins of the same protein if

an antibody whose epitope-binding properties are not modified by the mutagenesis is available.

14. Immunoaffinity matrix can be prepared by coupling the ligand (in this case, the specific mAb) to a CNBr-activated Sepharose matrix, following the protocol suggested by the manufacturer. The mAb can be obtained by culture of hybridomas or by ascites production in mice, and then purified by protein A chromatography. A ligand density of about 5 mg mAb per ml medium is recommended. For proteins of molecular masses ranging from 25 to 50 kDa, a theoretical capacity of about 0.8–1.6 mg of POI per ml of resin is expected.
15. Conditions for elution of bound substances will depend on the predominant type of interaction with the coupled mAb. Different approaches can be followed to recover the maximum quantity of the POI: pH change (elution is generally affected by a decrease of pH), ionic strength increase (NaCl gradient), reduced polarity of the eluent (such as dioxane up to 10% (v/v) or ethylene glycol up to 50% (v/v)), deforming agents (such as chaotropic salts, guanidine or urea). In all cases, care should be taken in order to use conditions that are compatible with the stability of the matrix, the ligand and the adsorbed substances.
16. According to the stability conditions of the POI, an appropriate vehicle should be selected in order to preserve its properties along time. Consider the buffer species (phosphate, citrate, etc.), the use of albumin or other stabilizing agents, and formulate the POI for physicochemical and in vivo characterization.
17. A biological characterization must also be performed in order to evaluate if the modifications have altered the protein's bioactivity. This might be carried out through in vitro bioassays according to the bioactivity displayed by the POI. For example, antiviral bioassays can be performed to measure the bioactivity of IFN derivatives, while proliferation bioassays may be used to evaluate hyperglycosylated products from colony-stimulating factors.
18. Commercial precast IEF gels can be obtained. They are offered at different pH ranges according to the application.
19. The elimination of salt traces is imperative in order to avoid band distortions during the migration, separation, and focusing of the glycoforms according to their p*I*.
20. The complexity of the pharmacokinetic model that is chosen to describe the behavior of a drug varies with the administration route, the duration of its tissue distribution and the elimination process. Compartmentally based models are useful tools in pharmacokinetics. One-compartment models offer the simplest way to describe the process of drug distribution and

elimination on the body. It assumes that the drug can enter or leave the body, which acts as a single, uniform compartment. Two-compartment models assume that the drug distributes in two compartments: a central compartment (blood and highly perfused tissues) in which the drug distributes rapidly, and a peripheral compartment, which includes tissues in which the drug equilibrates more slowly. Multicompartment models divide the different tissues in more compartments. Numerous disadvantages are associated with compartmental models, including their lack of biologic relevance, leading to the development of noncompartmental models or physiological models.

References

1. Elliott S, Lorenzini T, Asher S, Aoki K, Brankow D, Buck L, Busse L, Chang D, Fuller J, Grant J, Hernday N, Hokum M, Hu S, Knudten A, Levin N, Komorowski R, Martin F, Navarro R, Osslund T, Rogers G, Rogers N, Trail G, Egrie J (2003) Enhancement of therapeutic protein in vivo activities through glycoengineering. Nat Biotechnol 21(4):414–421. doi:10.1038/nbt799nbt799 [pii]
2. Sinclair AM, Elliott S (2005) Glycoengineering: the effect of glycosylation on the properties of therapeutic proteins. J Pharm Sci 94(8):1626–1635. doi:10.1002/jps.20319
3. Egrie JC, Browne JK (2001) Development and characterization of novel erythropoiesis stimulating protein (NESP). Br J Cancer 84(Suppl 1):3–10. doi:10.1054/bjoc.2001.1746S000709200191746X [pii]
4. Egrie JC, Dwyer E, Browne JK, Hitz A, Lykos MA (2003) Darbepoetin alfa has a longer circulating half-life and greater in vivo potency than recombinant human erythropoietin. Exp Hematol 31(4):290–299
5. Perlman S, van den Hazel B, Christiansen J, Gram-Nielsen S, Jeppesen CB, Andersen KV, Halkier T, Okkels S, Schambye HT (2003) Glycosylation of an N-terminal extension prolongs the half-life and increases the in vivo activity of follicle stimulating hormone. J Clin Endocrinol Metab 88(7):3227–3235. doi:10.1210/jc.2002-021201
6. Fares FA, Suganuma N, Nishimori K, LaPolt PS, Hsueh AJ, Boime I (1992) Design of a long-acting follitropin agonist by fusing the C-terminal sequence of the chorionic gonadotropin beta subunit to the follitropin beta subunit. Proc Natl Acad Sci U S A 89(10):4304–4308
7. Ceaglio N, Etcheverrigaray M, Kratje R, Oggero M (2008) Novel long-lasting interferon alpha derivatives designed by glycoengineering. Biochimie 90(3):437–449. doi:S0300-9084(07)00307-0 [pii]10.1016/j.biochi.2007.10.013
8. Ceaglio N, Etcheverrigaray M, Conradt HS, Grammel N, Kratje R, Oggero M (2010) Highly glycosylated human alpha interferon: an insight into a new therapeutic candidate. J Biotechnol 146(1-2):74–83. doi:S0168-1656(10)00025-8 [pii]10.1016/j.jbiotec.2009.12.020
9. Ceaglio N, Gugliotta A, Tardivo MB, Cravero D, Etcheverrigaray M, Kratje R, Oggero M (2016) Improvement of in vitro stability and pharmacokinetics of hIFN-alpha by fusing the carboxyl-terminal peptide of hCG beta-subunit. J Biotechnol 221:13–24. doi:S0168-1656(16)30019-0 [pii]10.1016/j.jbiotec.2016.01.018
10. Samoudi M, Minuchehr Z, Harcum SW, Tabandeh F, Omid Yeganeh N, Khodabandeh M (2017) Rational design of glycoengineered interferon-beta analogs with improved aggregation state: experimental validation. Protein Eng Des Sel 30(1):23–30. doi:gzw058 [pii] 10.1093/protein/gzw058
11. Samoudi M, Tabandeh F, Minuchehr Z, Ahangari Cohan R, Nouri Inanlou D, Khodabandeh M, Sabery Anvar M (2015) Rational design of hyper-glycosylated interferon beta analogs: a computational strategy for glycoengineering. J Mol Graph Model 56:31–42. doi:S1093-3263(14)00204-6 [pii]10.1016/j.jmgm.2014.12.001
12. Song K, Yoon IS, Kim NA, Kim DH, Lee J, Lee HJ, Lee S, Choi S, Choi MK, Kim HH, Jeong SH, Son WS, Kim DD, Shin YK (2014) Glycoengineering of interferon-beta 1a improves its biophysical and pharmacokinetic properties. PLoS One 9(5):e96967. doi:10.1371/journal.pone.0096967PONE-D-14-02849. [pii]

13. Croxtall JD, McKeage K (2011) Corifollitropin alfa: a review of its use in controlled ovarian stimulation for assisted reproduction. BioDrugs 25(4):243–254. doi:4 [pii]10.2165/11206890-000000000-00000
14. Aebi M (2013) N-linked protein glycosylation in the ER. Biochim Biophys Acta 1833 (11):2430–2437. doi:S0167-4889(13) 00132-8 [pii]10.1016/j.bbamcr.2013.04.001
15. Schwarz F, Aebi M (2011) Mechanisms and principles of N-linked protein glycosylation. Curr Opin Struct Biol 21(5):576–582. S0959-440X(11)00140-0 [pii]10.1016/j.sbi.2011.08.005
16. Van den Steen P, Rudd PM, Dwek RA, Opdenakker G (1998) Concepts and principles of O-linked glycosylation. Crit Rev Biochem Mol Biol 33(3):151–208. doi:10.1080/10409239891204198
17. Rudd PM, Dwek RA (1997) Glycosylation: heterogeneity and the 3D structure of proteins. Crit Rev Biochem Mol Biol 32(1):1–100. doi:10.3109/10409239709085144
18. Fares F, Havron A, Fima E (2011) Designing a long acting erythropoietin by fusing three carboxyl-terminal peptides of human chorionic gonadotropin beta subunit to the N-terminal and C-terminal coding sequence. Int J Cell Biol 2011:275063. doi:10.1155/2011/275063
19. Fares F, Guy R, Bar-Ilan A, Felikman Y, Fima E (2010) Designing a long-acting human growth hormone (hGH) by fusing the carboxyl-terminal peptide of human chorionic gonadotropin beta-subunit to the coding sequence of hGH. Endocrinology 151(9):4410–4417. doi:en.2009-1431 [pii]10.1210/en.2009-1431
20. Kasturi L, Eshleman JR, Wunner WH, Shakin-Eshleman SH (1995) The hydroxy amino acid in an Asn-X-Ser/Thr sequon can influence N-linked core glycosylation efficiency and the level of expression of a cell surface glycoprotein. J Biol Chem 270(24):14756–14761
21. Shakin-Eshleman SH, Spitalnik SL, Kasturi L (1996) The amino acid at the X position of an Asn-X-Ser sequon is an important determinant of N-linked core-glycosylation efficiency. J Biol Chem 271(11):6363–6366
22. Chauhan JS, Rao A, Raghava GP (2013) In silico platform for prediction of N-, O- and C-glycosites in eukaryotic protein sequences. PLoS One 8(6):e67008. doi:10.1371/journal.pone.0067008PONE-D-12-31390. [pii]
23. Chuang GY, Boyington JC, Joyce MG, Zhu J, Nabel GJ, Kwong PD, Georgiev I (2012) Computational prediction of N-linked glycosylation incorporating structural properties and patterns. Bioinformatics 28(17):2249–2255. doi:bts426 [pii]10.1093/bioinformatics/bts426
24. Chen YZ, Tang YR, Sheng ZY, Zhang Z (2008) Prediction of mucin-type O-glycosylation sites in mammalian proteins using the composition of k-spaced amino acid pairs. BMC Bioinformatics 9:101. doi:1471-2105-9-101 [pii]10.1186/1471-2105-9-101
25. Leung M-Y, Cardenas GA, Almeida IC, Gerken TA (2014) Isoform specific O-glycosylation prediction (ISOGlyP)Version 1.2.

Chapter 14

Expression of Glycosylated Proteins in Bacterial System and Purification by Affinity Chromatography

Carlos Alexandre Breyer, Marcos Antonio de Oliveira, and Adalberto Pessoa Jr

Abstract

The bacterial expression of glycoproteins has experienced significant progress in recent years, particularly in regard to the production of conjugate vaccines against pathogens. In this case, a protein carrier conjugated with glycosides is used to produce intense stimulation of the immune system. Glycoconjugate vaccines account for 35% of the global vaccine market, and consequently, several biotechnological companies have developed products for the purification of glycosylated proteins to attain homogeneity. In this chapter we present a general process for glycoprotein production in *Escherichia coli* and a practice method for purification of glycosylated proteins, using affinity chromatography.

Key words Glycosylated protein, Bacterial expression, Conjugate vaccines, Affinity chromatography

1 Introduction

Glycosylation is the most common posttranslational modification (PTM) of proteins [1] and it can occurs in several amino acid residues, but the most commonly modified residues are asparagine (*N*-glycosylation), threonine, and serine (*O*-glycosylation) [2]. In *N*-glycosylation, the glycan is attached to the amide nitrogen of Asn and in *O*-glycosylation, the glycosides are attached to the hydroxyl oxygen of the Ser or Thr residues [3]. For some time, it was believed that glycosylation occurred solely in eukaryotes; however, the process has been reported in other organisms such as archaea and bacteria. The human gastrointestinal pathogen *Campylobacter jejuni* is the best-characterized organism capable of performing protein glycosylation, with more than 60 N-glycosylated proteins described to date [2, 4–6]. Accordingly, it is used as a biological model in prokaryotes. Glycosylation in bacteria is related to pathogenicity, and the deletion of the genes involved in these processes results in a decrease in immunogenicity, invasion, and

Virgínia Picanço-Castro and Kamilla Swiech (eds.), *Recombinant Glycoprotein Production: Methods and Protocols*, Methods in Molecular Biology, vol. 1674, DOI 10.1007/978-1-4939-7312-5_14,

pathogen adherence, among other processes [2, 4]. The *C. jejuni* glycosylation pathway is encoded by the single *pgl-1* gene cluster, which encodes for glycosyltransferases, a flippase that translocates the oligosaccharide to the periplasm, and the oligosaccharyltransferase PglB. The latter is able to transfer the O-polysaccharide from a lipid carrier to an acceptor protein. The PglB enzyme covalently attaches the glycan to asparagine residues in the motif D/E-X-N-X-S/T, in which X can represent any amino acid except proline [5]. Walker and colleagues (2002) demonstrated that *C. jejuni pgl-1* gene cluster could be successfully transferred to the *Escherichia coli* with full functionality. The expression of the *C. jejuni* AcrA-1, a periplasmic protein that can carry two oligosaccharides, and the *pgl-1* gene cluster results in AcrA-1 glycosylation in *E. coli* [6]. It was demonstrated posteriorly that PglB oligosaccharyltransferase activity is sufficient to transfer the oligosaccharides from undecaprenyl phosphate to acceptor proteins containing glycosylation sites. This process is possible because *E. coli* can synthesize heterologous polysaccharides on their glycosyl carrier lipid [7]. To produce glycosylated protein that carries heterologous polysaccharides, Feldman and colleagues (2005) constructed the *E. coli* (K12) CLM24, a strain that does not synthetize the O-antigen, because it contains the deleted O-antigen ligase *waaL* gene (*ΔwaaL*). This enzyme is responsible for transferring the O antigen to a lipid A-core. The introduction of the gene cluster to express *E. coli* O7 antigen, *pglB* and *acra1* in *E. coli* CLM24 results in AcrA-1 covalently linked to O7 polysaccharides [4]. Similar results were obtained when polysaccharides O16 from *E. coli* and O11 from *Pseudomonas aeruginosa* were expressed heterologously in *E. coli* CLM24 [4].

Moreover, the introduction of N-glycan glycosylation sites (DQNAT) and the periplasm signal peptide was able to promote efficient glycosylation of several different proteins [8]. Therefore, the expression of *pglB*, the gene cluster for polysaccharide expression, the carrier protein, containing an N-glycan sites and periplasm signal peptide, in an *E. coli* CLM24 strain result in the efficient glycosylation of the carrier protein (Fig. 1).

The production of glycosylated protein in *E. coli* is primarily related to the production of glycoprotein conjugate vaccines. In this type of conjugate vaccine, the capsular or lipopolysaccharides antigens of the pathogenic bacteria are attached covalently to a carrier protein [9]. Glycoengineering technology was applied to the production of conjugate vaccines against *Shigella dysenteriae* type 1 and *Staphylococcus aureus* [9–11]. The use of modified *E. coli* strains (*ΔwaaL*), expressing a specific oligosaccharide, a carrier protein, and PglB, has been suggested as a simple, low-cost alternative to glycoprotein production for glycoconjugate vaccines [7, 9, 10].

This chapter provides a general protocol for glycoprotein expression in *E. coli* that exemplifies glycoprotein conjugate vaccine

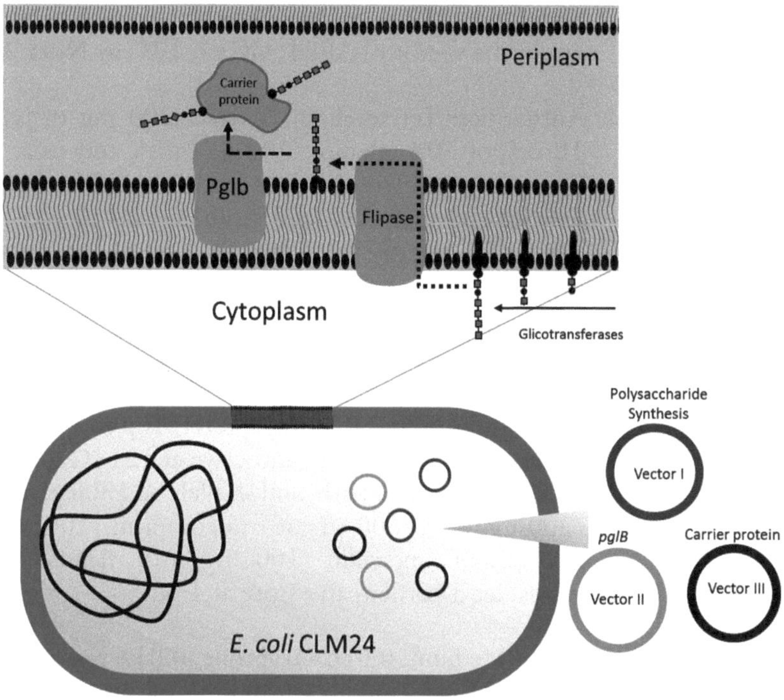

Fig. 1 Recombinant glycosylated protein expression in *E. coli* CLM24 strain. Vectors containing the gene cluster responsible for polysaccharides expression, *pglB* and carrier protein are transformed in *E. coli* CLM24. The transformation leads to polysaccharide production, expressed by a constitutive promoter. The polysaccharide is formed by expression of specific polymerase enzymes and the flippase responsible for transferring the polysaccharide to the periplasm. The PglB and carrier protein expression produces glycosylated carrier protein

production and a general method for glycosylated protein purification based on affinity chromatography, using immobilized lectin. The lectins are high affinity carbohydrate-binding proteins, to which diverse glycosides can bind. The elution is performed through a specific carbohydrate gradient [12]. Many lectins are commercially available, the most common including concanavalin A (*Canavalia ensiformis*), wheat germ (*Triticum vulgaris*), jacalin (*Artocarpus integrifolia*), and lentil (*Lens culinaris*).

2 Materials

2.1 Strains and Vectors

1. Bacteria strains: Electrocompetent *E. coli* CLM24: W3110 *ΔwaaL* [4].
2. Vectors: pMAF10 (pMLBAD), containing HA-tagged (human influenza hemagglutinin) and *pglB*, confers resistance trimethoprim [4, 8]. pTrc99A vector (GE Healthcare) confers resistence to ampicilin and is used to clone the carrier protein (CP) (pTrc99A_CP) (*see* **Note 1**). pLAFR1 confers resistance to

tetracycline and is used to insert the gene clusters (GC) resulting in the vector pLAFR1_GC [7, 13] (*see* **Note 2**).

2.2 Antibiotics and Selective Solid Media

1. Antibiotics: Tetracycline: Dissolve 200 mg of tetracycline in 10 mL of 70% ethanol (200 μg/mL), and store at −20 °C; Trimethoprim: Dissolve 500 mg of trimethoprim in 10 mL of DMSO (50 mg/mL), and store at −20 °C; Ampicillin: Dissolve 250 mg of ampicillin in 2.5 mL of ultrapure water (100 mg/mL), and store at −20 °C.
2. LB agar-plates: Add 200 mL of ultrapure water to a 1 L glass beaker. Weigh 5 g of tryptone, 2.5 g of yeast extract, 2.5 g of NaCl, 2.0 g of agar, and transfer to the beaker. Mix the solution and adjust pH to 7.0, using NaOH. Top to 1 L with deionized water, and sterilize in an autoclave at 121 °C for 20 min. Cool to 55 °C in a water bath, and sterilely add 500 μL of tetracycline (200 μg/mL), 500 μL of trimethoprim (100 mg/mL), and 500 μL of ampicillin (100 mg/mL). Transfer medium to plates, let it harden, and store at 4 °C.

2.3 Selective Media and Solutions to Protein Expression

1. 2XYT Medium: 0.016% tryptone, 0.01% yeast extract, 0.005% NaCl, pH 7.4. Add about 500 mL of ultrapure water to a 1 L glass beaker. Weigh 16 g of tryptone, 10 g of yeast extract, 5 g of NaCl, and transfer to the beaker. Mix and adjust pH to 7.0, using NaOH. Top to 1 L with deionized water, sterilize in autoclave at 121 °C for 20 min, and store at room temperature.
2. 1 M IPTG: Dissolve 1.19 g of IPTG in 5 mL of ultrapure water, and store at −20 °C.
3. 10% (w/v) arabinose: Dissolve 1 g of in arabinose in 9 mL in ultrapure water, and store at room temperature. Sterilize by filtration with a 0.22 μm filter.
4. Cracking buffer: 60 mM Tris–HCl, 1% (v/v) 2-mercaptoethanol, 1% (w/v) sodium dodecyl sulfate (SDS), 10% (v/v) glycerol, and 0.01% (w/v) bromophenol blue. Add about 50 mL of ultrapure water to a 100 mL glass beaker. Weigh 0.72 g of Tris, transfer to the beaker, mix and adjust pH to 7.0, using HCl. Add 1 mL of 2-mercaptoethanol, 1 g of SDS, 10 mL of glycerol, and 0.01 g of bromophenol blue. Mix and top to 100 mL with deionized water.

2.4 Solutions and Equipment to Protein Purification

1. 10% (v/v) Triton X-100: Dilute 1 mL of Triton X-100 in 9 mL of ultrapure water.
2. 100 mM PMSF: Dissolve 0.87 g of PMSF in 50 mL of ethanol, and store at 4 °C.
3. 10% (w/v) streptomycin sulfate: Dissolve 10 g of streptomycin sulfate in ultrapure water, and store at 4 °C.

4. HiTrap™ Con A 4B or HiTrap Capto lentil lectin (GE Healthcare Life Sciences) (*see* **Note 3**).
5. Peristaltic Pump.
6. Amicon® Ultra-15 (Millipore).
7. Binding Buffer: 20 mM Tris, 500 mM NaCl, 1 mM $MnCl_2$, and 1 mM $CaCl_2$. Add 500 mL of ultrapure water to a 1 L glass beaker. Weigh 2.24 g of Tris, 29.22 g of NaCl, 0.2 g of $MnCl_2$, and 0.11 g of $CaCl_2$, and transfer to the beaker. Mix and adjust pH to 7.4 using HCl. Top to 1 L using ultrapure water, and store at 4 °C.
8. Elution Buffer I: 20 mM Tris, 500 mM NaCl, 1 mM $MnCl_2$, 1 mM $CaCl_2$, and 100 mM methyl-α-D-glucopyranoside. Add about 500 mL of ultrapure water to a 1-L glass beaker. Weigh 2.24 g of Tris, 29.22 g of NaCl, 0.2 g of $MnCl_2$, 0.11 g of $CaCl_2$, and 19.41 g of methyl-α-D-glucopyranoside, and transfer to beaker. Mix and adjust pH to 7.4 using HCl. Top to 1 L with ultrapure water, and store at 4 °C (*see* **Note 2**).
9. Elution Buffer II: 20 mM Tris, 500 mM NaCl, 1 mM $MnCl_2$, 1 mM $CaCl_2$, and 200 mM methyl-α-D-glucopyranoside. Add 500 mL of ultrapure water to a 1 L glass beaker. Weigh 2.24 g of Tris, 29.22 g of NaCl, 0.2 g of $MnCl_2$, 0.11 g of $CaCl_2$, and 38.83 g of methyl-α-D-glucopyranoside, and transfer to the beaker. Mix and adjust pH to 7.4 using HCl. Top to 1 L with ultrapure water, and store at 4 °C.
10. Elution Buffer III: 20 mM Tris, 500 mM NaCl, 1 mM $MnCl_2$, 1 mM $CaCl_2$, and 300 mM methyl-α-D-glucopyranoside. Add about 500 mL of ultrapure water to a 1 L glass beaker. Weigh 2.24 g of Tris, 29.22 g of NaCl, 0.2 g of $MnCl_2$, 0.11 g of $CaCl_2$, and 58.25 g of methyl-α-D-glucopyranoside, and transfer to the beaker. Mix and adjust pH to 7.4 using HCl. Top to 1 L with ultrapure water, and store at 4 °C.
11. Elution Buffer IV: 20 mM Tris, 500 mM NaCl, 1 mM $MnCl_2$, 1 mM $CaCl_2$, and 400 mM methyl-α-D-glucopyranoside. Add approximately 500 mL of ultrapure water to a 1 L glass beaker. Weigh 2.24 g of Tris, 29.22 g of NaCl, 0.2 g of $MnCl_2$, 0.11 g of $CaCl_2$, and 77.67 g of methyl-α-D-glucopyranoside, and transfer to beaker. Mix and adjust pH to 7.4 using HCl. Top to 1 L with ultrapure water, and store at 4 °C.
12. Elution Buffer V: 20 mM Tris, 500 mM NaCl, 1 mM $MnCl_2$, 1 mM $CaCl_2$, and 500 mM methyl-α-D-glucopyranoside. Add 500 mL of ultrapure water to a 1 L glass beaker. Weigh 2.24 g of Tris, 29.22 g of NaCl, 0.2 g of $MnCl_2$, 0.11 g of $CaCl_2$, and 97.1 g of methyl-α-D-glucopyranoside, and transfer to beaker. Mix and adjust pH to 7.4 using HCl. Top to 1 L with ultrapure water, and store at 4 °C.
13. Standard Buffer: 10 mM Tris–HCl and 50 mM NaCl. Add about 500 mL of ultrapure water to a 1 L glass beaker. Weigh

1.21 g of Tris, and 2.92 g of NaCl, and transfer to beaker. Mix and adjust pH to 7.4 using HCl. Top to 1 L with ultrapure water, and store at 4 °C.

3 Methods

3.1 E. coli *CLM24 Transformation and Protein Expression*

1. Prepare *E. coli* CLM24 electrocompetent cells using basic protocols and transform them with 10 ng of pMAF10, pLAFR1_GC, and pTrc99A_CP vectors.
2. Plate culture in LB agar, containing trimethoprim (100 μg/mL), tetracycline (20 μg/mL), and ampicillin (100 μg/mL). Grow for 20 h at 37 °C.
3. Select a single *E. coli* CLM24 colony containing the three vectors with a sterile toothpick and inoculate it into 5 mL of 2XYT medium containing trimethoprim (100 μg/mL), tetracycline (20 μg/mL), and ampicillin (100 μg/mL). Grow overnight at 37 °C in an orbital shaker at 250 rpm.
4. Transfer culture to 500 mL of fresh LB, containing trimethoprim (100 μg/mL), tetracycline (20 μg/mL), and ampicillin (100 μg/mL). Grow for several hours at 37 °C in an orbital shaker at 250 rpm, until OD_{600} equals 0.6–0.8.
5. Add 0.25 mL of 1 M of IPTG (0.5 mM final) to induce the carrier protein expression and 10 mL of 10% (w/v) of arabinose (0.2% final) to induce PglB expression. Incubate overnight at 30 °C/250 rpm (*see* **Note 4**).
6. Check protein expression by SDS-PAGE. To this end, harvest 10 mL of culture, centrifuge for 5 min at 4000 × *g*/4 °C, and discard supernatant. Resuspend the pellet in an equivalent volume of cracking buffer and heat for 5 min at 100 °C. Load SDS–polyacrylamide gel with 20 μL of sample (*see* **Note 5**).
7. Centrifuge culture for 20 min at 4000 × *g*/4 °C, discard supernatant, and resuspend pellet in equivalent volume of water. Centrifuge for 20 min at 4000 × *g*/4 °C and discard the supernatant. At this stage, the pellet can be stored at −80 °C for later use or used directly in the procedure.
8. Unfreeze cell mass in bath ice and resuspend the pellet in 15 mL of binding buffer, containing 1.5 mL of 10% Triton X-100 (1% final) and 150 μL of 100 mM PMSF (final concentration = 1 mM). Disrupt cell in bath ice using three sonication cycles (amplitude 35%) of 30 s on and 1 min off. Add 150 μL 10% (w/v) of streptomycin sulfate, and shake for 30 min in ice bath.
9. Centrifuge 45 min at 20,000 × *g*/4 °C. Transfer the supernatant to a clean tube, and discard the pellet. Keep the sample on ice till purification.

3.2 Purification of Glycosylated Proteins

1. Wash HiTrap™ Con A 4B 5 mL with 50 mL (ten volumes) distillated water at a flow rate of 5 mL/min.
2. Equilibrate column with 50 mL (ten volumes) of binding buffer at a flow rate of 5 mL/min.
3. Load column with the sample at a flow rate of 1 mL/min. Collect flow-through for SDS-PAGE analysis.
4. Wash column with 50 mL of elution buffer I at flow rate of 5 mL/min. Collect flow-through.
5. Wash column with 50 mL of elution buffer II at flow rate of 5 mL/min. Collect flow-through.
6. Wash column with 50 mL of elution buffer III at flow rate of 5 mL/min. Collect flow-through.
7. Wash column with 50 mL of elution buffer IV at flow rate of 5 mL/min. Collect flow-through.
8. Wash the column with 50 mL of elution buffer V at flow rate of 5 mL/min. Collect flow-through (*see* **Note 6**).
9. Analyze the fractions for the presence of proteins, loading SDS–polyacrylamide gel with 20 μL of each sample collected in purification process. (*see* **Note** 7).
10. Combine fractions that contain the purified protein. Concentrate the sample, using an Amicon® Ultra-15 centrifugal filter at 4000 × *g*/4 °C for 2 mL final volume.
11. Complete Amicon® Ultra-15 centrifugal filter with 15 mL of standard buffer, and centrifuge at 4000 × *g*/4 °C to a final volume of 2 mL. Repeat this **step 4** times (*see* **Note 8**).
12. Determine protein concentration by the molar extinction coefficient, measuring absorbance at 280 nm (A_{280}) or using another protein quantification method as Bradford method [14, 15].
13. Load SDS–polyacrylamide gel with 20 μL of the sample to analyze glycosylated protein purity. Divide samples into aliquots, and freeze all samples at −80 °C until use.

4 Notes

1. pTrc99A vector (GE Healthcare) is used to clone the carrier protein (CP) (pTrc99A_CP). To obtain the carrier protein modified gene, the target gene can be synthesized in a expression vector containing in the upstream of the target gene sequence, the DNA sequence of the native signal peptide to periplasm from DsbA, two coding sequences for two N-glycan acceptors (glycosylation tag; GT) and at the downstream of the target gene another two GT sequences (Fig. 2).

Fig. 2 Schematic representation of a carrier protein gene containing spDsbA signal peptide, to enable protein transport to the periplasm, and the N-glycan acceptors peptide tags (GT), which will be glycosylated by PglB

2. pLAFR1 vector is used to carry the gene clusters (GC) (pLAFR1_GC) (Leibniz-Institute DSMZ–German Collection of Microorganisms and Cell Cultures) [12, 13].
3. Lectins are commonly used in glycoprotein purification using affinity purification chromatography. These proteins can bind to carbohydrates through noncovalent interactions, such as hydrogen bond, hydrophobic, electrostatic, and van der Waals interactions. The glycoprotein bond to the lectin can be resolved using a competitive substrate binding [16]. A commonly used method is based on lectin-Sepharose immobilization; the method is based on the commercially available column HiTrap Con A, which possesses a concanavalin A (Con A) isolated from *Canavalia ensiformis*, immobilized in Sepharose 4B. The Con A can bind two molecules containing α-D-mannopyranosyl, α-D-glucopyranosyl, and sterically related molecules and require the presence of C-3, C-4, and C-5 hydroxyl groups in the sugar. A similar protocol can be used in HiTrap Capto Lentil Lectin (GE Healthcare Life Sciences). This column possesses a lentil lectin from *Lens culinaris* immobilized to an agarose base matrix. The lentil lectin binds to molecules containing α-D-mannose, α-D-glucose, or sterically related carbohydrates. Compared to other methods, lentil lectin provides less distinction between glucosyl and mannosyl residues, and binds less strongly to them. It can also be used efficiently in the purification of detergent-solubilized membrane proteins.
4. Expression time and temperature may dependent of the protein solubility. In the case of insoluble carrier protein, the temperature and IPTG concentration can be lowered and the expression time increased.
5. To attain consistent results, expression control can be the performed using a similar culture without induction by IPTG or arabinose. Western blot analysis can also be performed, using specific antibodies to evaluate the protein expression over time as anti-HA or anti-spDsbA in the case of PglB and the carrier protein, respectively.
6. The elution condition can be optimized through the use of elution buffer supplemented with variable concentrations of methyl-α-D-glucopyranoside (100–500 mM).
7. To cleanse the column, rinse it with a buffer containing 50 mL of 20 mM of Tris–HCl, 500 mM of NaCl, pH 8.5, followed by

50 mL of 20 mM acetate, 500 mM of NaCl, pH 4.5. Repeat this process three times. Immediately wash the column with 25 mL of binding buffer. To store the column, wash it with 25 mL of 100 mM acetate, 1 M of NaCl, 1 mM of KCl, 1 mM of $MnCl_2$, 1 mM of $MgCl_2$, pH 6.0, containing 20% ethanol. Store the column at 4–8 °C.

8. This process removes the methyl-α-D-glucopyranoside of the sample. The Amicon® Ultra-15 used in concentration step must be compatible with protein molecular weight (e.g., Amicon 15 cutoff = 15 k Da; Amicon 30 cutoff = 30 kDa).

References

1. Valderrama-Rincon JD, Fisher AC, Merritt JH et al (2012) An engineered eukaryotic protein glycosylation pathway in *Escherichia coli*. Nat Chem Biol 8:434–436
2. Baker JL, Çelik E, DeLisa MP (2013) Expanding the glycoengineering toolbox: the rise of bacterial N-linked protein glycosylation. Trends Biotechnol 31:313–323
3. Li H, Debowski AW, Liao T et al (2017) Understanding protein glycosylation pathways in bacteria. Future Microbiol 12:59–72
4. Feldman MF, Wacker M, Hernandez M et al (2005) Engineering N-linked protein glycosylation with diverse O antigen lipopolysaccharide structures in *Escherichia coli*. Proc Natl Acad Sci 102:3016–3021
5. Nita-Lazar M, Wacker M, Schegg B et al (2005) The N-X-S/T consensus sequence is required but not sufficient for bacterial N-linked protein glycosylation. Glycobiology 15:361–367
6. Wacker M, Linton D, Hitchen PG et al (2002) N-linked glycosylation in *Campylobacter jejuni* and its functional transfer into E. coli. Science 298:1790–1793
7. Ravenscroft N, Haeuptle MA, Kowarik M et al (2016) Purification and characterization of a Shigella conjugate vaccine, produced by glycoengineering *Escherichia coli*. Glycobiology 26:51–62
8. Fisher AC, Haitjema CH, Guarino C et al (2011) Production of secretory and extracellular N-linked glycoproteins in *Escherichia coli*. Appl Environ Microbiol 77:871–881
9. Ihssen J, Kowarik M, Dilettoso S et al (2010) Production of glycoprotein vaccines in *Escherichia coli*. Microb Cell Fact 9:61
10. Garcia-Quintanilla F, Iwashkiw JA, Price NL et al (2014) Production of a recombinant vaccine candidate against *Burkholderia pseudomallei* exploiting the bacterial N-glycosylation machinery. Front Microbiol 5:381
11. Wacker M, Wang L, Kowarik M et al (2014) Prevention of *Staphylococcus aureus* infections by glycoprotein vaccines synthesized in *Escherichia coli*. J Infect Dis 209:1551–1561
12. Lam SK, Ng TB (2011) Lectins: production and practical applications. Appl Microbiol Biotechnol 89:45–55
13. Friedman AM, Long SR, Brown SE et al (1982) Construction of a broad host range cosmid cloning vector and its use in the genetic analysis of *Rhizobium mutants*. Gene 18:289–296
14. Simonian MH (2002) Spectrophotometric determination of protein concentration. Curr Protoc Cell Biol Appendix 3:Appendix 3B
15. Kruger NJ (2002) The Bradford method for protein quantitation. In: Protein protoc handbook. Humana Press, New Jersey, pp 15–22
16. Pohleven J, Trukelj B, Kos J (2012) Affinity chromatography of lectins. In: Affinity chromatography. InTech, Rijeka, Croatia, pp 49–74

Part II

Downstream Processes

Chapter 15

Purification Methods for Recombinant Factor VIII Expressed in Human Liver SK-Hep Cells

Vladimir Granovski, Mario Soares Abreu-Neto, and Dimas Tadeu Covas

Abstract

Coagulation factor VIII is one of the largest proteins attempted to be expressed in recombinant form. A very complex and labile protein which has a very short half-live and need a fast and efficient purification chain. Here, we describe a simple purification sequence using multimodal Capto MMC, affinity FVIII select and ion exchange SP-Fastflow chromatography steps without subjecting the target molecule to mechanical and temperature stress, separating impurities from rFVIII using net charge, hydrophobicity, and affinity of the molecules.

Key words Recombinant coagulation factor VIII, Affinity chromatography, Multimodal chromatography, Ion exchange

1 Introduction

Many proteins undergo post-translational modifications (PTM) soon after their synthesis. Without doubt the most common modification is glycosylation, which can represent over half of the human serum proteins [1]. The glycosylation is a covalent addition of a sugar to a specific site of a protein; after glycosylation the molecule become a glycoprotein [2]. The addition of sugars can be accomplished in two different organelles: an endoplasmic reticulum, which makes N-linked sugars, and Golgi apparatus making O-linked sugars [3]. Glycans can represent 1–80% of total mass and confer stability, activity, and half-life to the molecule [4]. Clearly the carbohydrates can also alter the structure, proper folding, interaction, pharmacokinetics and pharmacodynamics of proteins [5, 6]. However there are no clear difference in purifying a glycoprotein and its nonglycosylated form. Both molecules have a proper size, net charge, isoelectric point, hydrophobic patches and affinity properties to different ligands. The techniques that can be employed to purify glycoproteins are the same that purify simple, nonglycosylated forms. Evidently, there are specific characteristics

Virgínia Picanço-Castro and Kamilla Swiech (eds.), *Recombinant Glycoprotein Production: Methods and Protocols*, Methods in Molecular Biology, vol. 1674, DOI 10.1007/978-1-4939-7312-5_15, © Springer Science+Business Media LLC 2018

of sugar chains that can confer a specific selectivity to bind to a ligand in chromatography techniques [7].

One of the largest molecules attempted to be produced in the recombinant form is the clotting recombinant factor VIII (rFVIII). A very complex and highly glycosylated molecule, possessing 280 kDa in its mature form with six N-linked glycosylation and eight sulfation sites [8–10]. This molecule is extremely labile and has a very short half-life [11]. Thus, it is necessary to have a very efficient and rapid purification process in order to get a highly purified homogeneous product. Furthermore, the raw material in form of clarified supernatant have a large number of impurities that need to be removed. The removal consists in using several chromatography steps based on separation of ionizable molecules on the basis of differences in charge properties using Multimodal Capto MMC resin and SP-Fastflow resin, and for selectivity properties using FVIIISelect resin.

Here, we report a fast purification sequence for recombinant coagulation factor VIII with B-domain deleted produced in SK-HEP human cell line [12].

2 Materials

All chromatography columns (Capto MMC 5 mL Hitrap, FVIII Select 1.1 mL packed in a 5/50 Tricorn column and 1 mL IEX SP-FF Hitrap) and chromatograph system (Akta pure 25) were from General Electric Healthcare Life Sciences. The columns were washed and stored as instructed in provider manuals. All buffers were prepared using Milli-Q purified water and were filtered at 0.22 μm PVDF filters.

2.1 Supernatant Production

1. DMEM media—Add 800 mL of water in 1000 mL backer, 100 mL of inactivated fetal bovine serum (FBSi), and 13.5 g of Dulbecco's Modified Eagle's Medium (DMEM). Additionally, add 3.7 g of sodium bicarbonate and 2.5 g of HEPES. Make up to 1000 mL, adjust the pH to 7.0, and filter using 0.22 μm PVDF sterile filter.
2. PBS buffer—Add 900 mL of water to a 1000 mL backer, weight 8.0 g of NaCl, 0.2 g KCl, 1.2 g Na_2HPO_4, and 0.2 g KH_2PO_4 and transfer to the backer. Make up to 1000 mL, adjust pH to 7.4, and filter using 0.22 μm PVDF sterile filter.
3. Free Style 293 medium (Thermo Fisher Scientific).
4. Recombinant trypsin—animal origin-free, recombinant enzyme used for dissociating a wide range of adherent mammalian cells, including CHO, HEK293, A529, primary human keratinocytes, and embryonic stem cells.

5. T Flasks (72 cm^2).
6. Recombinant SK-HEP cells [12].

2.2 Purification

1. Tris Stocking solution pH 6.5: Add about 250 mL of water to a 500 mL graduated cylinder. Weight 77.2 g Tris–HCl and 1.15 g Tris base and transfer to a cylinder. Make up to 500 mL, filter using 0.22 μm membrane, and store at room temperature.
2. Tris Stocking solution pH 7.5: Add about 250 mL of water to a 500 mL graduated cylinder. Weight 65.9 g Tris–HCl and 9.87 g Tris base and transfer to a cylinder. Make up to 500 mL, filter using 0.22 μm membrane, and store at room temperature (*see* **Note 1**).

2.3 Multimodal Chromatography (MMC) Buffers

1. MMC Equilibration Buffer: 30 mM Tris and 300 mM sodium chloride, pH 6.5. Add about 50 mL of water to a 100 mL graduated cylinder. Add 3 mL of Tris pH 6.5 stock solution. Weight, 1.75 g sodium chloride and transfer to the cylinder. Add water to a 90 mL volume. Mix and adjust pH (if necessary) to 6.5. Make up to 100 mL with water, filter using 0.22 μm membrane, and store at 4 °C.
2. MMC Elution Buffer: 30 mM Tris, 1500 mM sodium chloride, 700 mM argenine, and 50% ethylene glycol, pH 6.5. Add about 30 mL of water to a 100 mL graduated cylinder. Add 3 mL of Tris pH 6.5 stock solution. Weight 8.76 g sodium chloride, 12.19 g of arginine and transfer to the cylinder. Add 50 mL ethylene glycol. Add water to a 90 mL volume. Mix and adjust pH (if necessary) to 6.5. Make up to 100 mL, filter using 0.22 μm membrane, and store at 4 °C.

2.4 FVIII Select Affinity Chromatography Buffers

1. Sample Dilution Buffer: 30 mM Tris, 20 mM histidine, 20 mM calcium chloride, and 0.02% Tween 80, pH 6.5. Add about 500 mL of water to a 1 L graduated cylinder. Add 30 mL of Tris pH 6.5 stock solution. Weight 3.1 g histidine, 2.21 g calcium chloride, and 200 μL of Tween 80, and transfer to the cylinder. Add water to a 900 mL volume. Mix and adjust pH (if necessary) to 6.5. Make up to 1 L with water, filter using 0.22 μm membrane, and store at 4 °C.
2. FVIII Select Equilibration Buffer: 30 mM Tris, 20 mM histidine, 20 mM calcium chloride, 30 mM sodium chloride, and 0.02% Tween 80, pH 6.5. Add about 50 mL of water to a 100 mL graduated cylinder. Add 3 mL of Tris pH 6.5 stock solution. Weight 0.31 g histidine, 0.22 g calcium chloride, 0.17 g sodium chloride, and 20 μL of Tween 80, and transfer to the cylinder. Add water to a 90 mL volume. Mix and adjust pH (if necessary) to 6.5. Make up to 100 mL with water, filter using 0.22 μm membrane, and store at 4 °C.

3. FVIII Select Elution Buffer: 30 mM Tris, 20 mM histidine, 20 mM calcium chloride, 1500 mM sodium chloride, 0.02% Tween 80, and 50% ethylene glycol, pH 6,5. Add about 30 mL of water to a 100 mL graduated cylinder. Add 3 mL of Tris pH 6.5 stock solution. Weight 0.31 g histidine, 0.22 g calcium chloride, 8.7 g sodium chloride, and 20 μL of Tween 80, and transfer to the cylinder. Add 50 mL of ethylene glycol. Add water to a 90 mL volume. Mix and adjust pH (if necessary) to 6.5. Make up to 100 mL with water, filter using 0.22 μm membrane, and store at 4 °C.

2.5 Ion Exchange Chromatography Buffers

1. Sample Dilution Buffer: 30 mM Tris, 20 mM histidine, 20 mM calcium chloride, and 0.02% Tween 80, pH 7.5. Add about 500 mL of water to a 1 L graduated cylinder. Add 30 mL of Tris pH 7.5 stock solution. Weight 3.1 g histidine, 2.21 g calcium chloride, and 200 μL of Tween 80, and transfer to the cylinder. Add water to a 900 mL volume. Mix and adjust pH (if necessary) to 7.5. Make up to 1 L with water, filter using 0.22 μm membrane, and store at 4 °C.
2. IEX Equilibration buffer: 30 mM Tris, 20 mM histidine, 20 mM calcium chloride, 30 mM sodium chloride, and 0.02% Tween 80, pH 7.5. Add about 50 mL of water to a 100 mL graduated cylinder. Add 3 mL of Tris pH 7.5 stock solution. Weight 0.31 g histidine, 0.22 g calcium chloride, 0.17 g sodium chloride, and 20 μL of Tween 80, and transfer to the cylinder. Add water to a 90 mL volume. Mix and adjust pH (if necessary) to 7.5. Make up to 100 mL with water, filter using 0.22 μm membrane, and store at 4 °C.
3. IEX Elution buffer: 30 mM Tris, 20 mM histidine, 20 mM calcium chloride, 1500 mM sodium chloride, and 0.02% Tween 80, pH 7.5. Add about 50 mL of water to a 100 mL graduated cylinder. Add 3 mL of Tris pH 7.5 stock solution. Weight 0.31 g histidine, 0.22 g calcium chloride, 8.7 g sodium chloride, and 20 μL of Tween 80, and transfer to the cylinder. Add water to a 90 mL volume. Mix and adjust pH (if necessary) to 7.5. Make up to 100 mL with water, filter using 0.22 μm membrane, and store at 4 °C.

2.6 Final Protein Detection

1. Protein analysis by SDS-PAGE: the SDS-PAGE analysis is performed using 10% Mini-Protean® TGX gels, 10 well, 30 μL and Mini-Protean® Tetra Cell—Bio-Rad accordingly to a provider manual (*see* **Note 2**). The purification progress gel can be observed in Fig. 1.
2. rFVIII detection method: the detection of rFVIII of each analyzed fraction is performed using a commercial chromogenic Chromogenix Coamatic® Factor VIII detection kit (Diapharma) accordingly to manufacturers manual (*see* **Note 3**).

3 Methods

3.1 Supernatant Production

1. The supernatant was generated after SKHep cell cultivation in T flasks initially (first inoculum) in DMEM medium containing 10% fetal bovine serum until a >95% of confluence. The metabolized medium was removed and the cells were washed with PBS. Freestyle medium (w/o serum) is added to the culture and incubated for 48 h. After incubation the culture is centrifuged and the supernatant filtered.

3.2 Multimodal Chromatography

1. To the filtered supernatant, add sodium chloride to a final concentration of 300 mM to achieve a conductivity of 20 mS/cm and adjust pH to 6.5.
2. Wash the Capto MMC column with MMC Equilibration Buffer until base line is established and all parameters such as pressure, UV, conductivity, and pH are constant.
3. Set UV auto-zero (*see* **Note 4**).
4. Inject the supernatant at the flow velocity of <600 cm/h.
5. Collect flow-through (*see* **Note 5**).
6. After injection, reequilibrate the system with MMC Equilibration Buffer until the UV base line is reached.
7. Elute the bound proteins with MMC Elution Buffer (*see* **Note 6**).

3.3 FVIII Select Affinity Chromatography

1. Dilute the eluted fraction from previous step 50× in Sample Dilution Buffer 1 (*see* **Note 7**).
2. Wash the FVIII select column with FVIII Select equilibration Buffer until base line is established and all parameters such as pressure, UV, conductivity, and pH are constant. Set UV auto-zero.
3. Inject the supernatant at the flow velocity of <300 cm/h.
4. Collect the flow-through.
5. After injection, reequilibrate the system with FVIII Select equilibration Buffer until the UV base line is reached.
6. Elute the bound proteins with FVIII Select Elution Buffer (*see* **Note 8**).

3.4 IEX Chromatography SP-FF

1. Dilute the eluted fraction from previous step 30× in Sample Dilution Buffer 2 (*see* **Note 9**).
2. Wash the SP-FF column with IEX Equilibration buffer until base line is established and all parameters such as pressure, UV, conductivity, and pH are constant.
3. Set UV auto-zero.

4. Inject the supernatant at the linear flow of 1 mL/min.
5. Collect the flow-through.
6. After injection, reequilibrate the system with IEX equilibration Buffer until the UV base line is reached.
7. Elute the bound proteins with IEX Elution buffer (*see* **Note 10**).

3.5 Detection

1. The samples are prepared by adding 20 μL of SB 5× to 80 μL of sample and placed in a hot bath at 90 °C for 15 min.
2. Apply 20–25 μL of boiled samples to the SDS-PAGE wells and run the electrophoreses at 100 V, 400 mA until the bromophenol blue reach the bottom of the gel.
3. Use silver stain method.
4. Use Chromogenix Coamatic® Factor VIII detection kit accordingly to manufacturers manual to detect the presence of recombinant factor VIII.

4 Notes

1. The Tris pH 6.5 and 7.5 are prepared balancing the Tris–HCl and Tris base accordingly to its thermodynamic pK_a/apparent pK_a and the molecular weight of each compound. In this way it is possible to add a very low amount of stocking solution to a buffer and have a very good control/balance over the pH.
2. Every eluate from the chromatography steps need to be evaluated for target protein presence applying 25 μL of the sample per Gel wells (Fig. 1).
3. All eluates and the initial supernatant are submitted to a chromogenic rFVIII detection to observe if the target protein is present. The test is empiric and visual, just to detect the protein and observe that the purification steps are working properly and there is no loss of product in the flow-through of each chromatography.
4. The auto zero is a standard option in all Akta equipment. If chromatography equipment do not have this option, the recommendation is to consider actual constant UV mean as zero.
5. It is highly recommended to collect the flow-through in order to analyze it for target protein content if there is any protein loss during the process. The flow-through is collected and stored at 4 °C for short period of time or at −80 °C for long term storage. This procedure must be applied for all chromatography steps.

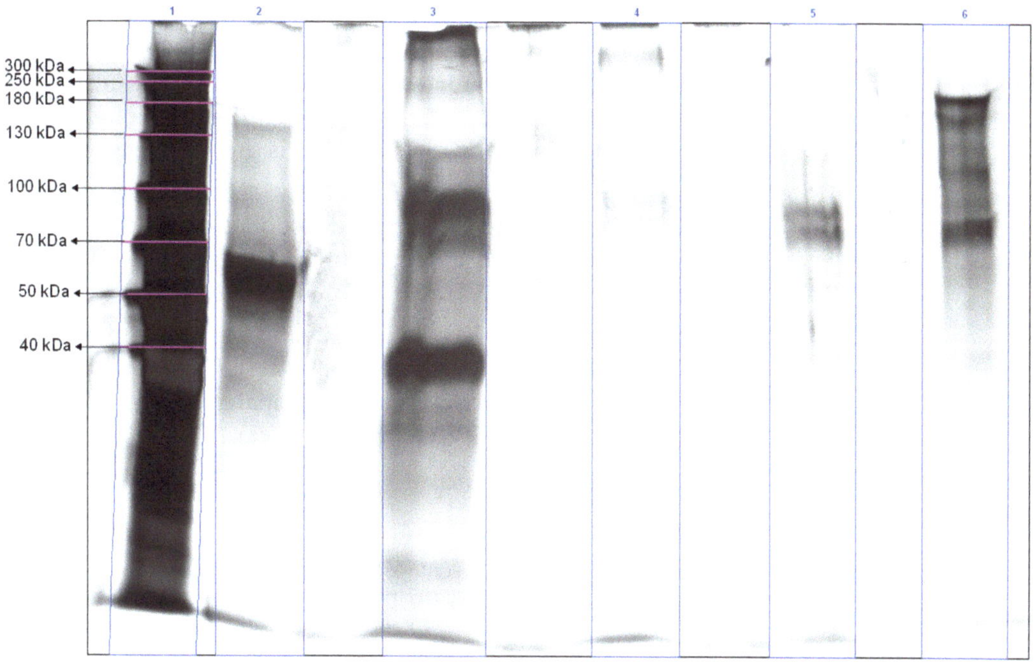

Fig. 1 SDS-PAGE gel of the retention samples from each chromatography. The wells represent (*1*) Molecular weight marker, Fisher 26625, (*2*) Raw supernatant, (*3*) Eluted fraction from Capto MMC, (*4*) Eluted fraction from FVIII select, (*5*) Eluted fraction from SP-sepharose, and (*6*) Commercial positive control (*see* **Note 11**)

6. The elution volume may vary depending on injection volume, resin type and column size. In our case, the elution volume was 10 mL from a 5 mL resin. Also, it is important to mention that the target protein elution is not always followed by UV spike. The eluate is collected in several fractions that are analyzed by chromogenic activity.
7. The sample dilution is needed due to the fact that the same 50% ethylene glycol and sodium chloride concentration used to elute the target protein from MMC columns, is used to elute the FVIII protein from the FVIII select. In order to avoid buffer interference and usage of dialysis system (can damage the target protein) the concentration of ethylene glycol is diluted to 1% and sodium chloride to 30 mM. In this way, it is possible to make the interaction between FVIII protein and the stationary phase of the column without buffer interference.
8. In the same way as mentioned in **Note 4**, the elution volume is strongly dependent on several factors. In the present case, the volume of eluate corresponded to 11 mL from a 1.1 mL of resin. Also, as mentioned before, the protein elution not always occurs with UV peak.

9. The buffer used to elute the FVIII protein from FVIII Select chromatography has a high concentration of sodium chloride. Since the high sodium chloride concentration is needed to elute the protein from the next step chromatography, the salt concentration must be lowered. This is the main reason for the 30× sample dilution.
10. The elution volume is strongly dependent of the column volume. Since we used a 1 mL Hitrap column the elution was performed in 5 mL.
11. It is possible to notice the progression of rFVIII purification (Fig. 1) during the chromatography tests. The Well 5 clearly shows the high purity even when compared to a commercially available control (ADVATE from Baxter).

References

1. Apweiler R, Hermjakob H, Sharon N (1999) On the frequency of protein glycosylation, as deduced from analysis of the SWISS-PROT database. Biochim Biophys Acta 1473(1):4–8
2. Berg JM, Tymoczko JL, Stryer L (2002) Section 11.3: Carbohydrates can be attached to proteins to form glycoproteins. In: Freeman WH (ed) Biochemistry, 5th edn. Available from: https://www.ncbi.nlm.nih.gov/books/NBK22521/
3. Nettleship JE (2012) Structural biology of glycoproteins, glycosylation. In: Petrescu S (ed) InTech. doi:10.5772/48154. Available from: http://www.intechopen.com/books/glycosylation/structural-biology-of-glycoproteins
4. Shylaja M, Seshadri HS (1989) Glycoproteins: an overview. Biochem Educ 17(4):170–178
5. Liu L, Gomathinayagam S, Hamuro L, Prueksaritanont T, Wang W, Stadheim TA, Hamilton SR (2013) The impact of glycosylation on the pharmacokinetics of a TNFR2:Fc fusion protein expressed in Glycoengineered Pichia Pastoris. Pharm Res 30(3):803–812. doi:10.1007/s11095-012-0921-3. Epub 8 Nov 2012
6. Higel F, Seidl A, Sörgel F, Friess W (2016) N-glycosylation heterogeneity and the influence on structure, function and pharmacokinetics of monoclonal antibodies and FC fusion proteins. Eur J Pharm Biopharm 100:94–100. doi:10.1016/j.ejpb.2016.01.005
7. Yang Z, Hancock WS (2005) Monitoring glycosylation pattern changes of glycoproteins using multi-lectin affinity chromatography. J Chromatogr A 1070(1–2):57–64
8. Kosloski MP, Miclea RD, Sathy V (2009) Role of glycosylation in conformational stability, activity, macromolecular interaction and immunogenicity of recombinant human factor VIII. AAPS J 11(3):424
9. Sandberg H, Almstedt A, Brandt J, Gray E, Holmquist L, Oswaldsson U, Sebring S, Mikaelsson M (2001) Structural and functional characteristics of the B-domain-deleted recombinant factor VIII protein, r-VIII SQ. Thromb Haemost 85(1):93–100
10. Santagostino E (2014) A new recombinant factor VIII: from genetics to clinical use. Drug Des Devel Ther 8:2507–2515
11. Gringeri A, Wolfsegger M, Steinz KN, Reininger AJ (2015) Recombinant full-length factor VIII (FVIII) and extended half-life FVIII products in prophylaxis new insights provided by pharmacokinetic modelling. Haemophilia 21(3):300–306. doi:10.1111/hae.12605
12. Rosa NG, Swiech K, Picanço-Castro V, Russo-Carbolante EMS, Neto MSA, Fernandes AC, Faça VM, Covas DT (2012) SK-HEP cells and lentiviral vector for production of human recombinant factor VIII. Biotechnol Lett 34(8):1435–1443

Chapter 16

Purification Method for Recombinant hG-CSF by Affinity Chromatography

Bruna Samham Archangelo and Elisa Maria de Sousa Russo

Abstract

The human granulocytic colony-stimulating factor (hG-CSF) acts mainly by promoting the maturation of granulocytes and stimulating their phagocytic and chemotactic activity. It has been used in the treatment of many diseases, in particular in neutropenic conditions. Here, we describe the purification process of the recombinant protein hG-CSF expressed in *Pichia pastoris*. The protein purification proved to be efficient using the nickel affinity chromatography method described in this chapter.

Key words Affinity chromatography, G-CSF, *Pichia pastoris*, Purification, Recombinant protein

1 Introduction

Colony stimulating factors (CSF) comprise a group of proteins that control the proliferation and differentiation of hematopoietic cells [1]. The human granulocyte colony stimulating factor (hG-CSF) is included in this group and it has a regulatory function in the growth, differentiation, survival, and activation of neutrophils and its precursors, being central for neutrophil-based immune defenses [2]. Its biological activity is accomplished through interaction with specific receptors, whose function is to transmit the extracellular signal of hG-CSF to the cytoplasm of cells expressing these specific receptors [3].

The study of this protein and its use as a drug has being possible once human G-CSF has been cloned and expressed as a recombinant protein. Nowadays it is an established and commercially available biological product and is one of the most commonly used recombinant proteins for clinical purposes, being employed in the treatment of neutropenia caused by immunosuppressant diseases or treatments [1, 4].

The hG-CSF has been produced in several expression systems like *Escherichia coli* [5–7], *Pichia pastoris* [8, 9], *Bacillus subtilis*

Virgínia Picanço-Castro and Kamilla Swiech (eds.), *Recombinant Glycoprotein Production: Methods and Protocols*, Methods in Molecular Biology, vol. 1674, DOI 10.1007/978-1-4939-7312-5_16, © Springer Science+Business Media LLC 2018

[10], *Hansenula polymorpha* [11], and also mammalian cells. Currently, the recombinant human granulocyte colony stimulating factor is commercially available in its nonglycosylated form, produced in prokaryotic organisms and in the glycosylated form, derived from expression in mammalian cells [12].

Some studies report different forms of rhG-CSF purification. The most used method is liquid chromatography on an anion exchange [9, 10] or cation exchange columns [6, 13, 14]. Size exclusion chromatography is also used to purify rhG-CSF produced in *E. coli* as described by Vemula et al. [15]. We describe here a specific method for the purification of a rhG-CSF expressed in *Pichia pastoris* system by affinity chromatography through the histidine tag, present in the N-terminus end of the protein. The purification process is performed in a Fast Protein Liquid Chromatography (FPLC).

2 Materials

All solutions should be prepared using ultrapure water and analytical grade reagents. Prepare and store all reagents at room temperature (unless indicated otherwise). Diligently follow all waste disposal regulations when disposing of waste material. We do not add sodium azide to reagents. All solutions used in the Akta FPLC System should be deaerated.

2.1 Sample Preparation

1. 0.45 μm filter.
2. 5 mL syringe.
3. 2 mL microtube.
4. Sample: recombinant *Pichia pastoris* expression supernatant containing rhG-CSF.
5. Centrifuge.

2.2 Purification of rhG-CSF

1. HisTrap™ HP 1 mL column (*see* **Note 1**).
2. AKTAPurifier UPC-900 Protein Purification System (*see* **Note 2**).
3. DEPC water.
4. Buffer A: 100 mM Tris, 100 mM NaCl, and 17.4 mM NaH_2PO_4. Prepare a 1 L solution and Adjust pH to 8.0. Store at 4 °C. Weight 1.21 g of Tris, 6.89 g of NaH_2PO_4, and 5.84 g of NaCl. Dissolve the reagents in 1 L of water. Adjust pH to 8.0 and store at 4 °C.
5. Buffer B: 100 mM Tris, 100 mM NaCl, 17.4 NaH_2PO_4, and 250 mM Imidazole. Prepare a 1 L solution. Adjust pH to 8.0. Store at 4 °C. Weight 1.21 g of Tris, 6.89 g of NaH_2PO_4,

5.84 g of NaCl, and 17.02 g imidazole. Dissolve the reagents in 1 L of water. Adjust pH to 8.0 and store at 4 °C.

6. Ethanol 20%.

2.3 SDS–Polyacrylamide Gel

1. Resolving gel buffer: 2 M Tris–HCl pH 8.8. Add 100 mL water to a 1 L glass beaker (*see* **Note** 3). Weight 242.3 g Tris–HCl and transfer to the glass beaker. Add water to a volume of 900 mL. Mix and adjust pH with HCl (*see* **Note 4**). Make up to 1 L with water. Filter on a filter paper (*see* **Note 5**). Store at 4 °C.
2. Stacking gel buffer: 2 M Tris–HCl pH 6.8. Prepare a 1 L solution as in previous step but adjust pH to 6.8 with HCl. Filter on a filter paper (*see* **Note 6**). Store at 4 °C.
3. Thirty percent acrylamide–bis solution (30:0.8) acrylamide-- bis: Weight 75 g of acrylamide monomer and 2 g bis-acrylamide (cross-linker) and transfer to a 250 mL glass beaker containing about 170 mL of water. Make up to 250 mL with water and filter through a number 2 Watchman filter. Store at 4 °C, in a bottle wrapped with aluminum foil (*see* **Note** 7).
4. Ammonium persulfate: 10% solution in water (*see* **Note 8**).
5. Sodium dodecyl sulfate (SDS) 10% solution: Dissolve 5 g SDS in 50 mL water.
6. SDS-PAGE running 10× concentrated pH 8.3: dissolve 30.28 g of Tris base, 144.12 g of glycine, and 10 g of SDS in 1 L of water.
7. SDS denaturation buffer (3×): 2 M Tris–HCl pH 6.8, 10% SDS, 87% glycerol, 25% β-mercaptoethanol, and 0.5% Bromophenol Blue. Leave one aliquot at 4 °C for current use and store the remaining aliquots at −20 °C.
8. Coomassie Blue Staining: 0.2% Coomassie Blue, 50% methanol, 10% acetic acid. Weight 1 g of Coomassie Brilliant Blue. Add 200 mL of methanol, 40 mL of acetic acid, and 160 mL of water for a solution of 400 mL final volume. First dissolve Coomassie Blue in methanol and after add the other reagents. Filter in filter paper number 1. Store at room temperature.
9. Decolorant solution: 80 mL ethanol 100%, 80 mL acetic acid and to fulfill for 1 L with purified water.

3 Methods

Carry out all procedures at room temperature unless otherwise specified.

3.1 Sample Preparation for Purification

1. After the induction of rhG-CSF with methanol for 144 h in *Pichia pastoris*, the culture must be centrifuged to separate the cell fraction from the supernatant at 3000 × *g*, 5 min at room temperature in a centrifuge.
2. The recombinant *Pichia pastoris* expression supernatant containing rhG-CSF must be filtered using a 0.45 μm filter and a syringe (*see* **Note 9**).
3. Store the prepared sample at 4 °C.

3.2 Purification of Protein in Akta FPLC

1. Fit the column HisTrap 1 mL to the equipment, check the looping used.
2. Pass 10 column volumes (10 mL in this case) of water to clean the column, which is stored in ethanol.
3. Balance the column with 20 column volumes with Buffer A.
4. Apply the sample (2 mL in this case).
5. Wash the column with 18 column volumes with Buffer A.
6. Elution gradient starting from 0% of Buffer B to 100% of Buffer B (250 mM of imidazole) in 20 column volumes (Fig. 1).
7. Apply 100% of Buffer B for 8 column volumes.
8. Pass 10 column volumes of water to wash the column.
9. Pass 10 column volumes of 20% of ethanol for refrigerated storage of the column HisTrap at 4 °C.

3.3 SDS-PAGE 12%

The SDS-PAGE will be used to analyze the fractions collected from the purification method.

1. Mix in a 50 mL conical flask 1.5 mL of resolving gel buffer, 3.2 mL of acrylamide mixture, 3.18 mL of water, 80 μL of SDS, 50 μL of ammonium persulfate, and 10 μL of TEMED and cast gel within a 7.25 cm × 10 cm × 1.5 mm gel cassette. Allow space for stacking gel and gently overlay with isobutanol or water.
2. For the stacking gel: mix 250 μL of stacking gel buffer, 600 μL of acrylamide mixture, 3.09 mL of water, 40 μL of SDS, 37.5 μL of ammonium persulfate, and 10 μL of TEMED. Insert a 10-well gel comb immediately without introducing air bubbles.
3. Wait for the gel polymerization (*see* **Note 10**).
4. In a micro tube add 20 μL of the samples (supernatant containing rhG-CSF) and 10 μL of SDS denaturation buffer. Heat at 100 °C for 10 min.
5. Add protein standard at the first lane of the gel e add the samples to be analyzed in the other lanes.

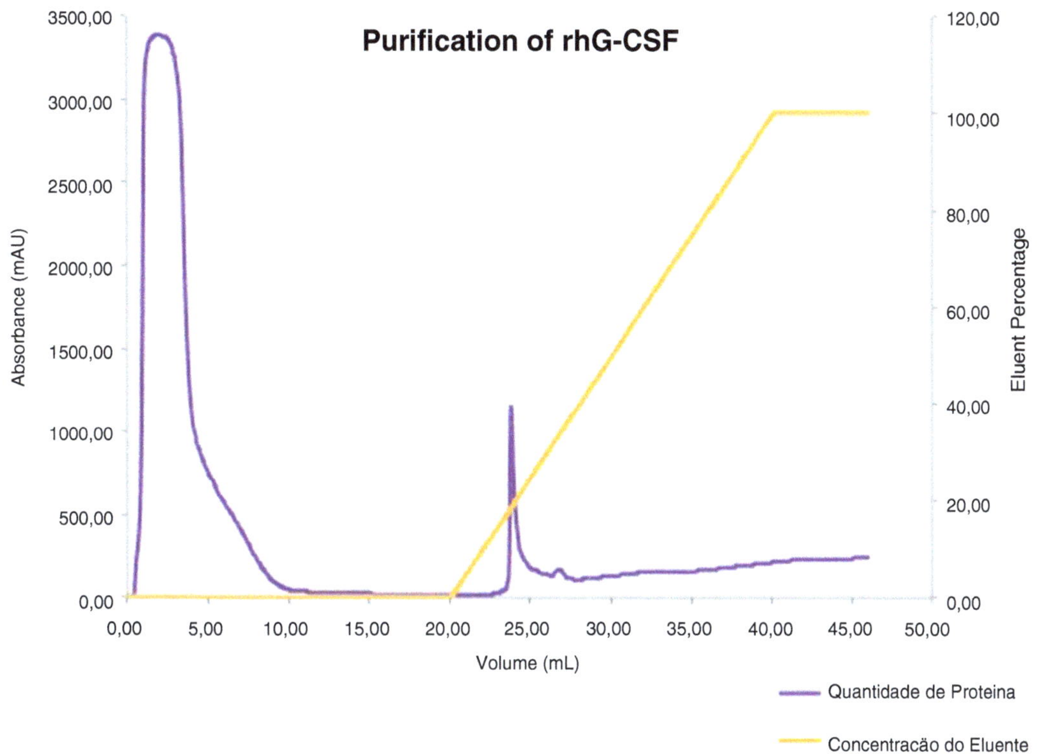

Fig. 1 Profile of Nickel affinity chromatography for the purification of rhG-CSF. Eluent B refers to the Imidazole Lysis Buffer

6. Electrophorese at 80 V until the samples has entered the gel and then continue at 110 V until the dye front (from the SDS denaturation buffer) reached the bottom of the gel (*see* **Note 11**).
7. After electrophoresis, remove the gel plates apart with the use of a spatula. The gel remains on one of the glass plates. Rinse the gel with water and transfer carefully to a container with Coomassie Blue staining. Place enough Coomassie Blue staining to cover the gel.
8. Put the container with the gel on a balance and let the gel color completely.
9. When the gel is very blue, begin to bleach with decolorant solution. Do it until the bands on the gel become visible (Fig. 2).

4 Notes

1. HisTrap columns are prepacked with Ni Sepharose High Performance and designed for simple, high-resolution purification of histidine-tagged proteins by immobilized metal ion affinity chromatography (IMAC).

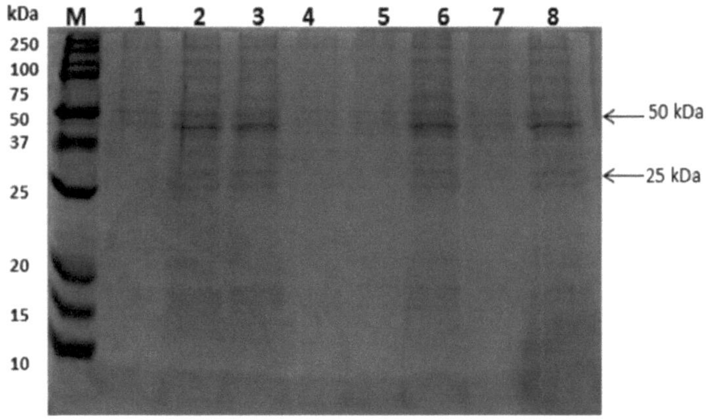

Fig. 2 Analysis of the purification of rhG-CSF by nickel affinity chromatography. 12% SDS-PAGE stained with Coomassie Blue. Run fractions, with approximately 2 mL of supernatant, showing in *M*: molecular weight marker; *1*: eluate; *2*: washing; *3*: washing; *4*: washing; *5*: sample eluted with 5% of imidazole; *6*: sample eluted with 10% of imidazole; *7*: sample eluted with 15% of imidazole; *8*: sample eluted with 20% of imidazole; *9*: sample eluted with 25% of imidazole. "Black arrows" indicate the possibly glycosylated protein, approximately 50 kDa in size and the nonglycosylated protein, of approximately 25 kDa

2. Akta FPLC has the ability to purify virtually any biomolecule. This system can handle the simplest and the toughest of challenges. The platform covers all major chromatographic and cross flow techniques, from research laboratory to process development and manufacturing. Akta systems are operated by the intelligent and scalable UNICORN software, which makes it easy to control every stage of your purification process.
3. If using a glass beaker, Tris can be dissolved faster if the water is warmed to about 37 °C. However, the downside is that care should be taken to bring the solution to room temperature before adjusting pH.
4. Concentrated HCl (12 N) can be used at first to narrow the gap from starring pH to the required pH. When it is close to the required pH, would be better to use a series of HCl (1–6 N) to avoid a sudden drop in pH below the required.
5. Use a filter paper number 0 to filter the solution to avoid contaminant particles.
6. Use a filter paper number 1 to filter the solution to avoid contaminant particles.
7. Wear a mask and gloves when weighing acrylamide. Read the instructions of the manufacturer. The acrylamide solution can be stored at 4 °C for 1 month.
8. It is better to use this solution freshly prepared each time or to use frozen aliquots that can be thawed at the time of use.

9. Filter to remove any residual material that may have remained in the sample. When applying, be careful not to form bubbles in the sample as it may disrupt the purification equipment.
10. Wait for about 30 min for the gel to completely polymerize. This amount of reagents is enough for two gels.
11. This step may take about 2 h. It will depend on the gel and the equipment used.

Acknowledgments

This work was supported by University of São Paulo (USP) and by CAPES.

References

1. Clark SC, Kamen R (1987) The human hematopoietic colony-stimulating factors. Science 236:1229–1237
2. Callard RE (1994) The cytokine facts book, vol 24. Academic Press, San Diego, p 265
3. Iwasaki H, Shimoda K, Okamura S, Otsuka T, Nagafuji K, Harada N et al (1999) Production of soluble granulocyte colony-stimulating factor receptors from myelomonocytic cells. J Immunol 163:6907–6912
4. Hubel K, Engert A (2003) Clinical applications of granulocyte colony-stimulating factor: an update and summary. Ann Hematol 87:207–213
5. Hammerling U, Kroon R, Sjodin L (1995) In vitro bioassay with bioassay with enhanced sensitivity for human granulocyte colony-stimulating factor. J Pharm Biomed Anal 13:9–20
6. Babaeipour V, Khanchezar S, Mofid MR, Abbas MPH (2015) Efficient process development of recombinant human granulocyte colony-stimulating factor (rh-GCSF) production in *Escherichia coli*. Iran Biomed J 2:102–110
7. Karimi Z, Nezafat N, Negahdaripour M, Aydin B, Hemmati S, Ghasemi Y (2015) The effect of rare codons following the ATG start codon on expression of human granulocyte-colony stimulating factor in Escherichia coli. Protein Expr Purif 114:108–114
8. Maity N, Thawani A, Sharma A, Gautam A, Mishira S, Sahai V (2015) Expression and control of codon-optimized granulocyte colony-stimulating factor in *Pichia pastoris*. Appl Biochem Biotechnol 178:159–172
9. Apte-Deshpande A, Somani S, Mandal G, Soorapaneni S, Padmanabhan S (2009) Over expression and analysis of O-glycosylated recombinant human granulocyte colony stimulating factor in Pichia Pastoris using Agilent 2100 Bioanalyzer. J Biotechnol 143:44–50
10. Bashir S, Sadaf S, Ahmad S, Akhtar MW (2015) Enhanced and secretory expression of human granulocyte colony stimulating factor by *Bacillus subtilis* SCK6. BioMed Res Int. ID 636249:9p
11. Talebkhan Y, Samadi T, Samie A, Barkhordari F, Azizi M, Khalaj V, Mirabzadeh E (2016) Expression of granulocyte colony stimulating factor (GCSF) in *Hansenula polymorpha*. Iran J Microbiol 8:21–28
12. Codevilla CF et al (2004) Biological potency evaluation and characterization of rhG-CSF in pharmaceutical products. Braz J Hematol Hemother 26:104–108
13. Gomes FR, Maluenda AC, Tapias JO, Oliveira FL et al (2012) Expression of recombinant human mutant granulocyte colony stimulating factors (Nartograstim) in *Escherichia coli*. World J Microbiol Biotechnol 28:2593–2600
14. Vans ALS, Renard G, Palma MS et al (2008) Human granulocyte colony stimulating factor (hG-CSF): cloning, overexpression, purification and characterization. Microb Cell Factories 7:13
15. Vemula S, Thunuguntla A, Dedaniya A, Kokkiligadda S, Palle C, Ronda SR (2015) Improved production and characterization of recombinant human granulocyte colony stimulating factor from *E. coli* under optimized downstream processes. Protein Expr Purif 108:62–72

Chapter 17

Microplate-Based Method for High-Throughput Screening (HTS) of Chromatographic Conditions Studies for Recombinant Protein Purification

Rimenys J. Carvalho and Thayana A. Cruz

Abstract

High-throughput screening (HTS) systems have emerged as important tools to provide fast and low cost evaluation of several conditions at once since it requires small quantities of material and sample volumes. These characteristics are extremely valuable for experiments with large number of variables enabling the application of design of experiments (DoE) strategies or simple experimental planning approaches. Once, the capacity of HTS systems to mimic chromatographic purification steps was established, several studies were performed successfully including scale down purification. Here, we propose a method for studying different purification conditions that can be used for any recombinant protein, including complex and glycosylated proteins, using low binding filter microplates.

Key words Microplate-based method, High-throughput screening (HTS), Liquid chromatography, Adsorption studies, Elution studies, Protein purification, Design of experiments (DoE), Protein analytical methods, Scale down

1 Introduction

The large increase of recombinant proteins titers in upstream processing has shifted the bottleneck of several biomanufacturing processes toward purification. Achieving the required level of purity, good quality attributes, and high yields usually requires the use of different types of liquid chromatographic steps [1, 2]. The development of chromatographic purification steps on columns is a laborious task and involves several operating parameters that can affect its performance. Consequently, optimization of purification processes can be time- and material-consuming [3, 4]. In this context, new rational methodologies like high-throughput screening (HTS) have emerged as efficient tools which can be applied on chromatographic studies for designing optimal protein purification processes [5, 6].

Virgínia Picanço-Castro and Kamilla Swiech (eds.), *Recombinant Glycoprotein Production: Methods and Protocols*, Methods in Molecular Biology, vol. 1674, DOI 10.1007/978-1-4939-7312-5_17, © Springer Science+Business Media LLC 2018

Small-scale batch chromatography formats based on microplates are becoming an integral component of purification process development workflow allowing the screening of a large number of conditions in a short period of time with minimal sample and material (media and buffers) requirements [7, 8]. The technique is usually based on parallel purification in 96-well plates filter devices containing chromatographic media for the investigation of several purification conditions such as, different buffers conditions (buffer excipients, salt concentration or pH values) and media (ligand and support matrices) on adsorption and elution studies. Once the media are applied into the wells, the adsorption of the target protein occurs by reaching equilibrium during agitation, also called static binding. All chromatographic steps (loading, washing, and elution) can be performed in different batches. Between each step, the desired fractions are collected for further direct or indirect analyses [3, 8]. Additionally, microplate platforms can be used as screening tools to simplify subsequent purification optimization on larger scales based on chromatography columns, adsorptive membranes or monoliths [7, 8].

A valuable approach to this kind of method is the combination of accumulated knowledge of biochemical properties of proteins and critical factors influencing process performance with the experimental planning or DoE, letting the expansion of the multivariate space considering each well in a microplate as an independent experiment [4, 5]. DoE is a technique used for planning a minimum number of screening experiments with two or more variables suitable to generate sufficient data to be analyzed. Based on the information of the studied process, mathematical models can be developed and used to understand the importance of different experimental variables and to predict an optimized process [9, 10]. Here, we describe the use of a microplate-based method for HTS of different chromatographic conditions for recombinant protein purification that provides a rapid analysis of multiple chromatographic experimental conditions in small scale.

2 Materials

It is recommended to prepare all the solutions using ultrapure water (18.2 MΩ cm at 25 °C) purified by an appropriate equipment and analytical grade reagents. The solutions can be stored at 4 °C for increasing their lifetime, but it is not necessary to add sodium azide. All buffers must be filtered using hydrophilic membranes of up to 0.45 μm pore size.

2.1 Adsorption and Elution Studies

1. Multiscreen equipment for vacuum of microplates (96 or more wells) with vacuum pump or microplate centrifuge.

2. 96-Well filter plates (*see* **Note 1**). The membrane cutoff depends on the main objective; usually 0.45 or 0.22 μm is used. Deep well plates must be used for working volumes higher than 250 μL.
3. 96-Well microplates. Deep well plates must be used for collecting volumes higher than 250 μL.
4. Microplate shaker able to perform rotation higher than 1000 rpm (*see* **Note 2**).
5. Desired chromatographic media: adsorptive resins, gels, or membranes.
6. Reagents for solutions of each assay step: equilibration, washing, elution, regeneration, sanitization and storage.

3 Methods

3.1 Adsorption Studies

1. First, the chromatographic media has to be prepared in order to add the exact amount into the well. Add the particles in a graduated tube with storage solution and let them sediment completely for a long period in order to obtain the exact particle volume (*see* **Note 3**). For example, agarose particles require at least 24 h to sediment; on the other hand, silica particles (e.g., controlled porous glass—CPG) require only 8 h, since silica is heavier than agarose.
2. Plan and design your experiment conditions in the plate (*see* **Note 4**). It is possible to separate the plate in different zones of chromatographic media, buffers or conditions as showed in Fig. 1.

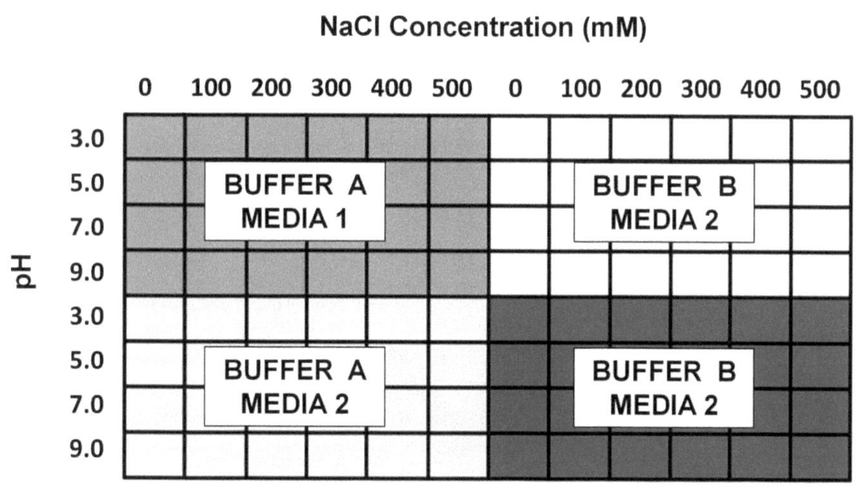

Fig. 1 Microplate layout for adsorption studies based on different conditions such as pH values, NaCl concentrations, buffers and chromatographic media. It is recommended to prepare a triplicate for each test condition

3. After sedimentation, the final volume of the particles will be obtained by total compaction by gravity. Add water or storage solution up to fourfold of the particle volume in order to get a known and more diluted suspension; high dilutions will facilitate the pipetting of the particle suspension into the wells (*see* **Note 5**). In case of adsorptive membrane, add a known membrane area instead of volume.
4. Define the volume ratio of particle:sample and calculate the volume of particles per well to be used in the experiment. (*see* **Note 6**). For example, when 1:10 (particle:sample) ratio is defined using 200 μL of sample, add 20 μL of particles per well. Thus, for a fourfold diluted particle suspension pipette 80 μL of the suspension per well. Before pipetting the particles, the suspension must be homogenized. If using heavier particles such as CPG, homogenize the suspension before each pipetting.
5. After applying the medium into the well, place the filter plate into the multiscreen device and remove the storage solution by either vacuum or centrifugation. For this procedure, a regular 96-well plate for waste is placed in the bottom of the multiscreen device or centrifuge to collect the storage solution while a filter plate stays exactly above it as shown in Fig. 2. For deep well plates, the filter plate is placed over the cover of the device. Close the device and connect it to a vacuum pump. Once the dry wells are seen after vacuum or centrifugation, move on to the next step to equilibrate the media.
6. For equilibration, add a buffer solution into the wells containing the media, the same volume as calculated for the sample must be added (*see* **Note 7**). Then, place the membrane plate in the plate shaker for 10 min at least. After agitating, vacuum or

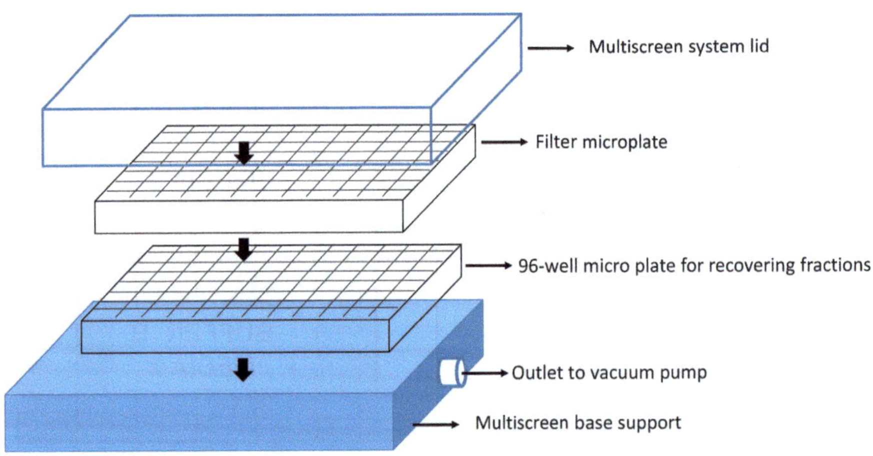

Fig. 2 Scheme of multiscreen HTS device for vacuum filter plates. System based on Merck Millipore product. Differences in the device structure can be found for different manufactures

centrifuge the plate as described in **step 4** in Subheading 3.1. Perform this procedure three times.

7. After equilibration, add the previously prepared sample to the well and shake the plate for the required time (*see* **Note 8**). Use the temperature required to keep your product stable and/or to obtain the best adsorption. Studies using different temperatures can be performed. It is recommended to add sample into the wells without chromatographic media to be used as the initial sample for adsorption calculations, since low adsorption, degradation, or denaturation of the proteins may occur during the process on the membrane filter.
8. Turn the shaker off when the adsorption time is reached. Put the filter plate on the 96-well microplate in the centrifuge or in the multiscreen device for further vacuum. Be sure that both plates are placed in the same lines and columns directions (*see* **Note 9**).
9. Reserve the 96-well micro plate with flow-through fractions for further analysis (*see* Subheading 3.3).
10. Before reusing the membrane plate, perform the elution, regeneration, and sanitization with the proper buffers/solutions and add the storage buffer into the wells in the end. If a well is used more than two times, it may present linking or medium compaction; in such case, do not reuse it.

3.2 Elution Studies

1. For elution studies, follow all steps described at adsorption studies section.
2. After recovering the flow-through fraction, add the same sample volume of washing buffer into the wells in order to remove all possible remaining samples.
3. Shake the filter plate for 10 min at least and place it into the multiscreen device or centrifuge over a regular 96-well microplate to recover the washing fraction. Collect the washing fraction if it is necessary for the experiment analysis. Perform the washing step twice.
4. Elute the target protein from the chromatographic media using the selected elution buffers by shaking 30 min at least and then vacuum or centrifuge the elution fraction. Save this fraction for further analysis (*see* Subheading 3.3). In order to guarantee the total recovery, it is recommended to perform the elution twice.
5. If reusing the filter plate, perform the regeneration and sanitization with proper buffers/solutions and add the storage buffer into the wells. If a well is used more than twice, it may present linking or media compaction; in such case, do not reuse it.

Table 1
Summary of the common analytical methods used for protein quantification and characterization

Subject of analysis	Method
Total protein concentration	Direct UV (OD 280 nm) Colorimetric assays (Bradford, BCA, Lowry)
Target protein concentration	ELISA, HPLC (RP, AC)
Electrophoretic patterns and homogeneity	SDS-PAGE
Peptide mapping	MS
Posttranslational modifications (PTMs)	HPLC (IEX, RPLC, AC, HILIC) and MS
Aggregation	SDS-PAGE, SEC
Critical contaminants	
• Residual DNA	Hybridization
• HCP	ELISA
Identity	*Western blot and dot-blot*
Biological activity	Specific biological assay (in vitro)

Adapted from Ref. 11. *AC* affinity chromatography, *BCA* bicinchoninic acid, *ELISA* enzyme-linked immunosorbent assay, *HCP* host cell protein, *HILIC* hydrophilic interaction liquid chromatography, *HPLC* high-performance liquid chromatography, *IEX* ion-exchange chromatography, *MS* mass spectrometry, *RPLC* reversed-phase liquid chromatography, *SDS-PAGE* sodium dodecyl sulfate polyacrylamide gel electrophoresis, *SEC* size-exclusion chromatography, *UV* ultraviolet

3.3 Analytical Methods

Different kinds of protein analytical assays can be employed to analyze the results generated by these HTS purification processes. Some of them can also be performed directly into the microplates where fractions were recovered facilitating the analyses throughput. For example, fractions can be directly collected into specific microplates to UV measurement at 280 nm wavelength and after being saved for other analytical assays. Differently, samples collect into ELISA microplates cannot be reused due to specific interactions with antibodies (*see* **Note 10**). On the other hand, other methods might demand indirect analysis since the recovered fraction in common 96-well plates will be taken to another device (*see* **Note 11**). Table 1 summarizes the main analytical protein quantification and characterization assays which can be used to investigate the results obtained in these HTS approaches (*see* **Note 12**). Some examples of results can be observed in Fig. 3.

A simplified workflow containing the main steps of HTS approach for protein purification studies can be observed in Fig. 4.

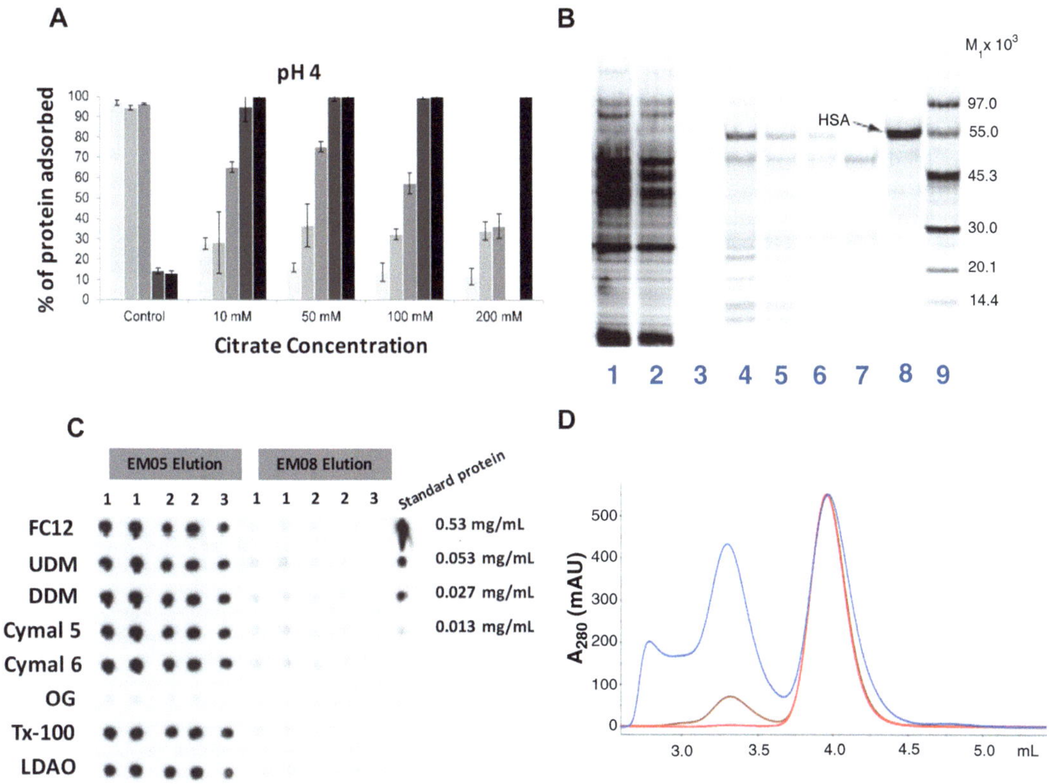

Fig. 3 Examples of analytical assays employed to analyze the results generated by HTS purification processes. (**a**) Adsorption study using phenylboronate as ligand on agarose matrix. Protein concentration was determined by UV absorbance at 280 nm. Different classes of proteins (glycosylated or not) with different p*I* values were used as feed. The graphic shows the effect of citrate buffer at pH 4 on the percentage of adsorption of amyloglucosidase (■); pepsin (■); cellulase (■); carbonic anhydrase (■); and conalbumin (■). The experiment was performed in the presence of different concentrations of citrate: 10, 50, 100, and 200 mM. The positive control was performed with 10 mM acetate at pH 4 (reproduced from Ref. 12 with permission from Wiley). (**b**) Capture of HSA (human serum albumin) from *E. coli* cell lysate by immobilizing a biotinylated polyclonal antibody to Streptavidin HP MultiTrap™ 96-well plates. Analysis of the collected fractions between chromatographic steps was performed by SDS-PAGE and the gel was post-stained with Deep Purple Total Protein Stain. The target protein was enriched 180-fold relative to the starting material. Lanes—*1*: Start material (diluted 1:30); *2*: Flow-through (diluted 1:30); *3*: Fifth wash; *4*: First elution; *5*: Second elution; *6*: Third elution; *7*: Antibody; *8*: HAS (Mr 65,000); *9*: Low molecular weight marker (reproduced from Ref. 13 with permission from GE Healthcare). (**c**) Screening study for optimizing solubility of membrane proteins. Eight detergents (FC12, UDM, DDM, Cymal™ 5, Cymal 6, OG, Triton™ X-100 and LDAO) were screened for their effect on the solubility of histidine tagged membrane proteins (EM05 and EM08) expressed in *E. coli*, on His MultiTrap™ FF 96-well plates. Data evaluation of elution fractions was performed by Dot-blot analysis on nitrocellulose membrane using INDIA HisProbe™-HRP Western blotting probe for the detection of target proteins (reproduced from Ref. 13 with permission from GE Healthcare). (**d**) Evaluation of aggregate content in purified fractions derived from clarified cell culture supernatant of Chinese hamster ovary (CHO) cells expressing a monoclonal antibody, using MabSelect SuRe™ (affinity chromatography) and Capto™ adhere (multimodal chromatography) plates. The concentration of monomers and aggregates was calculated in the starting material and the flowthrough fraction from the peak areas associated with SEC analysis: starting material (MabSelect SuRe™ pool, *brown*), flow-through/elution fraction (Capto™ adhere, *red*), and the discarded fraction (Capto™ adhere, *blue*); curves were normalized against the curve showing the flow-through/elution fraction (reproduced from Ref. 14 with permission from GE Healthcare)

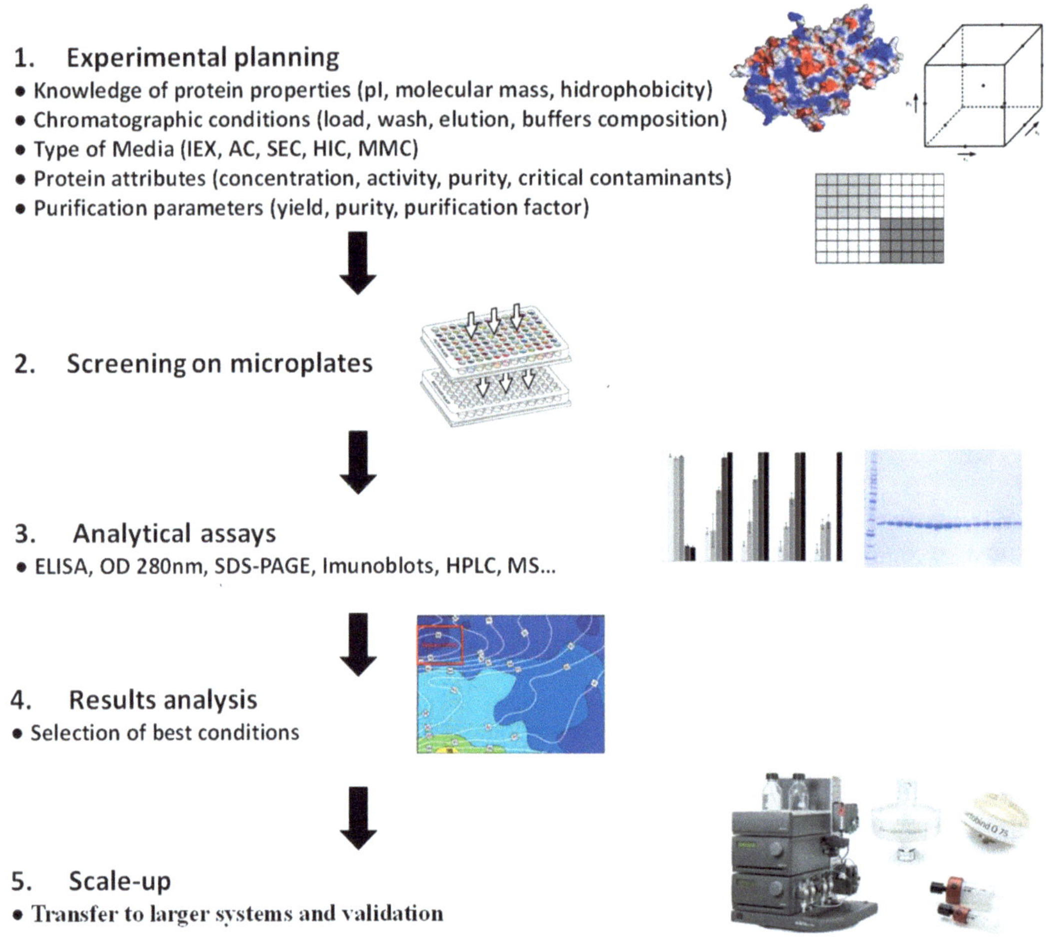

Fig. 4 Scheme of the workflow used in HTS methods for the optimization of chromatographic conditions. Adapted from Ref. 15

4 Notes

1. The membrane material must have hydrophilic low protein binding properties in order to avoid protein retention on the membrane surface. PVDF (polyvinylidene fluoride) is the most common membrane material used for low protein binding. Different manufacturers can be found in the market for both filter plate and device such as GE, Merck Millipore, Sartorius, and Pall.
2. Rotations higher than 1000 rpm are required in order to obtain a homogenized suspension for maximum adsorption. Otherwise, the particles may be stuck in the bottom of the well.

3. The robot workstation may be applied for this method, but it is not essential. The use of a robot system may reduce the operation time and possible errors during handling.
4. To plan your screening experiments use: (1) Specific DoE software to input and define the critical parameters to be studied using robust mathematical approaches or (2) use your previous knowledge about your target protein to select the main variables to be tested. It is recommended to have experimental triplicate data for each condition. If you have several conditions to investigate, use different plates for replicates, otherwise distribute your conditions in a single plate and use the different wells as your replicates.
5. Make sure to use a higher volume than that required for the experiment.
6. The dilution of the particles is important in order to avoid formation of lumps in the tip. The higher the concentration of the particles, the higher the probability for lump formation. The lump can deter the pipetting of exact volume and increase the error of particle volume per well.
7. The volume of particles applied into the wells will depend on the particles:solution working ratio. For instance, for 200 μL of sample volume per well and a ratio of 1:10, 20 μL of particles for each well is required. Different ratios can be used (e.g., 1:5, 1:10, or 1:20); however, the researcher must be aware about the differences in adsorption capacity.
8. The equilibration buffer must have similar conditions (pH, conductivity, etc.) as the sample in column chromatographic procedures.
9. To reach equilibrium and obtain maximum adsorption, it is recommended to obtain previously the kinetics of this interaction, and then determine how long the agitation should be maintained.
10. Small volumes can be retained into the wells; an average value can be calculated if the volume value is important for the result of the experiment. This retained volume can increase as long as the plate is being reused due to membrane blocking, impurities, contaminants, membrane damaged and compacted matrices.
11. Some samples may be out of range of the analysis if previous dilution is not performed. So it is recommended to use common 96-well plates as collection plates to treat the samples before analysing them.
12. Define the critical analytical assays to evaluate your results and divide the small volume of samples in aliquots for further use.

Acknowledgment

The authors would like to acknowledge the financial support from the Brazilian foundations FAPERJ, Capes, and CNPq.

References

1. Hanke AT, Ottens M (2014) Purifying biopharmaceuticals: knowledge-based chromatographic process development. Trends Biotechnol 32(4):210–220
2. Castilho LR, Moraes AM, Augusto EFP, Butler M (2008) Product purification processes. In: Castilho LR (ed) Animal cell technology: from biopharmaceuticals to gene therapy. Taylor & Francis, UK
3. Rege K, Pepsin M, Falcon B, Steele L, Heng M (2006) High-throughput process development for recombinant protein purification. Biotechnol Bioeng 93(4):618–630
4. Amadeo I, Mauro L, Ortí E, Forno G (2014) Establishment of a design space for biopharmaceutical purification processes using DoE. In: Labrou NE (ed) Protein downstream processing: design, development and application of high and low-resolution methods. Humana Press, Totowa, NJ, pp 11–27
5. Barker GA, Calzada J, Herzer S, Rieble S (2015) Adaptation to high throughput batch chromatography enhances multivariate screening. Biotechnol J 10:1493–1498
6. Wensel DL, Kelley BD, Coffman JL (2008) High-throughput screening of chromatographic separations: III. Monoclonal antibodies on ceramic hydroxyapatite. Biotechnol Bioeng 100:839–854
7. Sanaie N, Cecchini D, Pieracci J (2012) Applying high-throughput methods to develop a purification process for a highly glycosylated protein. Biotechnol J 7:1242–1255
8. Monié E (2006) Evaluation of the 96-well format for screening of chromatographic buffer conditions. Dissertation, Uppsala University School of Engineering
9. GE Healthcare (2014) Design of experiments in protein production and purification handbook. Available via: http://www.gelifesciences.com/file_source/doe/29103850AA.pdf. Accessed 27 Nov 2016
10. Olsson IM, Johansson E, Berntsson M, Eriksson L, Gottfries J, Wold S (2006) Rational DOE protocols for 96-well plates. Chemom Intell Lab Syst 83:66–74
11. Kálmán-Szekeres Z, Olajos M, Ganzler K (2012) Analytical aspects of biosimilarity issues of protein drugs. J Pharm Biomed Anal 69:185–195
12. Carvalho RJ, Woo J, Aires-Barros MR, Cramer SM, Azevedo AM (2014) Phenylboronate chromatography selectively separates glycoproteins through the manipulation of electrostatic, charge transfer, and cis-diol interactions. Biotechnol J 9:1250–1258
13. GE Healthcare (2009) MultiTrap 96-well plates applications and guidelines. Available via: https://www.gelifesciences.com/gehcls_images/GELS/Related%20Content/Files/1314787424814/litdoc28951127_20161014222744.pdf. Accessed 27 Nov 2016
14. Engstrand C, Rodrigo G, Forss A, Lacki K, Välimaa KN (2010) Rapid and scalable microplate development of a two-step purification process. Bioprocess Int 8:58–66. Available via: https://www.gelifesciences.com/gehcls_images/GELS/Related%20Content/Files/1314823637792/litdoc28987098_20161014202820.pdf. Accessed 27 Nov 2016
15. Pall Life Sciences (2011) Simple, efficient 96-well plate high throughput screening lab method to optimize the use of mixed-mode chromatography sorbents for the purification of a monoclonal antibody. Application Note USTR 2795. Available via: http://www.pall.com/pdfs/Biopharmaceuticals/10-4610_DoE%2096-WellPlate_AN_HR.pdf. Accessed 27 Nov 2016

Chapter 18

Purification and Autoactivation Method for Recombinant Coagulation Factor VII

Vladimir Granovski, Marcela C.C. Freitas, Mario Soares Abreu-Neto, and Dimas T. Covas

Abstract

Recombinant coagulation factor VII is a very important and complex protein employed for treatment of hemophiliac patients (hemophilia A/B) who develop inhibitors antibodies to conventional treatments (FVIII and FIX). The rFVII is a glycosylated molecule and circulates in plasma as zymogen of 50 kDa. When activated the molecule is cleaved to 20–30 kDa and has a half-life of about 3 h, needing to be processed fast and efficiently until freeze-drying. Here, we describe a very simple and fast purification sequence for rFVII using affinity FVII Select resin and a dialysis system that can be easily scaled up.

Key words Recombinant coagulation factor VII, Affinity chromatography, FVII autoactivation, rFVII purification methodology

1 Introduction

The inhibitor generation is very important issue in hemophilia treatment, where the repeated administration of coagulation factors VIII (FVIII) and FIX (FIX) generate antibodies that make the further treatment inefficacious [1]. The FVIII and FIX participate in the intrinsic coagulation pathway which lead to conversion of inactive factor X to its active form (Xa), forming a prothrombin activator and evolving a full thrombin burst [2]. One of the solutions to overcome the inhibition generation in hemophilia A/B patients is the administration of coagulation factor VII (FVII), which can be obtained from plasma or recombinant cultures (rFVII) [3]. The rFVII can bypass the FVIII and FIX needed for a thrombin burst acting on an extrinsic coagulation pathway activating the factor X [4].

The rFVII is a glycosylated molecule that is administered via injection and circulates in plasma as zymogen of 50 kDa [5]. When activated the molecule is cleaved to 20–30 kDa and has a half-life of

Virgínia Picanço-Castro and Kamilla Swiech (eds.), *Recombinant Glycoprotein Production: Methods and Protocols*, Methods in Molecular Biology, vol. 1674, DOI 10.1007/978-1-4939-7312-5_18,

about 3 h [6]. In vitro, the cleavage of the molecule occurs by auto activation and proximity [7]. Once produced in recombinant cell cultures, due to its very short half-life, rFVII needs to be processed fast and efficiently until freeze-drying.

Here, we describe the development of a simple and fast purification method using affinity FVIISelect resin and a simple dialysis process to purify and activate recombinant factor VII [8].

2 Materials

The chromatography column (Tricorn 5/50), resin (FVII Select) and the chromatographic system (Aktapure25) were from General Electric Healthcare Life Sciences. The 20 mL dialysis centrifuge tubes (Vivaspin 20) were from Millipore. All buffers were prepared using Milli-Q purified water and were filtered in 0.22 μm PVDF filters.

2.1 Supernatant Production

1. Recombinant rFVII producing cells (SK-Hep-1) [8].
2. Culture media Dulbecco's Modified Eagle Medium (DMEM)—In a 1000 mL reservoir add 500 mL of water; add DMEM powder as recommended in supplier manual. Supplement the culture media with 10 mM of HEPES and 44 mM of sodium bicarbonate. Inactivate 100 mL of fetal bovine serum by incubating it in 56 °C for 45 min and add to the culture. Make up to 1000 mL, adjust the pH to 7.0 (if necessary) and filter through 0.22 μm sterile filter.
3. Phosphate buffer saline (PBS).
4. T flasks (72 cm^2).
5. Animal-free recombinant trypsin.
6. Vitamin K solution.
7. Free Style 293 medium (Thermo Fisher Scientific).

2.2 Purification

1. FVII Select equilibration buffer: 50 mM Tris, 150 mM NaCl pH 7.5. Add about 50 mL of water to a 100 mL graduated cylinder. Add 0.66 g of Tris–HCl, 0.10 g of Tris base, and 0.877 g of NaCl. Make up to 100 mL, adjust pH to 7.5 if necessary, filter through 0.22 μm membrane and store at 4 °C.
2. FVII Select elution buffer: 50 mM Tris, 1500 mM NaCl, 50% (v/v) propylene glycol (PG) pH 7.5. Add about 40 mL of water to a 100 mL graduated cylinder. Add 0.66 g of Tris–HCl, 0.10 g of Tris base and 8.76 g of NaCl to a cylinder. Add 50 mL of PPG and make up to 100 mL, adjust pH to 7.5 in necessary, filter through 0.22 μm membrane and store at 4 °C.

3. Dialysis buffer (Freeze-Dry Buffer)—weight the following components; 292 mg of NaCl, 147 mg of calcium chloride dehydrate, 132 mg of glycyl glycine, 7.0 g of Tween 20, and 3.0 g of mannitol. Make up to 100 mL, adjust pH to 7.5 in necessary, filter through 0.22 μm membrane and store at 4 °C.
4. Protein analysis by SDS-PAGE: the SDS-PAGE analysis is performed using 4–20% gradient Mini-Protean® TGX gels, ten wells, 30 μL Mini-PROTEAN® Tetra Cell—Bio-Rad accordingly to a provider manual.
5. SDS-PAGE loading buffer—Mix the following components: 0.91 g Tris–HCl, 0.03 g of Tris base, 10 mL glycerol, 5 mL β-mercaptoethanol, 2.0 g sodium dodecyl sulfate, 4 mL of water, and 1 mL Bromophenol Blue (1% solution).
6. The detection of rFVII molecule can be performed using several commercially available methods such as chromogenic Chromogenix Coaset FVII kit, Factor VII Human ELISA Kit and finally a coagulation analyzer ACL-7000 with a Hemosil kit containing FVII calibration curve, positive and negative controls.

3 Methods

3.1 Supernatant Production

1. The supernatant was generated after SKHep cell cultivation in T flasks initially in DMEM medium containing 10% fetal bovine serum until a >95% of confluence. The metabolized medium was removed and the cells were washed with PBS. Free Style medium (w/o serum) and vitamin K (5 μg/mL) is added to the culture and incubated for 48 h. After incubation the culture is centrifuged and the supernatant filtered.

3.2 Affinity Chromatography

1. To each 100 mL of supernatant add 0.66 g of Tris–HCl and 0.10 g of Tris base, adjusting pH to 7.5 (if necessary).
2. Preequilibrate affinity chromatography column with FVIISelect affinity chromatography equilibration buffer until base line is established and all parameters such as pressure, UV, and conductivity are constant.
3. Inject the supernatant at the flow velocity recommended by the resin supplier (*see* **Note 1**).
4. Collect Flow-through (*see* **Note 2**).
5. After injection, reequilibrate the system with FVII Select equilibration buffer until the UV base line is reached.
6. Elute the bound proteins with FVII Select elution buffer (*see* **Note 3**).
7. If needed, the eluate can be stored at −20 °C until the dialysis (*see* **Note 4**).

3.3 Dialysis

1. Add 10 mL to a 20 mL Vivaspin 5 kDa of the eluate and complete until 20 mL with dialysis buffer (*see* **Note 5**).
2. Centrifuge the tube at a maximum of 4500 × *g* until half of tube volume is filtered.
3. Add 10 mL of dialysis buffer.
4. Repeat the last three steps for at least five more times (*see* **Note 6**).

3.4 Freeze-Drying

1. Aliquot 2 mL Eppendorf tubes with 500 μL of the dialyzed samples.
2. Cover the tubes with Parafilm and make small holes with a needle (*see* **Note 7**).
3. Freeze the tubes at −80 °C for 24 h.
4. Freeze-dry the samples.
5. In order to perform the characterization, resuspend the freeze-dried samples in 500 μL of cold ultrapure water (*see* **Note 8**).

3.5 rFVII Detection

1. Prepare the SDS-PAGE samples by mixing 10 μL of SDS-PAGE loading buffer and 40 μL of sample. Incubate the samples for 15 min in boiling water.
2. Apply boiled samples to the SDS-PAGE wells and run electrophoresis at 100 V, 400 mA until the Bromophenol Blue reach the bottom of the gel.
3. Use silver stain method (Fig. 1).
4. The detection of rFVII in the sample can be performed improving the methods cited in Subheading 2.

4 Notes

1. The resin provider supplies the information regarding maximum working flow velocity and maximum pressure. Do not exceed the pressure limits; it may impact on the target protein stability, protein binding to resin efficiency and elution properties. The working pressure in our experiments did not exceed 0.3 mPa.
2. The flow-through must be collected and analyzed for rFVII presence in order to monitor the rFVII biding efficiency to the resin. Once detected rFVII presence in flow-through, the resin must be discarded.
3. The rFVII elution can be observed by UV 280 nm peak. If no peak is noted, the eluate collection can be started when the conductivity signal pikes, since the elution buffer has high salt concentration. The eluate volume can be determined collecting

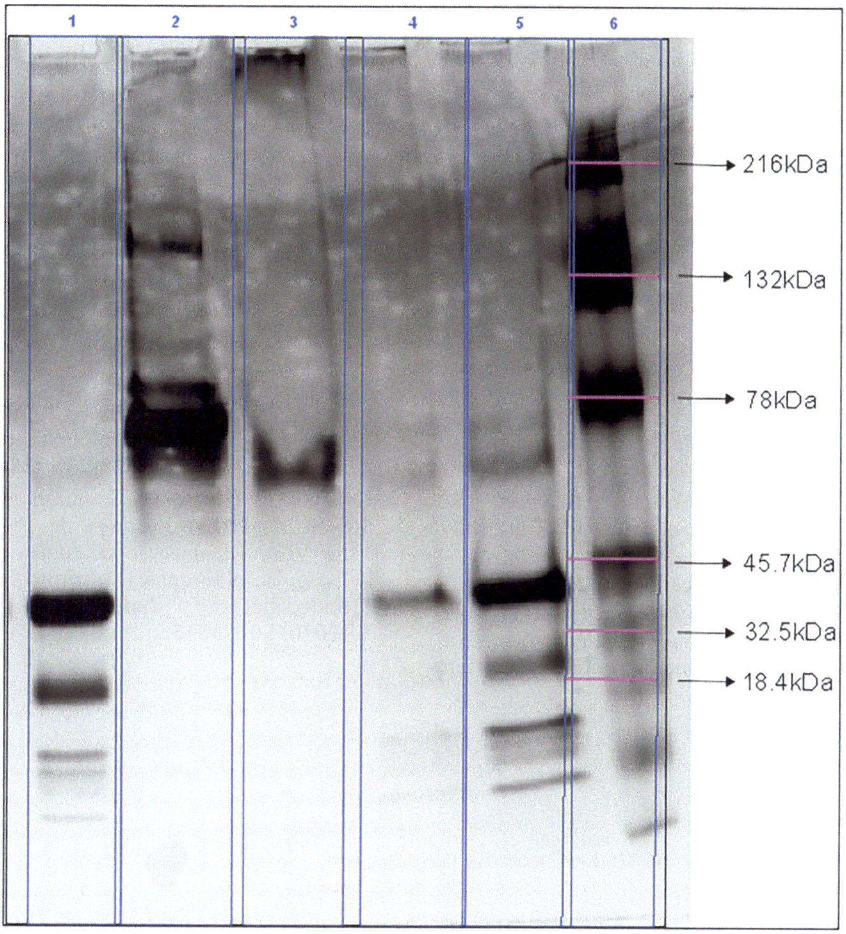

Fig. 1 SDS-PAGE gel of the retention samples of purification process. (*1*) Commercial control NovoSeven® (Novo Nordisk); (*2*) Cell supernatant; (*3*); Eluted fraction from FVII select; (*4*) Dialyzed fraction; (*5*) Solubilized freeze-dry sample (final product); (*6*) Commercial molecular weight marker

several fractions and detecting the rFVII presence. The Eluate volume must be determined empirically.

4. Since the rFVII is not cleaved at this point and autoactivation occurs only during the dialysis process, the sample can be stored up to 5 days at −20 °C. For long-term storage, freeze at −80 °C.
5. The Elution buffer contains 50% of propylene glycol (PG) and a high concentration of NaCl. It is very important to lower the concentration <1% of PG. Repeat the dialysis step six times.
6. After six dialysis buffer additions, the concentration of PG should be at 0.4–0.8%.
7. The freeze-drying methodology may vary for different machines.

8. After reconstitution the rFVII is highly unstable and has a half-life of 120 min. Any manipulation must be done on ice. rFVII can lose activity with many freeze–thaw cycles.

References

1. Franchini M, Mannucci PM (2012) Past, present and future of hemophilia: a narrative review. Orphanet J Rare Dis 7:24
2. Hedner U (2015) Recombinant activated factor VII: 30 years of research and innovation. Blood Rev Suppl 1:S4–S8
3. Yoshihiro M, Takamiya O (2012) The expression of recombinant factor VII variants is influenced by different cells for transfection using expression studies. Rinsho Byori 60(3):193–200
4. Eigenbrot C (2002) Structure, function, and activation of coagulation factor VII. Curr Protein Pept Sci 3(3):287–299
5. Bharadwaj D, Iino M, Kontoyianni M, Smith KJ, Foster DC, Kisiel W (1996) Factor VII central. A novel mutation in the catalytic domain that reduces tissue factor binding, impairs activation by factor Xa, and abolishes amidolytic and coagulant activity. J Biol Chem 271 (48):30685–30691
6. Tiede A (2015) Half-life extended factor VIII for the treatment of hemophilia A. J Thromb Haemost Suppl 1:S176–S179
7. Pedersen AH, Lund-Hansen T, Bisgaard-Frantzen H, Olsen F, Petersen LC (1989) Autoactivation of human recombinant coagulation factor VII. Biochemistry 28(24):9331–9336
8. Covas DT, Freitas MCC, Picanço-Castro V, Swiech K (2015) Processo de Produção de Fator VII de coagulação sanguínea e fator VII de coagulação sanguínea. Instituto Nacional da Propriedade Industrial (INPI) BR1020150123345

Part III

Formulation and Analytics

Chapter 19

Preparation of Immunoliposomes by Direct Coupling of Antibodies Based on a Thioether Bond

Raquel Petrilli, Josimar O. Eloy, Robert J. Lee, and Renata F.V. Lopez

Abstract

Drug delivery is of paramount importance, since the drug needs to be delivered to a specific site, in adequate concentration, avoiding degradation in order to provide therapeutic efficacy. Different nanocarriers have been used over the years for this purpose and liposomes are well-established systems due to the high biocompatibility and the possibility to vehiculate both hydrophilic and lipophilic drugs. In order to circumvent the rapid clearance by the reticuloendothelial system and to avoid the healthy cells exposure to the drug, long circulating liposomes containing polyethyleneglycol (PEG) and functionalized liposomes for targeted delivery have been developed. Immunoliposomes consist of liposomes containing antibodies or antibody fragments attached at the membrane surface. This attachment can be performed using PEG lipids, containing a reactive terminal group such as maleimide and thiolated antibodies. Additionaly, the use of PEG chains as spacers increases antibody–antigen affinity, since the antibody is not shielded by the steric hindrance of PEG and also due to the correct orientation of antibodies for interaction with receptors on cell surface. In this chapter, we describe and discuss in details the protocol to prepare anti-epidermal growth factor receptor (anti-EGFR) and anti-human epidermal growth factor receptor 2 (anti-HER2) liposomes using cetuximab and trastuzumab as antibodies. We present the direct coupling method based on the maleimide thioether reaction for these immunoliposomes preparation and present some characterization steps and in vitro studies in cell culture which can be used for better understanding these nanocarriers.

Key words Immunoliposomes, Targeting, Covalent coupling, Direct conjugation, Cetuximab, Trastuzumab

1 Introduction

The use of nanocarriers in drug delivery has been reported as advantageous compared to conventional formulation methods because it can prevent the degradation of therapeutic agents, enhance their in vivo efficacy due to the delivery to specific tumor sites and reduce the side effects [1]. Among them, liposomes have been extensively used in the clinic, and researched due to the high biocompatibility and the possibility to encapsulate both hydrophilic and lipophilic drugs. Nowadays, a variety of liposomal formulations are available in the market, such as Doxil®, Myocet®, Lipo-Dox®,

Virgínia Picanço-Castro and Kamilla Swiech (eds.), *Recombinant Glycoprotein Production: Methods and Protocols*, Methods in Molecular Biology, vol. 1674, DOI 10.1007/978-1-4939-7312-5_19, © Springer Science+Business Media LLC 2018

and Marqibo®, among others, which demonstrates the feasibility of the application of these nanocarriers in the clinic [2]. However, the rapid clearance and capture of liposomes by the reticuloendothelial system is one of the major drawbacks of conventional liposomes [3]. In order to circumvent these problems, different strategies can be used. PEGylation that is present in Doxil® and Lipo-Dox®, for example, is able to avoid macrophage escape and thus produces long circulating liposomes [4]. Different targeting moieties can also be conjugated to the surface of liposomes, both covalently attached or not, such as cell penetrating peptides [5], transferrin [6], folate [7], oligossacharides/polyssacharides [8], and antibodies [9], among others. These targeting moieties cause the active delivery of liposomes due to the recognition of overexpressed receptors at cell surface [10].

Liposomes that have antibodies attached at the surface are denominated immunoliposomes. Immunoliposomes can be produced by the coupling of whole antibodies or antibody fragments such as Fab or single-chain Fv fragments (scFv) using both non-covalent or covalent coupling methods [11, 12]. In the context of cancer treatment, different antibodies can be used both as combination therapy agents with conventional chemotherapeutic drugs or as targeting agents [13–15]. Cetuximab and trastuzumab are able to target epidermal growth factor receptor (EGFR) and human epidermal growth factor receptor 2 (HER2) and are currently used in the clinic for the treatment of squamous cell carcinoma and breast cancer, respectively [16, 17].

For targeted nanoparticles, the noncovalent coupling is performed using biotinylated antibodies that are linked to avidin-derivatized liposomes [18]. On the contrary, for covalently attachment of antibodies in liposomes, these can be both directly coupled to the surface using functional groups on liposomal surface or they can be linked using micelles of the antibody–lipid conjugate [12]. For covalent methods, different functionalized PEG-lipids can be used, containing a reactive terminal group such as hydrazine, N-3-(pyridyldithio)proprionate, maleimide, succinyl, p-nitrophenyl-carbonyl, and cyanuric choride [18]. Sterically stabilized immunoliposomes can be obtained by the attachment of PEG and antibodies in the liposomal surface. The use of PEG chains as spacers between the liposomes and antibodies increases antibody–antigen binding, since antibody is not shielded by the steric hindrance of PEG and due to the correct orientation of antibodies for interaction with receptors in cell surface [19].

Herein we describe in great detail a protocol to prepare both anti-EGFR and anti-HER2 liposomes using the direct coupling method. We discuss the covalent coupling method using Traut's reagent (Fig. 1) and the formulation requirements for the conjugation success. Characterization steps of the immunoliposomes and in vitro studies are presented and discussed.

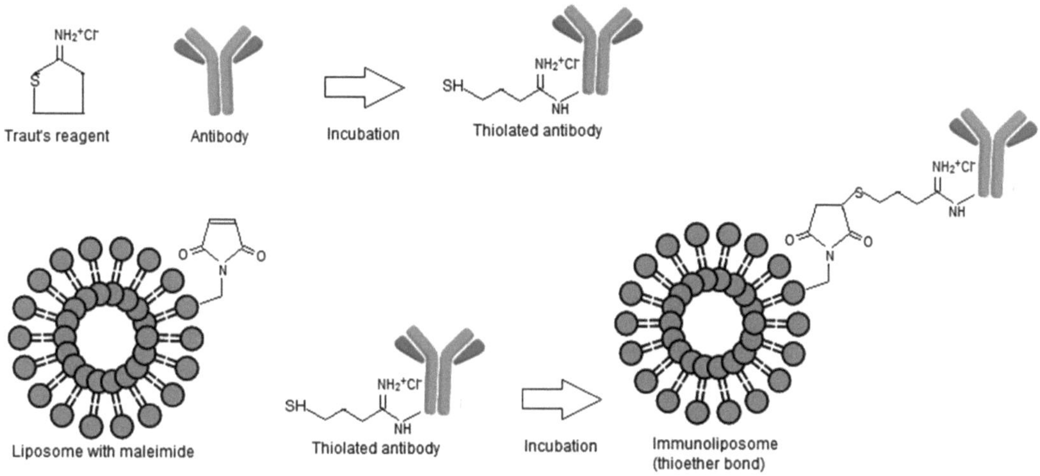

Fig. 1 Schematic representation of the immunoliposome reaction, adapted from [15] with authorization

2 Material

All solutions are prepared using ultrapure water (deionized water with sensitivity of 18.2 MΩ cm^2 at 25 °C) and stored at room temperature, unless specified otherwise.

2.1 Thiolation Reaction and Protein Quantification Assay

1. Antibodies to be attached: cetuximab or trastuzumab (*see* **Note 1**).
2. Traut's reagent.
3. Phosphate buffer saline (PBS), pH 7.4. Weigh 8 g NaCl, 0.2 g KCl, 0.2 g KH_2PO_4, and 1.15 g Na_2HPO_4. Prepare 1 L buffer solution and keep refrigerated until use (*see* **Note 2**). PBS buffer containing 5 mM EDTA, pH 8.0. Weigh 0.404 g EDTA, add to 200 mL of PBS buffer (pH 7.4) and adjust pH to 8.0.
4. 50 mM EDTA solution: Weigh 0.202 g EDTA and add to 10 mL of PBS buffer (pH 7.4) (*see* **Note 3**).
5. 0.1 N NaOH solution (*see* **Note 4**).
6. Disposable PD-10 desalting columns (*see* **Note 5**).
7. Bicinchoninic acid protein assay (BCA) protein assay kit.

2.2 Immunoliposomes Preparation

1. Lipids: L-α-Phosphatidylcholine from soybean (SPC), 1,2-distearoyl-*sn*-glycero-3-phosphocholine (DSPC), cholesterol (Chol), and 1,2-distearoyl-*sn*-glycero-3-phosphoethanolamine-*N*-[maleimide(polyethylene glycol)-2000] (ammonium salt) (DSPE-PEG-Mal) (*see* **Note 6**).
2. Drugs: Hydrophilic drugs should be added to the hydration buffer at the drug-to-lipid ratio of 0.1. Lipophilic drugs should be added to the organic phase at the drug-to-lipid ratio of 0.1.
3. Sepharose CL-4B column.

2.3 Protein Primary Structure Integrity

1. Ready Gel SDS-PAGE 4–20% mini gels.
2. Sample buffer 4×.
3. 2-Mercaptoethanol.
4. Tris–glycine–SDS buffer 10×, pH 8.3.
5. Precision plus dual color protein.

2.4 In Vitro Activity

1. The in vitro activity can be evaluated using cell viability studies using the 3-(4,5-dimethythiazol-2-yl)-2,5-diphenyl tetrazolium bromide (MTT).
2. Dulbecco's modified Eagle's medium (DMEM), 4500 mg/L glucose, containing 584 mg/L L-glutamine and 4 mg/L pyridoxine hydrochloride; RPMI-1640; and McCoy's 5A medium (Gibco) were used for A431 cells (EGFR+); MDA-MB231 (HER2−); and SK-BR-3 (HER2+) cell lines, respectively.
3. Heat-inactivated fetal bovine serum (FBS).
4. Antibiotic–antimycotic solution containing 10,000 IU of penicillin, 10 mg of streptomycin, and 25 μg of amphotericin B per mL.
5. 2.5% trypsin (10×) solution.
6. 2.5 mg/mL MTT solution.
7. 0.9% sterile saline.
8. Dimethyl sulfoxide (DMSO).

3 Methods

All solutions procedures should be carried out at room temperature, unless specified.

3.1 Thiolation Reaction

1. Mix 1 mg of antibody to PBS buffer containing 5 mM EDTA (pH 8.0). Add 50 mM EDTA solution until reach 5 mM EDTA in the final mixture. Adjust pH to 8.0 using 0.1 N NaOH solution.
2. Add 18.8 μL of a solution 2 mg/mL of Traut's reagent (Traut's reagent: antibody 40:1, mol:mol) dropwise, under continuous stirring at 37 °C. Keep the mixture under agitation for 3 min. This step should be carried out protected from light.
3. Protect the mixture antibody:Traut's reagent from light using aluminum foil and keep it for 1 h at 37 °C water bath.
4. Pass total 1 mL through a 10 mL PD-10 desalting column, collecting 14 fractions with 1 mL each using PBS containing EDTA buffer (pH 8.0).

3.2 Protein Quantification Assay

This procedure is applied to the quantification of antibody after conjugation to the liposomes and thus can be used to determine the conjugation efficiency.

1. For this step, the BCA protein assay kit (ThermoFisher) can be used.
2. Prepare the working reagent: mix 50 parts of BCA Reagent A with one part of BCA Reagent B.
3. Use dilution scheme for microplate procedure, as specified by the manufacturer.
4. Using 96-well microplates, apply 25 μL of the standard curve (25–750 μg/mL) and the samples (fractions obtained from the PD-10 column elution) in the wells.
5. Add 200 μL of the working reagent to each well.
6. Homogenize the microplates for 30 s on a microplate shaker and then incubate at 37 °C for 30 min.
7. Measure the absorbance at 562 nm.
8. Combine the fractions containing the antibody.

3.3 Immunoliposomes Preparation

1. Follow steps described in Subheadings 3.1 and 3.2 previous to the immunoliposomes preparation.
2. Weigh lipids DSPC–Chol–DSPE-PEG-Mal (55:45:2.5, molar proportion) or SPC–Chol–DSPE-PEG-Mal (10:2:0.5) for cetuximab and trastuzumab targeting, respectively (*see* **Note 7**). If intended for encapsulation into the immunoliposome, a lipophilic drug (e.g., rapamycin) should be added at an appropriate amount.
3. Add 2 mL chloroform and proceed solvent evaporation under rotary evaporation at 60 °C for 30 min (*see* **Note 8**).
4. Hydrate the thin lipid film using PBS buffer (pH 7.4). If intended for encapsulation in the immunoliposome, a hydrophilic drug (Example: 5-FU) should be added at an appropriate amount. Keep the formulation in the rotary evaporator without vacuum for 30–60 min to ensure complete film hydration.
5. Use a method to reduce particle size and lamellarity of the liposomes. High pressure homogenization for 30 min or extrusion five times through 0.2 μm pore size polycarbonate filters and three times through 0.1 μm pore size can be used (*see* **Note 9**).
6. Mix the thiolated antibody fractions (as described in Subheading 3.2, **step 8**) with the preformed liposomes. Allow the mixture to stand overnight at room temperature.
7. Add 1 mL of the immunoliposomes to a 10 mL Sepharose CL-4B column to remove unconjugated antibody. Collect 25 fractions eluted using PBS pH 7.4.

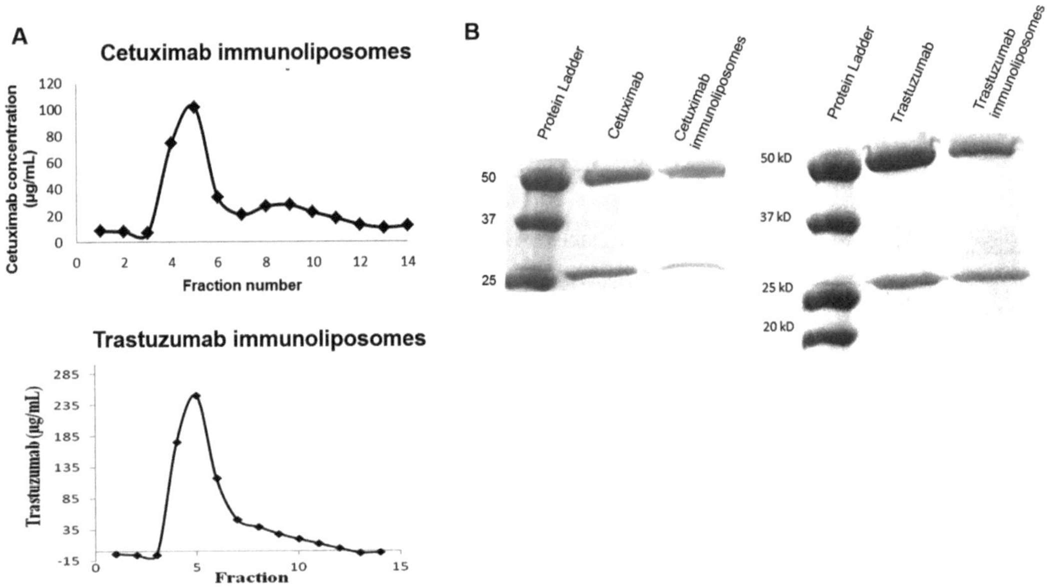

Fig. 2 (**a**) Chromatograms of cetuximab and trastuzumab immunoliposomes eluted in CL-4B columns with PBS pH 7.4. Fractions from 4–6 are attributed to the immunoliposomes. (**b**) Represents the antibodies integrity evaluation by SDS-PAGE electrophoresis under reducing conditions (adapted from [14, 15] with authorization)

8. Determine antibody concentration by the BCA method and generate a chromatogram (Fig. 2a). Calculate the targeting efficiency by comparing the amount of antibody conjugated to the liposomes and the amount of antibody added for the thiolation reaction.
9. Characterize the immunoliposomes using particle size, polydispersity, zeta potential, and antibody primary structure integrity.

3.4 Protein Primary Structure Integrity

1. Prepare the sample buffer using nine parts of Laemmli and one part of 2-mercaptoethanol.
2. Add 10 μL of sample buffer to 30 μL of the immunoliposomes (*see* **Note 10**).
3. Heat the samples at 60 °C for 10 min for denaturation prior to gel application.
4. Remove gel out of its pouch, cut the tape along the black line and mount for electrophoresis.
5. Apply 30 μL of each sample and the precision plus dual color protein for molecular weight confirmation.
6. Carry out electrophoresis at 100 V for 50 min in Tris–glycine–SDS buffer pH 8.3.

7. Perform staining using QC Colloidal Coomassie stain. Proceed the destaining with deionized water in three cycles of 30 min each (*see* **Note 11**).
8. Photograph gels for interpretation of results (Fig. 2b).

3.5 In Vitro Activity

1. Allow cells to reach 80–90% confluency.
2. Trypsinize using trypsin solution and sterile saline.
3. Seed 1×10^4 cells/well in 96-well plates for cell culture and incubate for 24 h at 37 °C in 5% CO_2 atmosphere.
4. Wash the cells using 200 μL saline solution and incubate with the samples diluted in incomplete medium (*see* **Note 12**).
5. Wash the cells twice using 200 μL saline solution to remove all samples.
6. Add 180 μL incomplete culture medium and 20 μL of a 2.5 mg/mL MTT solution.
7. Incubate for 4 h at 37 °C.
8. Remove all medium and solubilize the formazan crystals in 200 μL of DMSO.
9. Measure the absorbance values at 570 nm with a microplate reader.
10. Calculate the cellular viability as the percentage of viable cells compared to the control group (*see* **Note 13**).

4 Notes

1. Antibodies used were commercially available solutions, 2 mg/mL Erbitux and 21 mg/mL Herceptin® (after reconstitution of lyophilized powder) manufactured by Merck Serono and Roche, without further purification. These were kind gifts from Cancer Hospital Clinical Pharmacy.
2. A concentrated 10× PBS buffer was prepared and diluted at the moment of use. For this, simply use 10× the amount of salts indicated and dilute in the same amount of deionized water. Keep refrigerated.
3. The indicated amount of EDTA refers to ethylenediaminetetraacetic acid disodium salt dihydrate (MW = 372.24 g/mol).
4. Usual method to prepare a NaOH solution. Heat on the bottom of the flask is expected. Store using plastic flasks.
5. PD-10 desalting column preparation: the column is prepacked. Proceed equilibration using the elution buffer before use (approximately 30 mL buffer).
6. Lipids used with purity ≥99%.

7. Different lipid compositions can be used, but it is important to use a lipid conjugated to maleimide in order to allow the reaction between the thiolated antibody and the maleimide group. Despite the lower cost of other anchors, such as Chol [12], DSPE-PEG-Mal is the most frequently employed lipid for immunoliposome preparation, because it avoids the synthesis step. Also, the coupling of antibodies directly at the liposomal surface affects the targeting affinity due to the presence of steric hindrance caused by PEG coating. Thus, the use of linkers with long spacer arms is indicated to attach the antibodies distantly from the liposomal bilayer [20]. Additionally, for optimal formulation it is expected to have maximum drug loading, low leakage upon storage, and systemic circulation, with adequate release at the target site. For this, the incorporation of high-transition temperature lipids and/or Chol is suggested [2].
8. The amount of chloroform used varies depending on the amount of lipids used, based on the formulation developed. It is important to observe a clear solution in the solvent prior to rotary evaporation.
9. Different methods can be used for liposomal particle size and lamellarity reduction. It is important to study the methods influence on particle size, polydispersity, and morphology of the vesicles. This can be accomplished using particle size measurements and transmission electron microscopy or cryo-transmission electron microscopy.
10. In order to compare the gel bands, apply the same amount of protein for the different samples tested. The antibody solution should be applied and used as a control.
11. For the destaining step, immerse the gel in deionized water and place it on a microplate shaker platform. Change the destaining solution until the three cycles of washing or if necessary, until the background is clear.
12. For A431, the treatments were performed for 72 and 120 h, whereas for MDA-MB231 (HER2−) and SK-BR-3 (HER2+) the treatments were applied for 72 h.
13. It is necessary to have a control group (nontreated cells) in each microplate in order to calculate cell viability.

Acknowledgments

The authors would like to acknowledge grants #2012/23764-3, #2012/10388-3, #2012/21513-3, #2013/15134-2, and #2014/22451-7 from Sao Paulo Research Foundation (FAPESP) and CNPQ for grant #480962/2013-8.

References

1. Fomina N, Sankaranarayanan J, Almutairi A (2012) Photochemical mechanisms of light-triggered release from nanocarriers. Adv Drug Deliv Rev 64(11):1005–1020
2. Eloy JO, Claro de Souza M, Petrilli R, Barcellos JPA, Lee RJ, Marchetti JM (2014) Liposomes as carriers of hydrophilic small molecule drugs: strategies to enhance encapsulation and delivery. Colloids Surf B Biointerfaces 123:345–363
3. Torchilin VP (2005) Recent advances with liposomes as pharmaceutical carriers. Nat Rev Drug Discov 4(2):145–160
4. Chang HI, Yeh MK (2012) Clinical development of liposome-based drugs: formulation, characterization, and therapeutic efficacy. Int J Nanomedicine 7:49–60
5. Kwon SS, Kim SY, Kong BJ, Jin K, Noh GY, Im NR, Lim JW, Ha JH, Kim J, Park SN (2015) Cell-penetrating peptide-conjugated liposomes as transdermal delivery system of Polygonum aviculare L. extract. Int J Pharm 483:26–37
6. Mendonça LS, Firmino F, Moreira JN, De Lima MCP (2010) Transferrin receptor-targeted liposomes encapsulating anti- BCR-ABL siRNA or asODN for chronic myeloid leukemia treatment. Bioconjug Chem 21:157–168
7. Li H, Piao L, Yu B, Yung BC, Zhang W, Wang PG, Lee JL, Lee RJ (2011) Delivery of calf thymus DNA to tumor by folate receptor targeted cationic liposomes. Biomaterials 32 (27):6614–6620
8. Wijagkanalan W, Kawakami S, Higuchi Y, Yamashita F, Hashida M (2011) Intratracheally instilled mannosylated cationic liposome/NF-κB decoy complexes for effective prevention of LPS-induced lung inflammation. J Control Release 149(1):42–50
9. Loureiro JA, Gomes B, Fricker G, Cardoso I, Ribeiro CA, Gaiteiro C, Coelho MA, Pereira MC, Rocha S (2015) Dual ligand immunoliposomes for drug delivery to the brain. Colloids Surf B Biointerfaces 134:213–219
10. Petrilli R, Eloy JO, Marchetti JM, Lopez RFV, Lee RJ (2014) Targeted lipid nanoparticles for antisense oligonucleotide delivery. Curr Pharm Biotechnol 15(9):847–855
11. Kirpotin DB, Drummond DC, Shao Y, Shalaby MR, Hong K, Nielsen UB, Marks JD, Benz CC, Park JW (2006) Antibody targeting of long-circulating lipidic nanoparticles does not increase tumor localization but does increase internalization in animal models. Cancer Res 66(13):6732–6740
12. Pan X, Wu G, Yang W, Barth RF, Tjarks W, Lee J (2007) Synthesis of cetuximab-immunoliposomes via a cholesterol- based membrane anchor for targeted delivery of a neutron capture therapy (NCT) agent to glioma cells. Bioconjug Chem 18(1):101–108
13. Demarest SJ, Hariharan K, Dong J (2011) Emerging antibody combinations in oncology. MAbs 3(4):338–351
14. Petrilli R, Eloy J, Lopez RFV, Lee R (2017) Cetuximab immunoliposomes enhance delivery of 5-FU to skin squamous carcinoma cells. Anti Cancer Agents Med Chem 17:301–308
15. Eloy J, Petrilli R, Brueggemeier RW, Marchetti, JM, Lee RJ (2017) Rapamycin loaded immunoliposomes functionalized with trastuzumab: a strategy to enhance cytotoxicity to HER2 positive breast cancer cells. Anti Cancer Agents Med Chem 17:48–56
16. Chang HR (2010) Trastuzumab-based neoadjuvant therapy in patients with HER2-positive breast cancer. Cancer 116(12):2856–2867
17. Cohen MH, Chen H, Shord S, Fuchs C, He K, Zhao H, Sickafuse S, Keegan P, Pazdur R (2013) Approval summary: Cetuximab in combination with cisplatin or carboplatin and 5-fluorouracil for the first-line treatment of patients with recurrent locoregional or metastatic squamous cell head and neck cancer. Oncologist 18:460–466
18. Schnyder A, Krähenbühl S, Török M, Drewe J, Huwyler J (2004) Targeting of skeletal muscle in vitro using biotinylated immunoliposomes. Biochem J 377:61–67
19. Koning GA, Kamps JAAM, Scherphof GL (2002) Efficient intracellular delivery of 5-fluorodeoxyuridine into colon cancer cells by targeted immunoliposomes. Cancer Detect Prev 26(4):299–307
20. Nobs L, Buchegger F, Gurny R, Alle E (2004) Current methods for attaching targeting ligands to liposomes and nanoparticles. J Pharm Sci 93(8):1980–1992

Chapter 20

Polyester-Based Nanoparticles for the Encapsulation of Monoclonal Antibodies

Flávia Sousa, Pedro Fonte, Andreia Cruz, Patrick J. Kennedy, Inês Mendes Pinto, and Bruno Sarmento

Abstract

Aliphatic polyesters have been widely explored for biomedical applications (e.g., drug delivery systems, biomedical devices, and tissue engineering). Recently, polyesters have been used in nanoparticle formulations for the controlled release of monoclonal antibodies (mAbs) for the enhanced efficacy of antibody-based therapy. Polyester-based nanoparticles for mAb delivery provide decreased antibody dosage, increased antibody stability and protection and longer therapeutic action, ultimately translating to an increased therapeutic index. Additionally, nanoencapsulation holds the potential for the selective cellular recognition and internalization of mAbs, in the disease context when intracellular organelles and molecules (e.g., enzymes, transcription factors and oncogenic proteins) are the preferred target. We present here a detailed method to prepare mAb-loaded polyester-based nanoparticles and the various techniques to characterize the resulting nanoparticles and mAb structure. Finally, we highlight different biological approaches to assess the in vitro bioactivity of the antibody upon nanoparticle release.

Key words Polyester, Polymeric nanoparticles, Monoclonal antibodies, Controlled release, Antibody delivery, Antibody structure, Antibody bioactivity, Anti-VEGF, Angiogenesis, Endothelial cells

1 Introduction

Recombinant monoclonal antibodies (mAbs) are large biopharmaceuticals used in the treatment of several human diseases, namely, cancer, asthma, and immune and infectious diseases [1]. The number of FDA-approved mAbs in the market has been steadily increasing over the last few decades with approximately 50 mAbs available at the time of publication [2, 3]. Immunoglobulins, constituted by two light chains (~25 kDa each) and two heavy chains (~50 kDa each) linked by disulfide bonds, have the ability to bind with high affinity and specificity to almost infinite molecules. This targeted therapy helps to minimize secondary side effects [4]. However, significant production costs and issues associated with mAb administration, pharmacology, tissue penetration and mode of action

Virgínia Picanço-Castro and Kamilla Swiech (eds.), *Recombinant Glycoprotein Production: Methods and Protocols*, Methods in Molecular Biology, vol. 1674, DOI 10.1007/978-1-4939-7312-5_20, © Springer Science+Business Media LLC 2018

make antibody-based therapy very challenging [5]. Due to the large size of mAbs, pressure gradient inside tumor masses and the "binding site barrier effect," these biomolecules have shown low ability to penetrate tumors or to access inflammation sites [6, 7]. In addition, antibody-based therapy usually implies multiple and long (several hours) intravenous administrations reducing patient compliance [8]. In order to achieve a maximum response of mAb in the antibody-based therapy, a longer half-life, a higher penetration in the tissues, a minimal cytotoxic effect and a low immunogenicity is desired [9]. Thus, the nanoencapsulation of recombinant mAbs is a promising alternative to achieve a more effective and selective therapy.

Polyester-based nanoparticles, also known as polymeric nanoparticles, are constituted by aliphatic, semi-aromatic or aromatic polyesters in order to encapsulate several types of drugs including protein-based drugs. Linear polyesters can be composed of homopolymers, such as polylactic acid (PLA), polyglycolic acid (PGA), polycaprolactone (PCL) or by copolymers such as poly (lactic-co-glycolic acid) (PLGA) and poly (3-poly(3-hydroxybutyrate)-*co*-ε-caprolactone) [10]. Polymeric nanoparticles are made mainly by linear polyesters. Importantly, the US Food and Drug Administration (FDA) has approved PLGA, PGA and PLA for certain medical applications [11] due to their biocompatibility and the nontoxic degradation by-products that are eliminated via natural metabolic pathways. Particularly, PLGA nanoparticles are the most common type of polyester-based nanoparticles used for mAb delivery [12, 13]. Nanoencapsulation can enhance the treatment of several diseases because it may allow for dosage reduction leading to decreased toxicity and enhanced potency [14–16]. Another advantage of mAb nanoencapsulation is the possibility for controlled release of the mAb from the nanoparticles, keeping the plasma concentration of mAbs within a therapeutic range [14]. Intracellular delivery of mAbs may be achieved by nanoencapsulation when the intracellular components are intended to be targeted [17]. Despite all the advantages of encapsulation of mAbs into PLGA nanoparticles, antibody stability and antibody bioactivity must be assured. Thus, the physicochemical and biological characteristics of mAbs must be evaluated before and after nanoencapsulation, in order to understand if the formulation process affects the structure and bioactivity of the mAb.

In this chapter, we present detailed protocols for the preparation and characterization (structure and bioactivity) of PLGA nanoparticles containing a mAb (*see* Fig. 1). In the mAb bioactivity section, we use bevacizumab, a recombinant humanized mAb that acts against vascular endothelial growth factor type A (VEGF-A), as a molecular model system in order to provide a more detailed protocol about in vitro studies for mAb.

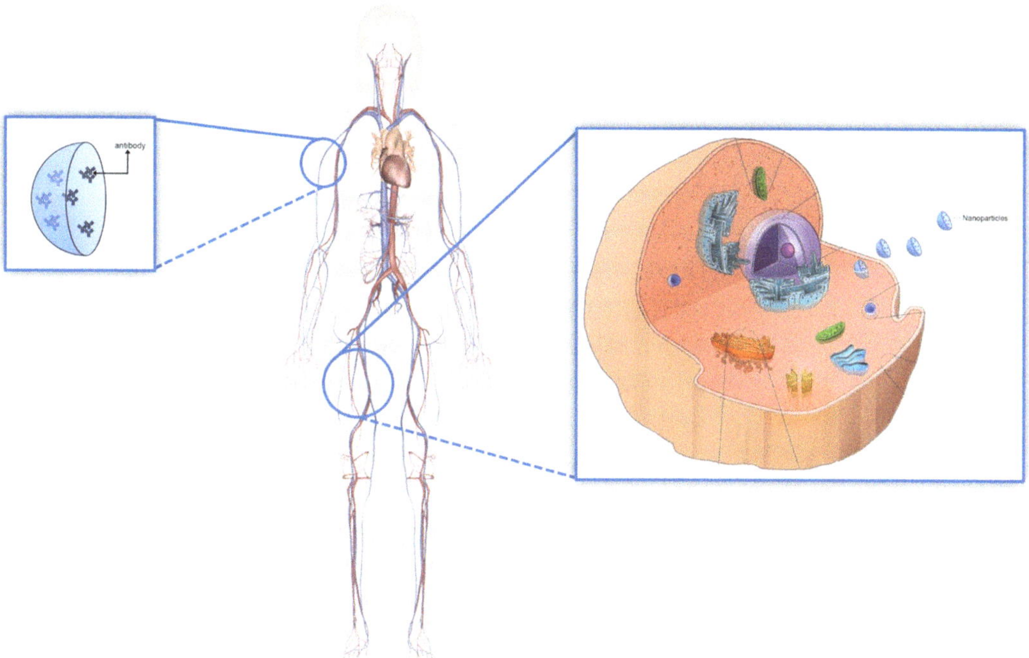

Fig. 1 Representative figure of this chapter purpose: polyester-based nanoparticles for recombinant mAb delivery

2 Materials

2.1 Reagents

1. Poly(D,L-lactide-*co*-glycolide) 50:50 (PLGA) (*see* **Note 1**).
2. Poly(vinyl alcohol) (PVA) (*see* **Note 2**).
3. 99.8% ethyl acetate (EA) (*see* **Note 3**).
4. Monoclonal antibody (~150 kDa) (*see* **Note 4**).
5. Coomassie Protein Assay Kit.
6. Phosphate buffered saline 10× (PBS): 1.37 M NaCl, 0.1 M Na_2HPO_4, 27 mM KCl, and 18 mM KH_2PO_4, pH 7.4.
7. Human umbilical vein endothelial (HUVEC) cell line cultured in M199 medium.
8. M199 medium supplemented with 20% v/v fetal bovine serum (FBS) and 1% v/v of penicillin–streptomycin (10,000 units/mL penicillin and 10,000 μg/mL streptomycin).
9. 2% gelatin solution (gelatin in ultrapure water).
10. 1% Triton X-100 in PBS buffer.
11. Triazolyl Blue Tetrazolium Bromide (MTT).
12. Dimethyl sulfoxide (DMSO).
13. Recombinant human $VEGF_{165}$.
14. Cell Proliferation ELISA, BrdU.

2.2 Equipment

1. Probe sonicator.
2. Beckman Optima TLM ultracentrifuge.
3. Eppendorf® LoBind microcentrifuge tubes.
4. DTS 1060 disposable zeta cell.
5. Zetasizer Nano ZS.
6. Quanta 400 FEG SEM (FEI).
7. JEOL JEM-1400 Electron Microscopy (JEOL Ltd).
8. ABB MB 3000 FTIR spectrometer (ABB, Zurich, Switzerland) with a MIRacle single reflection ATR accessory.
9. HorizonMB FTIR software (ABB, Zurich, Switzerland).
10. NanoDrop 2000c UV-Vis spectrophotometer (Thermo Scientific, Wilmington, DE, USA).
11. Jasco J815 CD Spectrometer.
12. Felix fluorescence RTC 2000 spectrometer.

3 Methods

3.1 PLGA Nanoparticles Preparation

PLGA nanoparticles can be made by several methods, such as solvent emulsification–evaporation, solvent displacement, interfacial deposition, solvent emulsification–diffusion, salting out, and nanoencapsulation [18]. The choice of nanoencapsulation method must take into account the method conditions (e.g., thermal or pressure stresses) and physical–chemical characteristics of the mAb (e.g., solubility in water and organic solvents, pH, and molecular weight). The method most commonly used to encapsulate mAbs is the modified solvent emulsification–evaporation method based on a water–oil–water (w/o/w) [19, 20]. For that reason, we describe the preparation of nanoparticles by this method (Fig. 2).

The production protocol comprises the following steps:

1. Dissolve 100 mg of PLGA in 1 mL of EA in a glass tube (*see* **Note 5**).
2. Add 2 mg of mAb (in aqueous solution) to the glass tube (*see* **Note 6**) and emulsify using a probe sonicator at 70% of amplitude for 30 s.
3. Transfer the primary emulsion to a PVA aqueous solution (2% w/v, 12.5 mL) and sonicate at 70% of amplitude for 30 s (*see* **Note 7**).
4. Transfer the double emulsion into a 50 mL glass baker and stir for 3 h at 300 rpm (*see* **Note 8**).
5. Wash the PLGA nanoparticles with milli-Q water three times by centrifugation using a Beckman Optima TLM ultracentrifuge at 40,000 × *g* for 20 min at 4 °C. Reconstitute the pellet with milli-Q water or phosphate-buffered saline (PBS) (*see* **Note 9**).

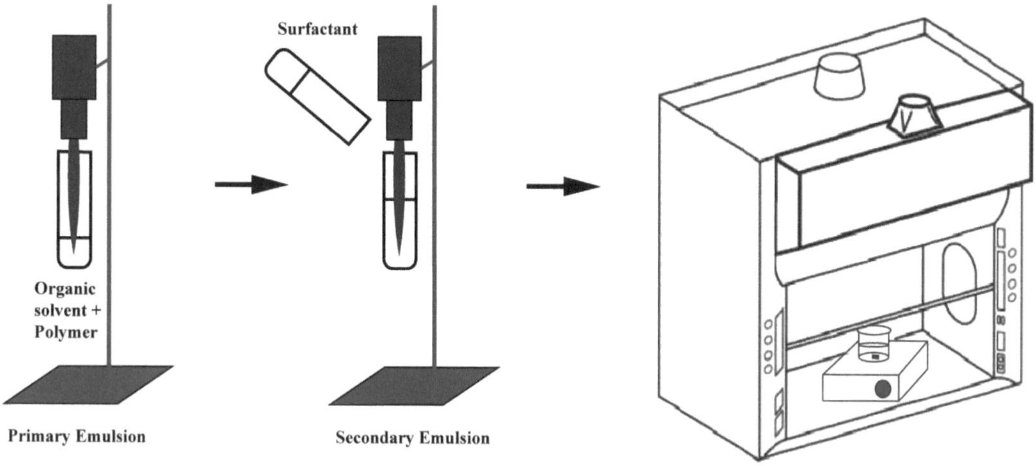

Fig. 2 Schematic procedure about modified solvent emulsification–evaporation method based on a water–-oil–water to prepare PLGA nanoparticles

3.2 PLGA Nanoparticle Characterization

The characterization of mAb-loaded PLGA nanoparticles can be made using several techniques such as determination of particle size and charge; determination of mAb association efficiency and drug loading; evaluation of nanoparticle morphology by transmission electron microscopy (TEM) and scanning electron microscopy (SEM); and the study of mAb release kinetics from PLGA nanoparticles. In this section, a brief list of these characterization methods is provided.

3.2.1 Particle Size and Potential Zeta Analysis

1. Dilute the nanoparticle formulation (1:150). Prepare 900 μL (6 μL of nanoparticles solution with 894 μL of mQ water) in Eppendorf® LoBind Microcentrifuge Tube: Protein (*see* **Note 10**).
2. Put 750 μL of sample in an activated and water-flushed DTS 1060 disposable zeta cell.
3. Determine nanoparticle size and zeta potential by dynamic light scattering (DLS) using a Zetasizer Nano ZS. Measure samples in triplicate at a scattering angle of 173° and 25°. The produced PLGA nanoparticles must present a mean particle size in the range of 200 nm and a zeta potential of around –30 mV.

3.2.2 mAb Association Efficiency (AE) and Drug Loading (DL)

1. The Bradford Protein Assay Kit is a colorimetric method used to quantify total protein levels and can be used to indirectly determine the amount of associated mAb (minimum detectable dose = 1 μg/mL).
2. Use the Bradford Protein Assay Kit according to manufacturer's instructions.

3. Centrifuge the PLGA nanoparticles as described in **step 5** of Subheading 3.1 and collect the supernatant. Use the whole or diluted supernatants in Bradford Protein Assay Kit.
4. Calculate the association efficiency and drug loading by the following equations (*see* **Note 11**):

$$\mathrm{AE} = \frac{\text{Total amount of mAb} - \text{Free mAb in supernatant}}{\text{Total amount of mAb}} \times 100$$

$$\mathrm{DL} = \frac{\text{Total amount of mAb} - \text{Free mAb in supernatant}}{\text{Total dry weight of nanoparticles}} \times 100$$

3.2.3 Nanoparticle Morphology

Nanoparticle morphology of mAb-loaded PLGA nanoparticles may be evaluated with TEM and SEM. TEM provides information about the shape of nanoparticles, whereas SEM provides information about the surface of nanoparticles. It is important to confirm the characteristic spherical shape with smooth surfaces typical of PLGA nanoparticles. These techniques also determine the particle size and, thus, can be compared to the results with DLS analysis. However, the technical procedures for both techniques are different. The protocol to evaluate the nanoparticle morphology by SEM comprises the following steps:

1. Dilute the nanoparticle formulation with ultrapure water (1:10).
2. Put 3 μL of each washed nanoparticle formulation on aluminum stubs and leave overnight to dry.
3. Before the microscopic observation, the samples should be coated with a layer of gold/palladium for 60 s using a 15 mA current.
4. Observe the nanoparticle morphology using an SEM with an accelerating voltage of 10 kV (*see* **Note 12**).

Nanoparticle morphology by TEM comprises the following steps:

1. Dilute the nanoparticle formulation (1:10).
2. Prior to microscopic observation, put the samples on a grid treated with uranyl acetate.
3. Observe the PLGA nanoparticles morphology using a TEM.

3.2.4 mAb In Vitro Release Study

1. Disperse 100 mg of nanoparticles in 20 mL of pH 7.4 PBS solution at 37 °C under magnetic stirring at 100 rpm. All samples should be run in triplicate (*see* **Note 13**).
2. At predetermined times (e.g., 1, 2, 4, 8, 24, and 48 h), collect 1 mL samples and separate the supernatant from PLGA nanoparticles by ultracentrifugation (the same velocity described in **step 5** of Subheading 3.1).

3. Replace with fresh buffer (1 mL) after each samples collection.
4. Determine the content of mAb in the supernatant using the Bradford Protein Assay Kit.

3.3 Evaluation of mAb Structure

Interactions between the mAb and the polyester-based nanoparticles (e.g., electrostatic, hydrogen bonds, van der Walls forces and hydrophobic) will occur during antibody nanoencapsulation [21]. These interactions, that are dependent on the type of carrier and the antibody itself, may be affect the mAb structure compromising its biological activity [22]. Thus, we recommend evaluating the secondary structure of the mAb using attenuated transmission reflectance-Fourier transform infrared spectroscopy (ATR-FTIR), circular dichroism (CD), and fluorescence spectroscopy (FS).

3.3.1 Attenuated Transmission Reflectance-Fourier Transform Infrared Spectroscopy (ATR-FTIR)

ATR-FTIR spectroscopy may provide information about the secondary structure (α-helices, β-sheets, and β-turns) of proteins in a noninvasive way [23]. mAb are constituted mainly by β-sheets and β-turns. The spectra region characteristic of evaluating the mAb secondary structure is the amide I region (1600–1700 cm^{-1}). mAbs usually are dominated by bands at 1615, 1635, and 1690 cm^{-1} which are representative of β-sheet structure [24]. On the other hand, β-turns can be represented by bands at 1660 and 1670 cm^{-1}. In order to quantify the mAb secondary structure, second derivative spectra are widely used [25]. Thus, mAb structural changes can be found overlapping the second derivative spectra of mAb native with the loaded mAb. The procedure to obtain FTIR spectra is composed of the following steps:

1. Prepare your sample and choose the correct support to place the sample (depending on physical state of samples) (*see* **Note 14**). Spectra of mAb-loaded PLGA nanoparticles, unloaded PLGA nanoparticles and the native mAb must be analyzed.
2. Obtain the FTIR spectra using ABB MB 3000 FTIR spectrometer. The spectrum collection should be made with 256 scans, at 4 cm^{-1} of resolution in the region of 4000–6000 $cm^{-1.}$ Triplicates of each sample should be analyzed.
3. After spectrum collection, spectral processing should be made using the HorizonMB FTIR software. Spectral processing starts with a double subtraction procedure (subtract the spectrum of mAb-loaded PLGA nanoparticles to the unloaded PLGA nanoparticles). After that, calculate a Savitzky-Golay second derivative with 13 points followed by a baseline correction with 3–4 point adjustment at the amide I region (1700–1600 cm^{-1}).
4. All spectra should be normalized using Origin 8 software in order to compare the second derivative spectra of mAb native with the loaded mAb.

3.3.2 Circular Dichroism (CD) Spectroscopy

CD spectroscopy is a simple, powerful and excellent technique for the assessment of the conformational properties of a protein [26]. CD spectroscopy rapidly determines the secondary and tertiary structure and the unfolding/folding properties. Contrary to ATR-FTIR spectroscopy, this noninvasive technique does not evaluate the secondary structure of the mAb within the nanoparticles. Therefore, we recommend using CD spectroscopy to characterize the secondary structure of mAb after its release from the nanoparticles. Similar to ATR-FTIR spectroscopy, structural changes may be identified with the overlap of the spectra of native and released mAb. Far-UV CD measurements may identify the secondary structural changes of mAbs, while Near-UV provides information about the tertiary structure of mAb [27]. The procedure to evaluate the secondary structure of mAb is described below:

1. Collect the mAb released from PLGA nanoparticles (*see* **Notes 15** and **16**) and prepare a 0.2 mg/mL mAb solution using the appropriate buffer (e.g., pH 7.4 PBS).
2. Put the sample in a 0.1 cm cell and determine the parameters for performing the reading. The Far-UV CD measurements are made from 190 to 250 nm. Set the following parameters: bandwidth—1 nm; data pitch—0.5 nm; data integration time (DIT)—2 s; scanning speed—50 nm/min; and accumulations—8. These parameters are dependent on the specific mAb molecule under evaluation.
3. Collect the buffer spectra and the mAb spectra under the same conditions using a Jasco J815 CD Spectrometer. A triplicate of each sample should be done (*see* **Note 17**).
4. Subtract the mAb spectra to the respective reference and export and convert data in order to express ellipticity. After that, all spectra should be normalized as described in **step 4** in Subheading 3.3.1.

3.3.3 Fluorescence Spectroscopy

Fluorescence spectroscopy is a rapid and robust method to study thermal conformation stability in several formulations and is able to detect slight changes in the tertiary structure of proteins using the second derivative [28]. This technique uses the fluorescence of aromatic amino acids in proteins to determine the denaturation temperature and folding/unfolding of proteins [29]. Thus, we recommend using fluorescence spectroscopy when the mAb is released from nanoparticles. The steps of this method are described below:

1. Collect the mAb released from PLGA nanoparticles and prepare a mAb solution using the PBS as reference buffer (*see* **Note 18**).
2. Put the buffer/sample in a 0.4 cm cell.

3. For each scan, fix the excitation wavelength at 280 nm (emission from tryptophan and tyrosine) and collect the emission spectra from 300 to 530 nm with an integration time of 0.1 s and 1 nm of step size. Collect all emission scans with an excitation slit width of 0.25 mm and an emission slit width of 6 mm. A set of triplicates of each formulation should be analyzed in the experiment.
4. Collect the buffer spectra and the mAb spectra under the same conditions using a Felix fluorescence RTC 2000 spectrometer, using a 75-W Zenon arc lamp coupled to a monochromator.
5. Subtract the spectrum of the released mAb from the respective reference and possible Raman peaks. After that, normalize and derive all spectra using a 15 points Savitzky–Golay second derivative.
6. Compare the second derivative spectra of native and released. A lower similarity of the spectra corresponds to higher mAb structural changes.

3.4 Evaluation of mAb Bioactivity

The encapsulation process may compromise mAb biological activity therefore, following structural analysis, it is important to access mAb bioactivity, efficiency and toxicity in vitro and in vivo models. In this section, we focus on mAbs that act against VEGF-A (e.g., Bevacizumab) in order to demonstrate a described protocol about mAb bioactivity in vitro. VEGF-A is a critical angiogenic factor [30] that promotes the growth of vascular endothelial cells; induces gene expression and endothelial migration; and regulates vascular permeability and vasodilatation [31]. However, VEGF has an important role as a regulator of tumor angiogenesis and metastatic spread [32]. Thus, recombinant, humanized anti-VEGF antibodies have been an important tool for antiangiogenesis therapy in cancer. The biological activity of an anti-VEGF mAb is characterized by the blocking and neutralization of the endogenous VEGF [33]. Human umbilical vein endothelial cells (HUVEC) are used mainly in the characterization of biological activity of these mAbs because they express VEGF and VEGF receptors [34].

3.4.1 In Vitro Cytotoxicity Assays

Anti-VEGF mAb binds to VEGF and inhibits the interaction between VEGF and its receptors, resulting in decreased vascularization and consequently decreasing cell viability [31]. In order to evaluate anti-VEGF bioactivity, VEGF is added to HUVEC cells and the cell viability is monitored. Here, we describe a detailed protocol using the MTT (3-(4,5-dimethylthiazolyl-2-)-2,5-diphenyltetrazolium bromide) assay, a colorimetric-based procedure for assessing cell metabolic activity. The principle of the method is based on the MTT reduction to purple formazan by active cells [35]. Next a solubilization solution is added to dissolve the insoluble purple formazan product into a colored solution, which is quantified by a spectrophotometer. Briefly, the absorbance is proportional to the purple formazan produced by active cells, reflecting the viability of cells.

1. Seed HUVEC cells without growth factors into 96-well tissue culture test plates with a cell density of 1×10^4 cells/well. Supplement M199 medium with 20% of FBS and 1% of Pen-Strep (*see* **Note 19**).
2. After 24 h, remove the M199 medium and replace with different concentrations of native mAb, unloaded PLGA nanoparticles, and mAb-loaded PLGA nanoparticles (0.1, 1, 10, and 100 μg/mL). Include a negative control (Triton X-100 1%) and a positive control (medium only after incubation of cells with VEGF). Replicates should be included (*see* **Note 20**).
3. After 24 h of incubation at 37 °C in 5% CO_2 atmosphere, remove the M199 medium, and add 150 μL of MTT solution per well (final concentration of 0.5 mg/mL). Incubate for 4 h at 37 °C in 5% CO_2 atmosphere.
4. Remove and discard the MTT solution, and add 150 μL of DMSO.
5. Shake for 10 min (or until all the formazan particles are dissolved) at room temperature in the dark to minimize photo damage.
6. Measure the absorbance using a microplate spectrophotometer at 590 and 630 nm.
7. Determine the viability of cells according to the following equation:

$$\%\ \text{viability} = \frac{\text{Experimental value} - \text{Negative control}}{\text{Positive control} - \text{Negative control}} \times 100$$

3.4.2 BrdU Assay

One of the biological activities of anti-VEGF mAb is the inhibition of endothelial cell proliferation. Therefore, to complement the results of MTT assay, the next steps are representatives of detailed protocol about HUVEC proliferation using a BrdU (5′-bromodeoxyuridine) incorporation assay [34]. The BrdU assay is a simple and fast colorimetric method to evaluate cell proliferation [36] based on the incorporation of 5-bromo-2′-deoxyuridine (BrdU) into cellular DNA during DNA synthesis in replicating cells (*see* Fig. 3).

1. Seed HUVEC cells without growth factors into wells of 96-well tissue culture test plates with a cell density of 1×10^4 cells/well. Supplement M199 medium with 20% of FBS and 1% of Pen-Strep.

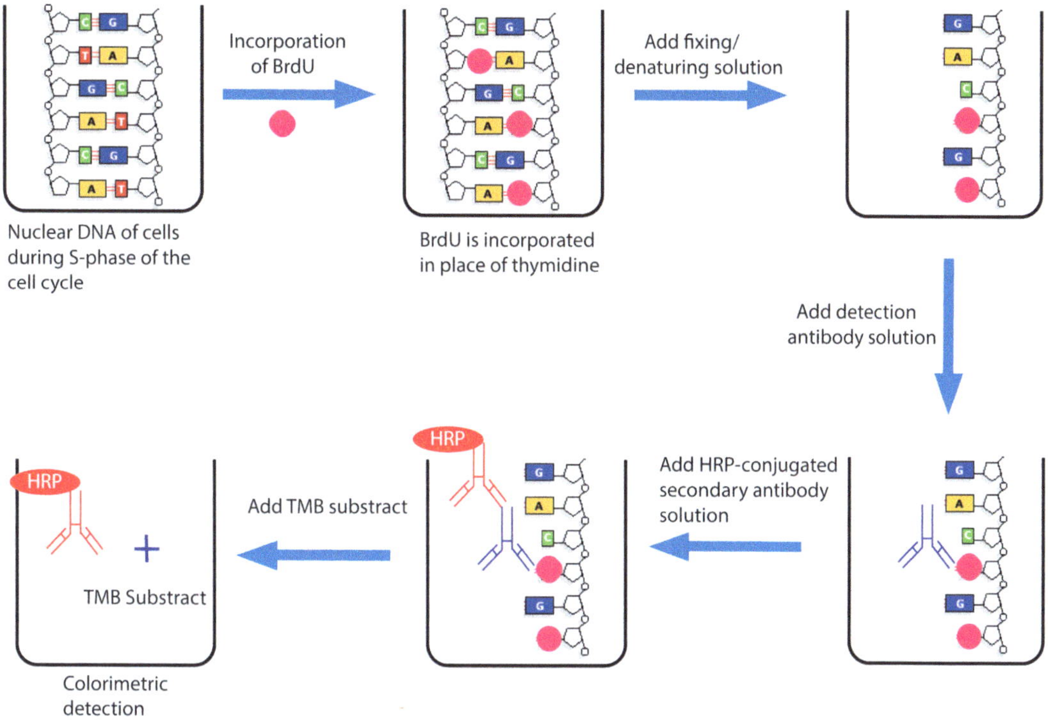

Fig. 3 Schematic procedure about BrdU cell proliferation assay to evaluate the bioactivity of mAbs

2. After 24 h, remove the M199 medium and replace with fresh warm medium supplemented with 50 ng/mL of recombinant VEGF and with different concentrations of mAb native, unloaded PLGA nanoparticles, and mAb-loaded PLGA nanoparticles (0.1, 1, 10, and 100 μg/mL). Include a negative control (cells and anti-BrdU-POD) (*see* **Note 21**).
3. Use the Cell Proliferation ELISA, BrdU Detection Kit according to the manufacturer's instructions (*see* **Note 22**).

4 Notes

1. PLGA must be stored at −20 °C. The suppliers may offer different types of PLGA which differ in lactic acid–glycolic acid ratio (25:75, 50:50, and 75:25), inherent viscosity midpoint, degree of crystallinity, and molecular weight. On other hand, all types of copolymers are available with different end groups, especially a free carboxylic end group or an ester end group. The choice of end groups is particularly important when the functionalization of PLGA nanoparticles is intended.
2. PVA was chosen because it is described in the literature that PLGA nanoparticles present a low polydispersity index when

PVA is used as surfactant (*see* Ref. 37). However, others types of surfactants can be used, such as Tween and Pluronic. The percentage of surfactant is also important when nanoparticles formulation is intended. The choice of the supplier of surfactant is very important because little variations of surfactants can originate different results in the preparation of nanoparticles.

3. Ethyl acetate was chosen as organic solvent because it can allow for the formation of smaller nanoparticles due to decreased surface tension of the polymer droplet in the primary emulsion (*see* Ref. 38). However, others organic solvents as dichloromethane and chloroform can be used.
4. The choice of mAb must take into account the physical-chemical properties (e.g., molecular weight, isoelectric point, solubility, or origin), storage, handling conditions and other added excipients.
5. The dissolution of PGLA in EA is very slow and for this reason, the dissolution should be left overnight followed by vortex the next morning. Avoid heating, once in the next step mAb will be added. Parafilm should be put in the top of the tube in order to avoid organic solvent evaporation.
6. The quantity of mAb added must take into account the drug loading intended and the mAb potency.
7. The dissolution of PVA in water is very difficult; thus, heating is needed. The pH of PVA solution can be changed as one way to improve the association efficiency, depending on the isoelectric point of the mAb.
8. The agitation should be made in a fume hood and under magnetic stirring in order to evaporate completely the organic solvent.
9. To determine the association efficiency of mAb-loaded nanoparticles by indirect method, ultracentrifugation is mandatory. However, in order to avoid the sedimentation of 150 kDa proteins speeds lower than 36,000 rpm are demanded.
10. The dilution of PLGA nanoparticles formulation must be guaranteed because the samples need to be optically clear for zeta potential measurements. If possible, the liquids used to dilute the samples should be filtered.
11. The total dry weight calculation can be obtained by freeze-drying an aliquot of hydrated nanoparticles.
12. As nanoparticles present a small size it is necessary to focus a small area in order to visualize the surface of nanoparticles. However, sometimes the strong electron beam can damage the nanoparticles and images with a good definition are impossible to obtain. Decreasing the intensity of the electron beam may be a good way to overcome this problem.

13. It should ensure that the quantity of mAb released is within the limits of detection and quantification of the Bradford Protein Assay Kit.
14. With ATR-FTIR it is possible to obtain spectra from a very wide range of solids, liquids and gases. If the sample is in a solid state, a small amount of sample is needed (~200 μg of mAb). However, when liquid samples are necessary, use higher quantities of sample (between 2 and 10 mg of mAb) to obtain a good FTIR spectra.
15. When performing CD measurements in a 0.1 cm cell, the maximum concentration of protein is 0.2 mg/mL due to signal saturation (make solutions between 0.05 and 0.2 mg/mL of mAb). The confirmation of the mAb concentration should be made by UV absorption at 280 nm using a NanoDrop 2000c UV-Vis spectrophotometer.
16. The choice of buffer should be done carefully because some buffers absorb strongly below 200 nm (*see* Ref. 39). Phosphate, Tris, and borate are suitable buffer ions for Far-UV CD measurements in a pH range from 6 to 10. Chloride ions should be avoided in high concentrations because they present absorption strongly below 200 nm.
17. As mAbs usually present a secondary structure rich in β-sheet, it is supposed that the CD spectrum shows one maximum at about 202 nm and a minimum at about 218 nm (*see* Refs. 40, 41).
18. The minimal concentration of protein to be analyzed is 1 μM.
19. Before seeding, each well should be coated with 1% of gelatin (dissolved in PBS). Allow surface to dry at least 30 min before introducing to the cells and medium in the incubator. After that, remove the excess gelatin from each well.
20. Sample preparation should be made with M199 medium supplemented with a recombinant human $VEGF_{165}$ (50 ng/mL).
21. Prepare the samples with M199 medium supplemented with a recombinant human $VEGF_{165}$ (50 ng/mL) and 5′-bromodeoxyuridine (BrdU) solution at a final concentration of 0.01 mM (according to the manufacturer's instructions).
22. When BrdU is incorporated into the cells, the pyrimidine analogue is integrated in place of thymidine into the new DNA of proliferating cells. After that, the procedure is to fix the cells and denature the DNA with fixing/denaturing solution. In order to detect the incorporated BrdU, mouse anti-BrdU mAb is added in the next step. To recognize the antibody, anti-mouse-HRP IgG is added. A subtract solution (tetramethyl-benzidine) is responsible for developing the color. Lastly, read the absorbance at 450 nm in an ELISA reader. The quantity of BrdU incorporated into cells is proportional to the absorbance (*see* Ref. 36).

Acknowledgments

Flávia Sousa would like to thank to Fundação para a Ciência e a Tecnologia (FCT), Portugal for financial support (Grant SFRH/BD/112201/2015). Patrick J. Kennedy gratefully acknowledges the BiotechHealth Programme (Doctoral Programme in Cellular and Molecular Biotechnology Applied to Health Sciences) and FCT for financial support (SFRH/BD/99036/2013) and beyond. This work was also financed by FEDER—Fundo Europeu de Desenvolvimento Regional funds through the COMPETE 2020—Operational Programme for Competitiveness and Internationalisation (POCI), Portugal 2020, and by Portuguese funds through FCT—Fundação para a Ciência e a Tecnologia/Ministério da Ciência, Tecnologia e Inovação in the framework of the projects "Institute for Research and Innovation in Health Sciences" (POCI-01-0145-FEDER-007274) and UID/BIM/04293/2013.

References

1. Scott AM, Wolchok JD, Old LJ (2012) Antibody therapy of cancer. Nat Rev Cancer 12 (4):278–287
2. Li J, Zhu Z (2010) Research and development of next generation of antibody-based therapeutics. Acta Pharmacol Sin 31(9):1198–1207
3. Ecker DM, Jones SD, Levine HL (2015) The therapeutic monoclonal antibody market. mAbs 7(1):9–14
4. Chames P, Van Regenmortel M, Weiss E, Baty D (2009) Therapeutic antibodies: successes, limitations and hopes for the future. Br J Pharmacol 157(2):220–233
5. Modjtahedi H, Ali S, Essapen S (2012) Therapeutic application of monoclonal antibodies in cancer: advances and challenges. Br Med Bull 104(1):41–59
6. Beckman RA, Weiner LM, Davis HM (2007) Antibody constructs in cancer therapy: protein engineering strategies to improve exposure in solid tumors. Cancer 109(2):170–179
7. Thurber GM, Schmidt MM, Wittrup KD (2008) Antibody tumor penetration: transport opposed by systemic and antigen-mediated clearance. Adv Drug Deliv Rev 60(12):1421
8. Samaranayake H, Wirth T, Schenkwein D, Raty JK, Yla-Herttuala S (2009) Challenges in monoclonal antibody-based therapies. Ann Med 41(5):322–331
9. Arruebo M, Valladares M, González-Fernández Á (2009) Antibody-conjugated nanoparticles for biomedical applications. J Nanomater 2009:24
10. Ravikumar M (2016) Handbook of polyester drug delivery systems. Pan Stanford, USA
11. Marin E, Briceño MI, Caballero-George C (2013) Critical evaluation of biodegradable polymers used in nanodrugs. Int J Nanomedicine 8:3071–3091
12. Varshochian R, Jeddi-Tehrani M, Mahmoudi AR, Khoshayand MR, Atyabi F, Sabzevari A, Esfahani MR, Dinarvand R (2013) The protective effect of albumin on bevacizumab activity and stability in PLGA nanoparticles intended for retinal and choroidal neovascularization treatments. Eur J Pharm Sci 50(3–4):341–352
13. Son S, Lee WR, Joung YK, Kwon MH, Kim YS, Park KD (2009) Optimized stability retention of a monoclonal antibody in the PLGA nanoparticles. Int J Pharm 368(1–2):178–185
14. Grainger DW (2004) Controlled-release and local delivery of therapeutic antibodies. Expert Opin Biol Ther 4(7):1029–1044
15. Yadav SC, Kumari A, Yadav R (2011) Development of peptide and protein nanotherapeutics by nanoencapsulation and nanobioconjugation. Peptides 32(1):173–187
16. Cardoso MM, Peca IN, Roque AC (2012) Antibody-conjugated nanoparticles for therapeutic applications. Curr Med Chem 19 (19):3103–3127
17. Srinivasan AR, Lakshmikuttyamma A, Shoyele SA (2013) Investigation of the stability and cellular uptake of self-associated monoclonal antibody (MAb) nanoparticles by non-small lung cancer cells. Mol Pharm 10(9):3275–3284

18. Pinto Reis C, Neufeld RJ, Ribeiro AJ, Veiga F (2006) Nanoencapsulation: I. Methods for preparation of drug-loaded polymeric nanoparticles. Nanomedicine 2(1):8–21
19. Li F, Hurley B, Liu Y, Leonard B, Griffith M (2012) Controlled release of bevacizumab through nanospheres for extended treatment of age-related macular degeneration. Open Ophthalmol J 6:54–58
20. Gdowski A, Ranjan A, Mukerjee A, Vishwanatha J (2015) Development of biodegradable nanocarriers loaded with a monoclonal antibody. Int J Mol Sci 16(2):3990–3995
21. Almeida A, Souto E (2007) Solid lipid nanoparticles as a drug delivery system for peptides and proteins. Adv Drug Deliv Rev 59 (6):478–490
22. Shoyele SA, Slowey A (2006) Prospects of formulating proteins/peptides as aerosols for pulmonary drug delivery. Int J Pharm 314(1):1–8
23. Fonte P, Soares S, Sousa F, Costa A, Seabra V, Reis S, Sarmento B (2014) Stability study perspective of the effect of freeze-drying using cryoprotectants on the structure of insulin loaded into PLGA nanoparticles. Biomacromolecules 15(10):3753–3765
24. Hawe A, Kasper JC, Friess W, Jiskoot W (2009) Structural properties of monoclonal antibody aggregates induced by freeze-thawing and thermal stress. Eur J Pharm Sci 38(2):79–87
25. Sarmento B, Ferreira DC, Jorgensen L, van de Weert M (2007) Probing insulin's secondary structure after entrapment into alginate/chitosan nanoparticles. Eur J Pharm Biopharm 65 (1):10–17
26. Greenfield NJ (2006) Using circular dichroism spectra to estimate protein secondary structure. Nat Protoc 1(6):2876–2890
27. Chelius D, Ruf P, Plöscher M, Liedtke R, Gansberger E, Hess J, Wasiliu M, Lindhofer H (2010) Structural and functional characterization of the trifunctional antibody catumaxomab. mAbs 2(3):309–319
28. Garidel P, Hegyi M, Bassarab S, Weichel M (2008) A rapid, sensitive and economical assessment of monoclonal antibody conformational stability by intrinsic tryptophan fluorescence spectroscopy. Biotechnol J 3(9–10):1201–1211
29. Oliveira Silva C, Petersen SB, Pinto Reis C, Rijo P, Molpeceres J, Vorum H, Neves-Petersen MT (2015) Lysozyme photochemistry as a function of temperature. The protective effect of nanoparticles on lysozyme photostability. PLoS One 10(12):e0144454
30. Ellis LM, Hicklin DJ (2008) VEGF-targeted therapy: mechanisms of anti-tumour activity. Nat Rev Cancer 8(8):579–591
31. Wang Y, Fei D, Vanderlaan M, Song A (2004) Biological activity of bevacizumab, a humanized anti-VEGF antibody in vitro. Angiogenesis 7(4):335–345
32. Goel HL, Mercurio AM (2013) VEGF targets the tumour cell. Nat Rev Cancer 13 (12):871–882
33. Ferrara N, Hillan KJ, Gerber H-P, Novotny W (2004) Discovery and development of bevacizumab, an anti-VEGF antibody for treating cancer. Nat Rev Drug Discov 3(5):391–400
34. Carneiro A, Falcao M, Azevedo I, Falcao Reis F, Soares R (2009) Multiple effects of bevacizumab in angiogenesis: implications for its use in age-related macular degeneration. Acta Ophthalmol 87(5):517–523
35. Stockert JC, Blázquez-Castro A, Cañete M, Horobin RW, Villanueva Á (2012) MTT assay for cell viability: intracellular localization of the formazan product is in lipid droplets. Acta Histochem 114(8):785–796
36. Cell Proliferation ELISA, BrdU (colorimetric). Roche Diagnostics, Mannheim, Germany. http://www.sigmaaldrich.com/catalog/product/roche/11647229001?lang=en&region=NO. Accessed 10 July 2016
37. Shi K, Cui F, Yamamoto H, Kawashima Y (2009) Optimized formulation of high-payload PLGA nanoparticles containing insulin–lauryl sulfate complex. Drug Dev Ind Pharm 35(2):177–184
38. Cohen-Sela E, Chorny M, Koroukhov N, Danenberg HD, Golomb G (2009) A new double emulsion solvent diffusion technique for encapsulating hydrophilic molecules in PLGA nanoparticles. J Control Release 133 (2):90–95
39. Kelly SM, Price NC (2000) The use of circular dichroism in the investigation of protein structure and function. Curr Protein Pept Sci 1 (4):349–384
40. Tetin SY, Linthicum DS (1996) Circular dichroism spectroscopy of monoclonal antibodies that bind a superpotent guanidinium sweetener ligand. Biochemistry 35 (4):1258–1264
41. Joshi V, Shivach T, Yadav N, Rathore AS (2014) Circular dichroism spectroscopy as a tool for monitoring aggregation in monoclonal antibody therapeutics. Anal Chem 86 (23):11606–11613

Chapter 21

Polyester-Based Nanoparticles for Delivery of Therapeutic Proteins

Pedro Fonte, Flávia Sousa, and Bruno Sarmento

Abstract

Therapeutic proteins are widely used to treat severe conditions, such as oncologic and metabolic diseases, and also for vaccination. They are recognized for its high biopotency, but their hydrophilicity and high molecular weight make its transport through membranes difficult to achieve. They may also suffer in vivo structural instability caused by a proteolytic cleavage, turning them biological inactive. Polyester-based nanoparticles are useful tools to overcome such problems, and deliver therapeutic proteins in a controlled and localized manner, decreasing the number of administrations needed and enhancing the therapeutic outcome. The biodegradable and biocompatible nature of such carriers makes them feasible to be administered by invasive and noninvasive routes. Since nanoparticles are produced in suspension form, they are usually lyophilized to increase its long-term storage stability. Therefore, the aim of this work is to propose an intuitive protocol for production and lyophilization of polyester-based nanoparticles for therapeutic proteins delivery. The characterization techniques to evaluate the nanoparticle and lyophilisate features, as well as the structural stability of the loaded protein are also described. For these purposes, PLGA nanoparticles and human insulin are used as models.

Key words Polyester, Nanoparticles, Protein, Protocol, Lyophilization, Stability, Drug delivery, Insulin, PLGA

1 Introduction

Therapeutic proteins are macromolecular drugs produced by biotechnology using live organisms and their active compounds, being useful in vaccination and in the treatment and management of severe conditions, such as oncologic and metabolic diseases. Comparatively to synthetic drug molecules, they present higher effect due to its similarity to biologic molecules, selectivity and ability to have a potent action, with fewer side effects [1]. Besides the benefits, their bioavailability may be reduced, mainly because of its hydrophilic nature and high molecular weight, hampering the ability to cross membranes [2]. Additionally, they present a limited in vivo stability, due to proteolytic cleavage and degradation, leading to its removal from the bloodstream. The parenteral route is the

Virgínia Picanço-Castro and Kamilla Swiech (eds.), *Recombinant Glycoprotein Production: Methods and Protocols*, Methods in Molecular Biology, vol. 1674, DOI 10.1007/978-1-4939-7312-5_21,

most used due to the instability and low permeability of proteins through biological membranes [3]. However, its invasive nature leads to reduced patient compliance and therapeutic inefficacy, because of the associated pain and discomfort. The oral, nasal, transdermal, ocular, and other noninvasive routes have been proposed to deliver therapeutic proteins, but their success as real alternatives remains a challenge.

The encapsulation of therapeutic proteins into nanoparticles is a useful strategy to protect them from degradation, and promote the uptake through biological membranes increasing its bioavailability [4]. Thus, nanoparticles can be used to deliver therapeutic proteins both by invasive and noninvasive routes. Aliphatic polyesters such as polycaprolactone (PCL), polylactic acid (PLA), poly (lactic-co-glycolic) acid (PLGA) and their copolymers are synthetic polymers used to produce nanoparticles and deliver therapeutic proteins [5–7]. They are recognized by its biodegradability, biocompatibility and ability to deliver drugs in a sustained, controlled, localized, and targeted manner [8].

Polyester-based nanoparticles loading therapeutic proteins are produced by different techniques, but the conditions of formulation, such as pH, ionic strength, temperature and others may influence the structural stability of proteins [1]. During formulation, proteins can suffer adsorption, aggregation and denaturation, limiting their in vivo concentration and bioactivity. Nanoparticles are usually produced in the suspension form, so stability problems during storage may occur mainly due to the hydrolytic activity of water on nanoparticles polymer and loaded protein. Lyophilization is widely used in the pharmaceutical industry to increase the long-term stability of products, and simultaneously allow its easy storage and handling by removing water from formulations [9]. The water removal occurs by freezing the sample, and removing water by sublimation and desorption using vacuum [10]. This dehydration process overcomes the instability in suspension of polyester nanoparticles, obtaining a solid product with good reconstitution and cake appearance, and low residual moisture content [11, 12]. After lyophilization, the nanoparticles should also maintain the initial physicochemical characteristics. However, the harsh conditions of the process may lead to protein stability, so an adequate lyophilization cycle needs to be used.

Overall, this work aims to propose an intuitive methodology to produce polyester-based nanoparticles for therapeutic proteins delivery, describing the most important techniques to characterize the nanoparticles features. Also, it is suggested an optimized lyophilization cycle to obtain a good lyophilized product. Finally, some techniques to assess the structural stability and bioactivity of the loaded therapeutic protein are also described. To fulfill all these proposes, PLGA nanoparticles and insulin are used as models.

2 Materials

2.1 Reagents

1. Recombinant human insulin.
2. 0.1 M HCl solution.
3. PLGA 50:50 Resomer® RG 503 H.
4. Dichloromethane.
5. Polyvinyl alcohol (PVA).
6. 2% (w/v) PVA solution.
7. Milli-Q water.
8. Trehalose.
9. Trifluoroacetic acid (TFA).
10. 0.1% (v/v) TFA solution.
11. Acetonitrile HPLC gradient grade.
12. Uranyl acetate.
13. Phosphate-buffered saline (PBS) buffer pH 7.4.
14. Chloroform.
15. 0.01 M HCl solution.
16. Enzyme-linked immunosorbent assay (ELISA) kit.

2.2 Equipment

1. Sonicator.
2. Magnetic stirrer.
3. Bench top lyophilizer.
4. Moisture analyzer,
5. HPLC instrument.
6. LiChrospher 100 RP-18 guard column, 5 μm particle size.
7. XTerra RP 18 column, 5 μm particle size, 4.6 mm internal diameter × 250 mm length.
8. Beckman Optima TL ultracentrifuge.
9. Malvern Zetasizer and Particle Analyzer 5000.
10. DTS1060 disposable zeta cell.
11. Transmission electron microscope.
12. Scanning electron microscope.
13. Medium Bench Centrifuge.
14. FT-IR spectrometer.
15. Triple reflection ATR accessory.
16. Horizon MB FTIR software (ABB, Zurich, Switzerland).
17. CD spectrometer.
18. Spectrophotometer.

19. Spectrofluorometer.
20. Laboratory refrigerator.
21. Laboratory incubator.

3 Methods

In this work, the production and characterization of PLGA nanoparticles loading insulin are used as models of polyester-based nanoparticles loading hydrophilic therapeutic proteins. Thus, the described techniques and protocols could be used with minor adjustments to obtain other polyester-based nanoparticles loading therapeutic proteins, considering their physicochemical characteristics.

3.1 Production of Nanoparticles

The nanoparticles are produced using a water-in-oil-in-water (W/O/W) double-emulsion technique, characterized for being a simple, robust and mild production method. The protocol was previously optimized by selecting the best one from a range of formulations with different PLGA ratios and surfactant types with distinct concentrations [13]. Possible modifications in process parameters and formulation composition may influence the characteristics of nanoparticles, namely its size, polydispersity index, zeta potential and drug association efficiency. The following steps describe the production protocol of insulin-loaded PLGA nanoparticles. The formulation should be produced in triplicate for further characterization.

1. 30 mg of insulin are poured into 0.6 mL of a 0.1 M HCl solution, which is gently hand swirled until complete solubilization—this is the inner aqueous phase (*see* **Notes 1** and **2**).
2. 200 mg of polymer are poured into 2 mL of dichloromethane in a glass test tube, which is hand shaken or vortexed until complete solubilization—this is the oily phase (*see* **Note 3**). In this and the next steps, the operation is performed in a fume hood to avoid the exposure of the operator to the organic solvent.
3. 0.2 mL of the inner aqueous phase are poured into the oily phase, and emulsified by sonication for 30 s at 70% of amplitude using a sonicator—the primary emulsion is obtained in this step (*see* **Notes 4** and **5**).
4. The primary emulsion is poured into a glass test tube with 10 mL of a 2% (w/v) PVA solution, which is the outer aqueous phase, and sonicated again for 30 s at 70% amplitude and poured into a 50 mL glass beaker containing the remainder

15 mL of 2% (w/v) PVA—this step produces the w/o/w double emulsion (*see* **Notes 4–6**).

5. The double emulsion is stirred in a magnetic stirrer for 3 h at 300 rpm for organic solvent evaporation and nanoparticles hardening—this step produces the nanoparticles suspension.

After production, the nanoparticles need to be characterized using analytical techniques, or lyophilized for further long-term storage. Lyophilization is frequently used to increase the storage stability of nanoparticles, or even to formulate dosage forms for pulmonary or oral delivery.

3.2 Lyophilization of Nanoparticles

The lyophilization is a dehydration process occurring in three steps: freezing, primary drying and secondary drying [9]. Besides the advantages of lyophilization on increasing the stability of nanoparticle formulations, the thermal and pressure conditions required may lead to possible aggregation or fusion of nanoparticles. Therefore, cryoprotectants and lyoprotectants can be added to formulations to preserve nanoparticles from freezing and drying stresses, respectively [11, 12]. Most of cryoprotectants act as lyoprotectants and vice versa, and the most commonly used are sugars. Some examples are trehalose, glucose, sucrose, fructose, sorbitol, and mannitol [7, 13–17].

The lyophilization cycle has to be adjusted considering the physicochemical properties of the formulations, to guarantee the obtainment of an amorphous cake with good redispersibility properties and low residual moisture content [12]. Therefore, the freezing step should occur at a temperature below the glass transition temperature of the frozen sample (T_g'), and the primary drying below the collapse temperature (T_c) [11]. The T_g' and T_c of formulations are obtained by differential scanning calorimetry (DSC) and lyophilization microscopy, respectively. The lyophilization cycle suggested in this section is adjusted regarding those parameters of the formulation described in a previous work [17]. Therefore, when using another polyester-based nanoparticle system or drug, the T_g' and T_c should be inferred and the lyophilization cycle adjusted accordingly.

Overall, the use of cryoprotectants and lyoprotectants and the lyophilization cycle adjustment regarding the physicochemical characteristics of formulations are tools to guarantee the obtainment of a good lyophilisate, the stability of nanoparticles and more important the structural and bioactivity maintenance of the loaded protein. The nanoparticle suspensions are lyophilized as follows:

1. The nanoparticle suspension is placed at a maximal height of 10 mm, into semi-stoppered glass vials with slotted rubber closures. Trehalose is used as protectant and added to the suspension and homogenized to achieve a final concentration

of 5–10% (w/v). Samples without protectant are used as controls, and each batch is lyophilized in triplicate.

2. The samples are placed inside the chamber of a benchtop lyophilizer, and frozen at −40 °C during 4 h.
3. Then, the primary drying occurs at −32 °C and 150 mTorr for 24 h.
4. At the end of primary drying, the secondary drying begins by rising the temperature to 20 °C and at 50 mTorr for 6 h.
5. Finally, the glass vials are closed inside the lyophilizer chamber and the closures are covered with Parafilm, to avoid the penetration of moisture into the vials during storage.

3.3 Macroscopic Evaluation and Residual Moisture Content of Lyophilisates

Immediately after lyophilization, the lyophilisates are visually inspected to verify any cake collapse or shrinkage. The collapse of the cake causes high reconstitution times and high residual moisture content of lyophilisates, which is detrimental to the stability of the formulation [12]. When performing the macroscopic evaluation of the lyophilisates, two main steps are to be regarded:

1. The color of the lyophilisate is evaluated. The lyophilisate color of a PLGA nanoparticles is white, so any modification can indicate the occurrence of an undesired physicochemical reaction during lyophilization.
2. The shape of the cake is evaluated. The cake should occupy the same volume and height of the initial nanoparticle suspension. Any modification indicates cake shrinkage or collapse.

The application of an optimized cycle, as used in Subheading 3.2, avoids the cake collapse and helps to achieve a low residual moisture content, leading to a good storage stability perspective. Figure 1 illustrates the difference between a noncollapsed and collapsed cake of insulin-loaded PLGA nanoparticles. For other

Fig. 1 Lyophilisates of insulin-loaded nanoparticles with a non-collapsed (*left vial*) and a collapsed cake (*right vial*). Reprinted with permission from [12]

polyester-based nanoparticles or drugs, the T_g' and T_c of the formulation should be evaluated, and the lyophilization cycle adjusted accordingly. The residual moisture content of lyophilisates is easily evaluated following the steps:

1. An equal amount of each lyophilisate is placed in a moisture analyzer. All samples are analyzed in triplicate.
2. The residual moisture content is quantified as percentage of the containing moisture in each lyophilisate. For pharmaceutical products, it is recommended a residual moisture content around 1% [9].

3.4 Resuspension of Nanoparticles

One of the more important characteristics of a good lyophilisate is its easiness to get resuspended after rehydration. The resuspension should be almost immediately, so to compare the resuspension of different hardly resuspendable lyophilisates, the time needed to obtain a complete resuspension can be recorded. Usually, collapsed cakes require long reconstitution times, and its complete resuspension can be difficult even using vortexing or sonication. The resuspension of lyophilisates can be easily performed as follows:

1. Milli-Q water is added on the inside wall of vials at the same volume of the initial nanoparticle suspension.
2. To guarantee the cake wetting, the vials are kept for 5–10 min.
3. Finally, to achieve a complete homogenization the samples are softly hand swirled.

3.5 Characterization of Nanoparticle Formulations

Several techniques can be used to evaluate the properties of nanoparticles loaded with therapeutic proteins after production and lyophilization. In this section, the critical characterization procedures are discussed.

3.5.1 Protein Quantification by High-Performance Liquid Chromatography-Ultraviolet (HPLC-UV) Detection Method

The evaluation of protein concentration in nanoparticle formulations is crucial, so it should be used a rapid and accurate method as HPLC-UV to evaluate important parameters such as the protein association efficiency (AE), loading capacity (LC), retention efficiency (RE) and in vitro release of the drug. In this section, it is described a rapid reverse-phase HPLC-UV method for determination of insulin in nanoparticulate systems, previously developed and validated [18]. Thus, to quantify other therapeutic proteins by HPLC, another method may be found in the literature or developed and validated.

The analysis is performed at room temperature in a Merck-Hitachi LaChrom HPLC instrument equipped with a LiChrospher® 100 RP-18, 5 μm particle size guard column; and an XTerra RP-18, 5 μm particle size, 4.6 mm internal diameter × 250 mm length column. The mobile phase is composed by a 0.1% (v/v) trifluoroacetic acid (TFA) aqueous solution and acetonitrile at a

70:30 (v/v) ratio, linearly changing to 60:40 (v/v) during 5 min. Then, the ratio is maintained from 5 to 10 min. The mobile phase has a flow rate of 1 mL/min, the injection volume is 20 μL and the detection wavelength is 214 nm. The retention time of insulin peak is around 6 min. Briefly; the insulin quantification by HPLC is performed using the following steps:

1. The insulin standard solutions at 1, 3, 6, 12.5, 25, and 100 mg/mL are prepared using a stock solution similar to that prepared to obtain the inner aqueous phase in the production of nanoparticles protocol (*see* Subheading 3.1), by rigorously diluting it with water to the concentration of each standard solution.
2. Each standard is analyzed in triplicate and the chromatograms are obtained. The total area of the peak of each standard solution is used to obtain the calibration curve.
3. Prior to the analysis of the samples, it may be necessary to dilute them to guarantee they fit in the linearity of the calibration curve.
4. Each test sample is analyzed in triplicate and the chromatograms are obtained (*see* **Note 7**). The total area of the peak of each sample is used to infer the concentration of insulin by interpolation using the calibration curve. If samples were previously diluted prior HPLC analysis, the dilution factor needs to be taken in consideration in the calculations.

3.5.2 Association Efficiency, Loading Capacity, and Retention Efficiency

The AE and LC are parameters evaluated after nanoparticles production, to evaluate the efficiency of the production protocol and nanoparticles ability to load the drug. The RE is evaluated after lyophilization to assess the amount of insulin that remains loaded, being an important indicator of the feasibility of the lyophilization process on preserving nanoparticles integrity [16, 17].

The AE represents the amount of drug associated with the nanoparticles, and is determined indirectly by the difference between the total amount of drug used to prepare the nanoparticles, and the amount of drug remaining in the supernatant after nanoparticles centrifugation, according to the equation:

$$\text{AE}\ (\%) = \frac{\text{Total amount of drug} - \text{Free drug in supernatant}}{\text{Total amount of drug}} \times 100$$

Similarly, the LC represents the mass percentage of containing drug of the total mass of nanoparticles, and is obtained using the equation:

$$\text{LC}\ (\%) = \frac{\text{Total amount of drug} - \text{Free drug in supernatant}}{\text{Total weight of nanoparticles}} \times 100$$

The total weight of nanoparticles is determined by lyophilization of hydrated nanoparticles obtained upon centrifugation. To evaluate both the AE and LC it is necessary to assess the amount of drug free in supernatant, using the next steps:

1. An aliquot of the formulation is centrifuged at 40,000 × *g* at 4 °C for 30 min in a Beckman Optima TL ultracentrifuge.
2. The supernatant is collected and the drug in the supernatant is quantified by HPLC-UV (*see* Subheading 3.5.1) taking into consideration the volume of the aliquot used in **step 1** (*see* **Note 8**). Samples are analyzed in triplicate.
3. About 50 mg of nanoparticles are collected and hydrated after centrifugation and lyophilized, to assess the total weight of nanoparticles.
4. The AE and LC are calculated using the described equations.

The occurrence of drug leakage after lyophilization indicates that the nanoparticles integrity was affected during lyophilization. The drug RE is evaluated after lyophilization using the equation:

$$\mathrm{RE}(\%) = \frac{\text{Amount of entrapped drug} - \text{Free drug in supernatant after lyophilization}}{\text{Amount of entrapped drug}} \times 100$$

The amount of entrapped drug was previously assessed, and represents the amount of drug loaded into the nanoparticles after production. The free drug in supernatant after lyophilization is obtained by quantification of drug in supernatant collected after resuspension of the lyophilisate and centrifugation of nanoparticles suspension. The RE is calculated following the next steps:

1. Using the previous quantification of supernatant obtained after centrifugation of nanoparticles suspension, the amount of drug entrapped into the nanoparticles is calculated by the arithmetic difference between the amount of insulin used to produce the nanoparticles and the amount of drug in the supernatant.
2. After resuspension of nanoparticles (*see* Subheading 3.4), an aliquot of the formulation is centrifuged at 40,000 × *g* and 4 °C for 30 min
3. The supernatant is collected and the drug in the supernatant is quantified by HPLC-UV (*see* Subheading 3.5.1), taking into consideration the volume of the aliquot used in **step 1** (*see* **Note 8**). Samples are analyzed in triplicate.
4. Calculate the RE using the described equation.

3.5.3 Nanoparticles Size, Polydispersity Index, and Zeta Potential

The nanoparticles size, polydispersity index and zeta potential are the features to always evaluate when developing nanocarrier systems for drug delivery. The determination of particle size is performed by photon correlation spectroscopy, also named dynamic light scattering (DLS), which is based on the scattering of a laser light by submicron particles in a time-dependent way [19]. Such technique is based on the idea that larger particles move with lower velocity than smaller ones. The variation between light intensity and time is correlated with the diffusion coefficient and converted to particle size, as an average of the hydrodynamic diameter and the polydispersity index, which represents the width of the size distribution [19].

The zeta potential is a measure of the charge of particles surface and is determined by assessing the electrophoretic mobility, which is the velocity of particles movement after application of an electric field [20]. The electrophoretic mobility may be determined by electrophoretic light scattering, phase analysis light scattering (PALS), and laser Doppler anemometry. Upon application of the electric field, particles move toward the opposite charge electrode, and a laser beam crosses the sample so the modifications in the intensity of light scattering by particles are related to its velocity. Thus, the zeta potential is calculated as function of the electrophoretic mobility. The zeta potential is a quantitative measure of the stability of colloidal systems. It is considered that colloidal systems with zeta potential higher than +30 mV and lower than −30 mV are stable due to particle-particle repulsion, avoiding particles aggregation, and particles with zeta potential between −10 mV and 10 mV are considered almost neutral [20].

The nanoparticles size, polydispersity index, and zeta potential are obtained following the next steps:

1. The hydrodynamic radius, polydispersity index, and zeta potential of nanoparticles are determined using a Malvern Zetasizer and Particle Analyzer 5000 (*see* **Note 9**).
2. Use an aliquot of diluted nanoparticles suspension in the determinations (*see* **Note 10**).
3. The samples are analyzed in triplicate, and size and polydispersity index of nanoparticles are determined with a detection angle of 90° at 25 °C.
4. The zeta potential is determined in triplicate at a scattering angle of 173° at 25 °C, using a DTS1060 zeta cell.

3.5.4 Morphological Analysis of Nanoparticles

The morphological analysis of nanoparticles may be performed by transmission electron microscopy (TEM) and scanning electron microscopy (SEM). Both techniques can give information about the size and shape of nanoparticles, and the SEM is useful to characterize the nanoparticles surface. The microscopy analysis can also help to corroborate the nanoparticles size evaluated by DLS (*see* Subheading 3.5.3). Both TEM and SEM are used to evaluate the nanoparticles in suspension form, and SEM can be also used to assess the morphology of the lyophilisates (Fig. 2).

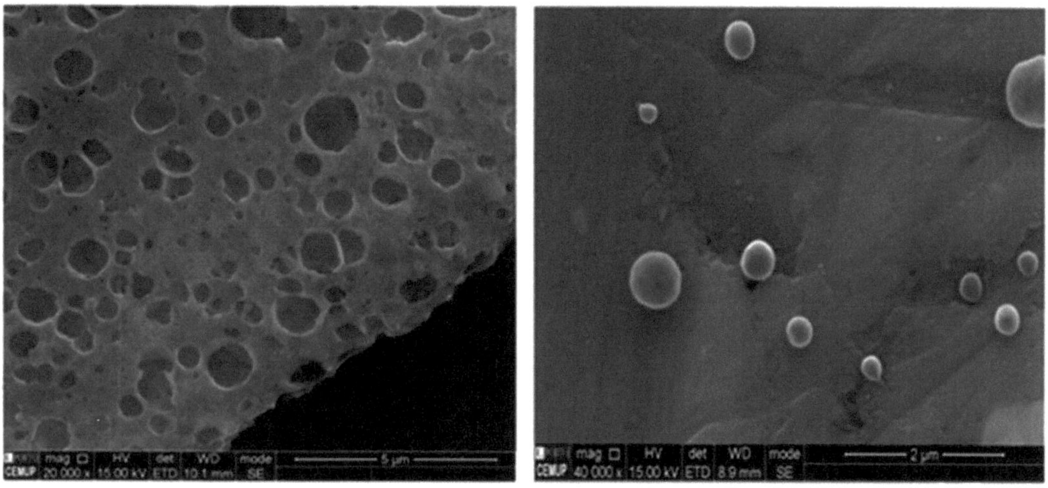

Fig. 2 SEM microphotographs of the lyophilisate of insulin-loaded PLGA nanoparticles with a cryoprotectant (*left*), and of nanoparticles after resuspension (*right*). Reprinted with permission from [12]

The TEM analysis is performed following the steps:

1. Ten microliters of nanoparticles suspension is placed on a grid during 60 s, to promote the adhesion of nanoparticles.
2. The grid is carefully dried with a paper tissue to remove the excess of nanoparticles suspension, without removing the nanoparticles on the grid surface.
3. The sample on the grid is treated with uranyl acetate during 30 s.
4. Similar to **step 2**, the excess of uranyl acetate is removed.
5. The nanoparticles on the grid are observed in a JEOL JEM-1400 transmission electron microscope, and the images are captured by an attached camera.

On its turn, the SEM analysis takes place by the following steps:

1. 10 μL of nanoparticles suspension is placed on a metal stub.
2. The samples are left overnight to dry (*see* **Note 11**).
3. For the nanoparticles lyophilisates, a carbon adhesive tape is placed on the metal stub and the lyophilisate is placed on the surface of the tape to promote its adhesion.
4. Using dry air, the excess of lyophilisate is carefully removed from the metal stub.
5. Prior observation, the samples are coated under vacuum, with a layer of gold/palladium at 15 mA current for 60 s.
6. The samples on the metal stub are observed in a FEI Quanta 400 FEG scanning electron microscope, and images are captured by an attached camera.

Overall, the PLGA nanoparticles present a spherical shape with a smooth surface, which is typical of these colloidal systems [13, 15].

3.5.5 In Vitro Release Study

The polyester-based nanoparticles are recognized to have a sustained and/or controlled release of loaded drugs. These properties are important to prolong the therapeutic effect of drugs and reduce the number of administrations. Thus, the release of the loaded drug in the bloodstream can be evaluated by an in vitro release study, performed following the steps:

1. For nanoparticle suspension samples, nanoparticles are collected by centrifugation in a Heraeus Megafuge 1.0 R centrifuge at 23,000 × *g* and 4 °C for 30 min.
2. For nanoparticle lyophilisates, the amount needed for the study is directly weighed.
3. 100 mg of the sample is placed into screw-capped cylindrical bottles containing 20 mL of PBS buffer at pH 7.4 and 37 °C (±1 °C). This step is performed in triplicate to obtain an acceptable statistical correlation.
4. The samples in the closed bottles are incubated under magnetic stirring at 100 rpm for 24 h and 37 °C using a thermostatic water bath.
5. At predetermined time points, 0.4 mL aliquots of samples are collected and replaced by the same volume of PBS fresh medium (*see* **Note** 12).
6. The samples from the previous step are centrifuged at 40,000 × *g* and 4 °C for 30 min, and the supernatant is collected (*see* **Note 8**).
7. The protein in the supernatant is quantified by HPLC-UV (*see* Subheading 3.5.1).

3.5.6 Analysis of Protein Integrity

During nanoparticles production, proteins are exposed to possible stress conditions such as pH, temperature, agitation and sonication, leading to possible denaturation and bioactivity problems. Also, during lyophilization the harsh conditions of the process can cause conformational changes in proteins that can be detrimental to its bioactivity. Therefore, it is crucial to ensure the loaded protein maintain its conformational structure prior nanoparticles administration. The assessment of proteins integrity is performed by Fourier transform infrared (FTIR) spectroscopy, circular dichroism (CD) and fluorescence spectroscopy.

3.5.7 Fourier Transform Infrared Spectroscopy

FTIR spectroscopy is a noninvasive technique that allows the direct analysis of a protein loaded into nanoparticles. Additionally, it can analyze samples both in liquid or solid form. For this purpose, the coupling of an attenuated total reflectance (ATR) accessory to the FTIR equipment facilitates the analysis, since it dispenses the use of KBr disks bearing the solid sample, or the use of specific cells with CaF_2 crystal windows for liquid samples analysis. Another advantage is the possibility to recover the sample under analysis, which is important for samples with a small amount.

The secondary structure of proteins, dominated by α-helices or β-sheets can be assessed by FTIR [21]. The amide I absorption is related with the amide bonds linking the amino acids, so the shape of the amide I band, occurring around 1700–1600 cm^{-1}, has been correlated to the secondary structure of proteins. The principle behind this technique is to obtain the spectra of the native protein and the loaded protein, and by spectral treatment focus in the amide I region and overlap both spectra. The degree of similarity between the two spectra is given by the percentage of area of overlap (AO), and represents the percentage of secondary structure maintenance after protein processing. The steps for FTIR analysis are:

1. The FTIR spectra are collected in the 4000–600 cm^{-1} region, with 256 scans and at 4 cm^{-1}, in an ABB MB3000 FTIR spectrometer equipped with a MIRacle triple reflection ATR accessory. The spectral treatment is performed using the Horizon MB FTIR software.
2. The nanoparticles suspension or the lyophilisates are placed on the surface of the ATR accessory crystal, and the spectra are collected (*see* **Note 13**). Specifically, the lyophilisates are compressed on the crystal surface using the ATR accessory. Both for liquid and solid samples, the same amount of sample of each formulation is used. All samples are analyzed in triplicate.
3. The spectra of native insulin are collected. These spectra are also subjected to the procedures of **steps** 7 and **8**.
4. The spectra of nanoparticles loading insulin are collected.
5. The spectra of unloaded nanoparticles and a water vapor spectrum are collected.
6. A double subtraction procedure of the spectra collected in **step 4** from the spectra collected in **step 5** is performed, accordingly to the literature [22].
7. The second derivative spectra are obtained using a 15-points Savitzky–Golay derivative function, and the spectra are truncated to 1710–1590 cm^{-1}.
8. A baseline correction using a 3–4 points adjustment is performed, and the spectra are area normalized for further comparison.
9. The area-normalized second-derivative spectra of native insulin is compared with the spectra of the loaded protein (Fig. 3), and the degree of similarity is obtained using the AO algorithm [23].

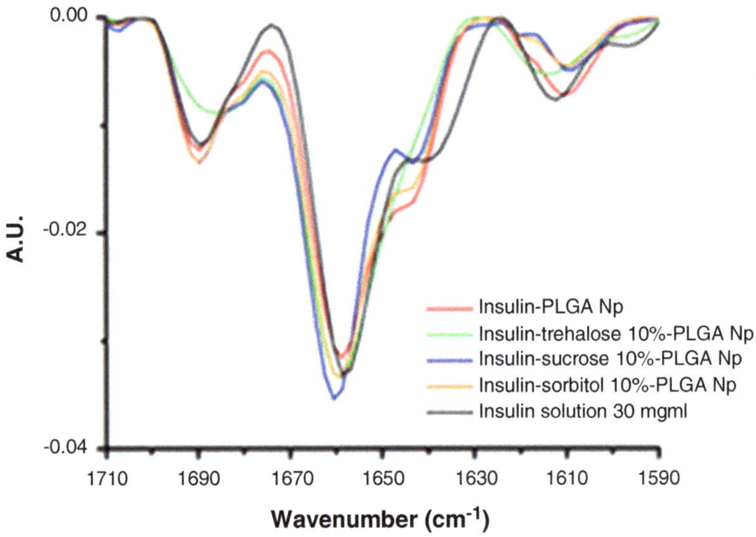

Fig. 3 Area-normalized second-derivative amide I FTIR spectra of insulin-loaded PLGA nanoparticles after lyophilization. Np stands for nanoparticles. Reprinted with permission from [17]

3.5.8 Circular Dichroism Spectroscopy

The CD spectroscopy is useful to study the conformation of asymmetrical molecules using the circularly polarized light. The tertiary and secondary structure of proteins may be analyzed using near (240–340 nm) or far (190–250 nm) UV radiation, respectively [24]. When compared with the spectra of the native protein, possible modifications in the characteristic bands of the CD spectra of the protein being tested indicate changes in its conformational structure that may lead to protein denaturation or inactivity [25].

In this work, the CD is used as a complimentary technique to evaluate the secondary structure of insulin. Contrary to FTIR, for CD analysis it is recommended the extraction of the loaded protein from nanoparticles to avoid detrimental interferences. Therefore, both nanoparticles in suspension and the lyophilisates are treated with a mixture of chloroform, to dissolve the PLGA polymer, and 0.01 M HCl to dissolve insulin (*see* **Note 14**). After this treatment, there is a separation between the organic and aqueous phases, and the latter dissolving insulin is collected and centrifuged at 23,000 × *g* and 4 °C for 30 min. Then, the supernatant is used for CD analysis following the next steps (*see* **Note 15**):

1. The analysis is performed at 25 °C using a Jasco J-815 CD Spectrometer. The CD spectra are recorded in the far-UV region using a 0.1 cm cell at 190–250 nm, with a step size of 0.5 nm, a bandwidth of 1.5 nm, and an averaging time of 5 s (*see* **Note 16**). The lamp housing is continuously purged with nitrogen. All samples are analyzed in triplicate and with an average of five scans.

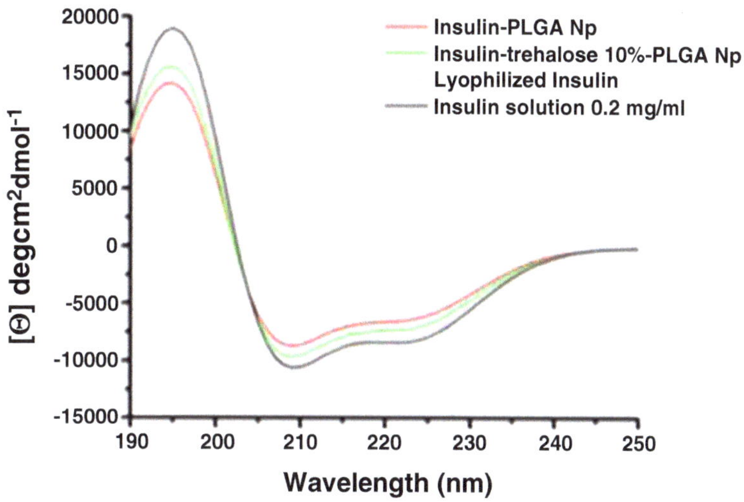

Fig. 4 Far-UV CD spectra of insulin extracted from insulin-loaded PLGA nanoparticles after lyophilization. Np stands for nanoparticles. Reprinted with permission from [16]

2. The spectra of a 0.2 mg/mL native insulin solution are collected.
3. The spectra of each insulin sample are collected, as well as the spectra of a 0.01 M HCl solution to be used as reference.
4. The CD signal of the reference spectrum is subtracted from the test samples spectra.
5. Then, the molar ellipticity of insulin in each sample is calculated using the formula: CD signal × MRW (mean residual weight of each insulin residue, 116 Da) [insulin concentration (mg/mL) × cell path length (0.1 cm)]. The insulin concentration is determined by UV absorption at 280 nm in a NanoDrop 2000c UV-Vis Spectrophotometer, using a molar extinction coefficient of 6200 M^-1 cm^-1 for 1.0 mg/mL.
6. The spectra of each sample are compared with the spectrum of native insulin, to evaluate conformational changes (Fig. 4).

3.5.9 Fluorescence Spectroscopy

The fluorescence spectroscopy is also used to assess the conformation of proteins, namely its tertiary structure. Lacking in tryptophan, the insulin fluorescence depends in its four tyrosine residues [26]. Modifications in the maximum or intensity of emission spectra of intrinsic protein fluorescence indicate changes in insulin structure [27].

Similar to CD, for fluorescence spectroscopy analysis the protein should be extracted from nanoparticles to avoid undesired interferences. Thus, the procedure to extract insulin from nanoparticles described in Subheading 3.5.6 can also be used prior to fluorescence analysis. The analysis is performed following the steps:

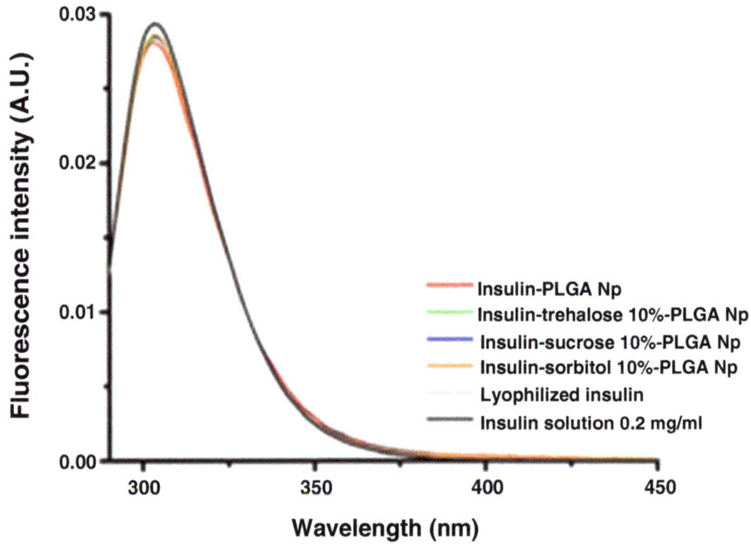

Fig. 5 Fluorescence spectra of insulin extracted from insulin-loaded PLGA nanoparticles after lyophilization. Np stands for nanoparticles. Reprinted with permission from [17]

1. The spectra are collected in a Jasco FP-6500 spectrofluorometer at 25 °C, in the range of 290–450 nm with 1 nm step. The excitation occurs at 280 nm, with the excitation and emission slits set to 3 nm, and an integration time per data point of 0.1 s. All samples are run in triplicate with an average of five scans.
2. The spectra of a 0.2 mg/mL native insulin solution are collected.
3. The spectra of each insulin sample are collected, as well as the spectra of a 0.01 M HCl solution to be used as reference.
4. The signal of the reference spectrum is subtracted from the test samples spectra, and normalized based in insulin concentration in each sample determined by UV absorption at 280 nm in a NanoDrop 2000c UV-Vis Spectrophotometer, using a molar extinction coefficient of 6200 M^{-1} cm^{-1} for 1.0 mg/mL.
5. The spectra of each sample are compared with the spectrum of native insulin, to evaluate conformational changes (Fig. 5).

3.6 Analysis of Insulin Bioactivity

After structural characterization, the confirmation of insulin bioactivity is evaluated using an ELISA commercial kit, consisting in a solid phase two-site enzyme immunoassay (*see* **Note 17**). The scientific principle is based in the direct sandwich technique, in which two monoclonal antibodies are targeted against different antigenic determinants in the insulin molecule. Aliquots of samples after insulin release (*see* Subheading 3.5.5), and extraction from nanoparticles (*see* Subheading 3.5.6) are diluted with water to the concentration range of the method (1–200 mU/l) given by the manufacturer protocol. The samples are analyzed in triplicate following the steps:

1. The samples are incubated and insulin reacts with peroxidase-conjugated anti-insulin antibodies.
2. A washing step is performed to remove unbound enzyme-labeled antibody, and the bound conjugate is detected upon reaction with 3,3′,5,5′-tetramethylbenzidine.
3. Then, the reaction is stopped upon addition of an acid solution giving a colorimetric endpoint read at 450 nm.

3.7 Long Term Stability of Nanoparticles

When nanoparticle formulations are intended to reach the market, their long-term stability needs to be verified. Thus, following the ICH recommendations both nanoparticle suspensions and lyophilisates are stored at three different storage conditions: 4 °C/60% RH in a laboratory refrigerator, 25 °C/60% RH and 40 °C/75% RH in a laboratory incubator [15, 28]. After 1, 3, 6, and 12 months the formulations are characterized using the aforementioned techniques in terms of the nanoparticles features and protein stability and bioactivity.

4 Notes

1. The amount of insulin dissolved in a 0.1 M HCl solution and the solvent volume may be changed regarding the final concentration of drug needed. Nevertheless, the total solubilization of the protein should be always guaranteed. In every case, 0.2 mL of insulin solution should be used to prepare the nanoparticles, so when preparing the insulin solution, the number of batches to be produced has to be known, and the amount of drug and solvent volume needed should be calculated accordingly. For instance, in the described protocol, it is recommended to have 10 mg of insulin in each batch, so 30 mg of the protein is dissolved in 0.6 mL of 0.1 M HCl solution. Once the nanoparticles are produced in triplicate, 0.2 mL from the previous insulin solution are used in each batch production.
2. The 0.1 M HCl solution is used to dissolve insulin, since it does not dissolve in water. When using other hydrophilic drugs, water or other solvent immiscible with the oily phase may be used.
3. To avoid the organic solvent evaporation, the tubes should be covered with Parafilm. Other organic solvents may be used, but in this choice a balance between its volatility, toxicity, and immiscibility with the aqueous phases has to be performed.

4. To avoid the solvent evaporation during sonication steps, the glass test tube may be placed in an ice bath.
5. Sonication is recommended to promote the emulsification of the mixture, because allows the formation of nanoparticles with lower diameters comparatively to mechanical homogenization. Even though, the latter can still be used in the production of nanoparticles using the described protocol.
6. To obtain nanoparticles with a more homogenous size distribution and a more efficient sonication, it should be used about half of the volume of the aqueous outer phase in the secondary sonication step. As described in the protocol, after sonication the mixture is poured into the remainder aqueous phase.
7. It should be guaranteed that the samples to be quantified contain insulin only. Thus, interfering substances such as polymer residues are removed by centrifugation or filtration prior HPLC analysis.
8. The supernatant must be collected as much limpid as possible, without nanoparticles or polymer residues.
9. During the analysis, it is important to ensure the cell walls are unscratched and clean, and the sample has no bubbles. These factors may interfere with the laser beam, causing misleading results.
10. The dilution of concentrated samples is advised both to avoid multiple scattering from particle aggregates, and to standardize the analyses.
11. During the overnight drying of samples, it should be guaranteed that no dust or other particles can reach the metal stub where the samples are placed to avoid interferences during the microscopy observation.
12. The replacement of the collected sample volume by PBS fresh medium is important to maintain the sink conditions.
13. When there is no ATR accessory available for FTIR analysis, the liquid samples are placed in a cell with CaF_2, whereas the solid samples are mixed with a KBr powder and compressed into disks using a manual tablet press. In the latter, it is important to guarantee the solid sample and KBr mixture is homogeneous, and all disks have similar thickness and without cracks.
14. For extraction of other drugs from different polyester-based nanoparticles, a different organic solvent can be used to dissolve the nanoparticles, but it cannot damage the protein structure and needs to be immiscible with the aqueous phase in which the protein is dissolved. Similarly, the aqueous phase needs to dissolve the drug to be extracted from nanoparticles, and has to be immiscible with the organic solvent.

15. The volume of nanoparticle suspension or amount of lyophilisate necessary to be treated with chloroform needs to be calculated, in order to obtain a theoretical concentration of insulin below 0.2 mg/mL in the supernatant. More concentrated protein samples are detrimental to the CD signal.
16. In CD experiments, it is important to use clean cells without scratches, to avoid undesired interferences.
17. The ELISA commercial kit described is specific to assess the bioactivity of insulin. Therefore, for other therapeutic proteins a specific ELISA kit should be used.

Acknowledgments

Flávia Sousa would like to thank the Fundação para a Ciência e a Tecnologia (FCT), Portugal for financial support (Grant SFRH/BD/112201/2015). This work was also financed by FEDER—Fundo Europeu de Desenvolvimento Regional funds through the COMPETE 2020—Operacional Programme for Competitiveness and Internationalisation (POCI), Portugal 2020, and by Portuguese funds through FCT—Fundação para a Ciência e a Tecnologia/Ministério da Ciência, Tecnologia e Inovação in the framework of the project "Institute for Research and Innovation in Health Sciences" (POCI-01-0145-FEDER-007274).

References

1. Frokjaer S, Otzen DE (2005) Protein drug stability: a formulation challenge. Nat Rev Drug Discov 4(4):298–306
2. Mahmood I (2008) Methods to determine pharmacokinetic profiles of therapeutic proteins. Drug Discov Today Technol 5(2–3):e65–e69
3. Antosova Z, Mackova M, Kral V, Macek T (2009) Therapeutic application of peptides and proteins: parenteral forever? Trends Biotechnol 27(11):628–635
4. de la Fuente M, Csaba N, Garcia-Fuentes M, Alonso MJ (2008) Nanoparticles as protein and gene carriers to mucosal surfaces. Nanomedicine 3:845–857
5. Ibraheem D, Agusti G, Elaissari A, Fessi H (2014) Preparation and characterization of albumin-loaded polycaprolactone nanoparticles for in vivo applications. J Colloid Sci Biotechnol 3(2):160–166
6. Gao X, Zhang X, Zhang X, Cheng C, Wang Z, Li C (2010) Encapsulation of BSA in polylactic acid–hyperbranched polyglycerol conjugate nanoparticles: preparation, characterization, and release kinetics. Polym Bull 65(8):787–805
7. Fonte P, Araújo F, Seabra V, Reis S, van de Weert M, Sarmento B (2015) Co-encapsulation of lyoprotectants improves the stability of protein-loaded PLGA nanoparticles upon lyophilization. Int J Pharm 496(2):850–862
8. Kumari A, Yadav S, Yadav S (2010) Biodegradable polymeric nanoparticles based drug delivery systems. Colloids Surf B Biointerfaces 75(1):1–18
9. Williams NA, Polli GP (1984) The lyophilization of pharmaceuticals: a literature review. J Parenter Sci Technol 38(2):48–59
10. Franks F (1998) Freeze-drying of bioproducts: putting principles into practice. Eur J Pharm Biopharm 45(3):221–229
11. Abdelwahed W, Degobert G, Stainmesse S, Fessi H (2006) Freeze-drying of nanoparticles: formulation, process and storage considerations. Adv Drug Deliv Rev 58(15):1688–1713
12. Fonte P, Reis S, Sarmento B (2016) Facts and evidences on the lyophilization of polymeric

nanoparticles for drug delivery. J Control Release 225:75–86

13. Fonte P, Soares S, Costa A, Andrade J, Seabra V, Reis S, Sarmento B (2012) Effect of cryoprotectants on the porosity and stability of insulin-loaded PLGA nanoparticles after freeze-drying. Biomatter 2(4):329–339
14. Soares S, Fonte P, Costa A, Andrade J, Seabra V, Ferreira D, Reis S, Sarmento B (2013) Effect of freeze-drying, cryoprotectants and storage conditions on the stability of secondary structure of insulin-loaded solid lipid nanoparticles. Int J Pharm 456(2):370–381
15. Fonte P, Soares S, Sousa F, Costa A, Seabra V, Reis S, Sarmento B (2014) Stability study perspective of the effect of freeze-drying using cryoprotectants on the structure of insulin loaded into PLGA nanoparticles. Biomacromolecules 15(10):3753–3765
16. Fonte P, Lino PR, Seabra V, Almeida AJ, Reis S, Sarmento B (2016) Annealing as a tool for the optimization of lyophilization and ensuring of the stability of protein-loaded PLGA nanoparticles. Int J Pharm 503(1–2):163–173
17. Fonte P, Andrade F, Azevedo C, Pinto J, Seabra V, van de Weert M, Reis S, Sarmento B (2016) Effect of the freezing step in the stability and bioactivity of protein-loaded PLGA nanoparticles upon lyophilization. Pharm Res 33(11):2777–2793
18. Sarmento B, Ribeiro A, Veiga F, Ferreira D (2006) Development and validation of a rapid reversed-phase HPLC method for the determination of insulin from nanoparticulate systems. Biomed Chromatogr 20(9):898–903
19. Photon Correlation Spectroscopy (2002) In: Scarlett B (ed) Particle characterization: light scattering methods. Springer Netherlands, Dordrecht, pp 223–288. doi:10.1007/0-306-47124-8_5
20. Clogston JD, Patri AK (2011) Zeta potential measurement. Methods Mol Biol 697:63–70
21. Van de Weert M, Hering J, Haris P (2005) Fourier transform infrared spectroscopy. In: Jiskoot W, Crommelin D (eds) Methods for structural analysis of protein pharmaceuticals. AAPS Press, Arlington, TX, pp 131–166
22. Dong A, Huang P, Caughey W (1990) Protein secondary structures in water from second-derivative amide I infrared spectra. Biochemistry 29(13):3303–3308
23. Kendrick B, Dong A, Allison S, Manning M, Carpenter J (1996) Quantitation of the area of overlap between second-derivative amide I infrared spectra to determine the structural similarity of a protein in different states. J Pharm Sci 85(2):155–158
24. Kelly S, Jess T, Price N (2005) How to study proteins by circular dichroism. Biochim Biophys Acta 1751(2):119–139
25. Greenfield NJ (2006) Using circular dichroism spectra to estimate protein secondary structure. Nat Protoc 1(6):2876–2890
26. Bekard IB, Dunstan DE (2009) Tyrosine autofluorescence as a measure of bovine insulin fibrillation. Biophys J 97(9):2521–2531
27. Lakowicz JR (2006) Principles of fluorescence spectroscopy. Springer US, New York
28. ICH Topic Q 1 A (R2), Stability testing of new drug substances and products (2003) European Medicines Agency, vol CPMP/ICH/2736/99. London, UK

Chapter 22

Quantification of Coagulation Factor VIII by Selective Reaction Monitoring

Edson Galdino do Nascimento Filho, Guilherme Pauperio Lanfredi, Mário Soares Abreu-Neto, Dimas Tadeu Covas, and Vitor Marcel Faça

Abstract

Coagulation factor VIII (FVIII) is an important glycoprotein involved in the extrinsic coagulation cascade. Mutations in FVIII gene results in hemophilia A, a recessive coagulation disorder that is clinically managed by administration of purified FVIII from blood donors or recombinant FVIII. Because of its fundamental therapeutic application, biotechnological production of FVIII requires rigid quality control and monitoring in patients and clinical trials. Here, we describe a protocol for a mass spectrometry based approach termed selective reaction monitoring (SRM) as an important alternative tool for accurate and sensitive quantitation of purified or recombinant FVIII.

Key words Mass spectrometry, Selective reaction monitoring (SRM), Coagulation factor VIII, Hemophilia A

1 Introduction

Biotechnological products require strict quality control during all the stages of development and therapeutic application. Mass spectrometry (MS) has been increasingly used in biotechnological industry, as an important tool for accurate and sensitive quantification of small molecules, peptides and intact proteins, also being able to show molecule modifications and presence of minor contaminants [1]. The MS tool becomes even more powerful in the field when coupled with liquid chromatography (LC-MS), allowing analysis of more complex mixture such as patient samples during clinical studies of new drugs. More recently, a multiplex, highly specific and sensitive MS approach termed selective reaction monitoring (SRM), has emerged as an important strategy to obtain accurate quantification of peptides and proteins present in complex matrix [2, 3].

SRM is based on selecting and quantifying specific sets of peptides derived from a single or multiple proteins, using liquid

Virgínia Picanço-Castro and Kamilla Swiech (eds.), *Recombinant Glycoprotein Production: Methods and Protocols*, Methods in Molecular Biology, vol. 1674, DOI 10.1007/978-1-4939-7312-5_22, © Springer Science+Business Media LLC 2018

chromatography tandem mass spectrometry (LC-MS/MS) analysis [2]. The method is based on the ion filtering capabilities of quadrupole-based MS instruments for selection and detection of precursor peptides and the corresponding fragment ions produced during the collision-induced dissociation stage of the process. Because of the rapid tandem quadrupole duty cycle and online chromatographic separation of peptides, multiple precursor–fragment transitions can be monitored simultaneously.

The peptides monitored by SRM are produced by proteolytic hydrolysis of target proteins. The selection of such peptides, also referred to as proteotypic peptides [4], is the first and most important step to obtain an efficient SRM method. Currently, this method development process can be aided by databases, such as the PeptideAtlas [5], and open source tools, such as Skyline [6], but also depends on the use of the synthetic proteotypic peptides for improved accuracy and sensitivity. Additional requirements in terms of peptide stability, lack of potential modifications, and good ionization and chromatographic properties are also important for SRM methods.

Plasma-derived Factor VIII (pdFVIII) or recombinant Factor VIII (rFVIII) is a key protein in the extrinsic coagulation cascade and represent the first line of treatment used for the recessive genetic disorder Hemophilia A. FVIII is a plasma glycoprotein synthesized mainly in hepatocytes and endothelial cells, which presents a complex structure formed by multiple chains having points of post-translational modifications. Moreover, during the coagulation activation, FVIII suffers post-translational proteolysis, turning it into an active protease that processes other downstream members in the cascade. Such structure complexity is fundamental do develop appropriate methods for quality control of pdFVIII or rFVIII biotechnological production as well as to monitoring its efficacy as therapeutic agent.

Here, we present a detailed protocol used to develop the SRM approach for accurate detection and quantitation of human pdFVIII or rFVIII. The protocol also indicates detailed information about the proteotypic peptides chosen and used as references for quantitative analysis. Overall, the presented method can improve and streamline the quantification of FVIII in all its stages of development and use as therapeutic agent.

2 Materials

2.1 Samples Containing FVIII Used for Method Development

1. Specialty Assayed Reference Plasma (S.A.R.P.) was used in this study as reference for normal levels of FVIII in human plasma.
2. A human cell line SK-Hep-1 producing proprietary recombinant FVIII [7, 8] was cultivated to generate protein material for method development (*see* **Note 1**).

2.2 Sample Preparation Components for SRM Analysis

1. VivaSpin 10 kDa MWCO (GE Healthcare).
2. Bradford protein quantitation kit.
3. Denaturation buffer: 8 M urea and 0.15 M Tris–HCl (pH 8.5).
4. Trypsin solution: 100 μg/mL sequencing grade modified trypsin in 100 mM ammonium bicarbonate (pH 8.0).
5. Reducing solution: 1% DTT in 0.15 M Tris–HCl (pH 8.5) (*see* **Note 2**).
6. Alkylation solution: 2% iodoacetamide (IAA) in 0.15 M Tris–HCl (pH 8.5) (*see* **Note 2**).
7. Dilution buffer: 0.15 mM Tris–HCl (pH 8.5).
8. Solid phase purification: OASIS HLB solid-phase-extraction columns (Waters Corporation; Milford, MA, USA).
9. Equilibration solution: 5% acetonitrile and 0.1% formic acid.
10. Elution solution: 50% acetonitrile and 0.1% formic acid.
11. Reconstitution buffer: 3% acetonitrile and 0.1% formic acid.

2.3 Solvents and Chromatographic Conditions for LC-MS/MS Analysis

1. Aqueous solvent (A): 94.9% MilliQ water, 5% acetonitrile, and 0.1% formic acid.
2. Organic solvent (B): 99.9% acetonitrile–0.1% formic acid.
3. Gradient was developed at 100 μL/min from 5% to 35% of solvent B in 25 min, at 35 °C.

2.4 Equipment and Software

1. SpeedVac concentrator.
2. Tandem quadruplole Xevo-TQs mass spectrometer coupled to Class I Ultra Performance Chromatographic System (UPLC) (Waters Corporation).
3. Acquity UPLC T3 Column, 1.8 μm particle size, 1 mm I. D. × 150 mm (Waters Corporation).
4. Software Skyline (v3.5) [6].

3 Methods

3.1 Sample Preparation for SRM Proteomic Analysis

1. Quantitate total proteins from SARP solution or conditioned media from FVIII producing cell by Bradford protein assay and use 50 μg aliquots for SRM analysis (*see* **Note 3**).
2. Reduce protein cysteine residues by addition of 5 μL reduction solution and maintain at 37 °C for 1 h.
3. Alkylate protein sulfhydryl residues by addition of 10 μL alkylating solution and maintain at room temperature for 1 h.
4. Dilute samples so urea is approximately 0.6 M (*see* **Note 4**).

5. Perform in-solution trypsin digestion by addition of 10 μL trypsin solution (1:50 enzyme–protein ratio) and incubate for 2 h at 37 °C.
6. Add an additional 5 μL trypsin solution and incubate for a further 16 h or overnight at 37 °C.
7. Desalt samples using solid-phase extraction in OASIS columns as follows:
   (a) Condition column with 1 mL of acetonitrile.
   (b) Equilibrate column with 1.6 mL equilibration solution.
   (c) Apply sample through the column.
   (d) Wash column with 1.6 mL of equilibration solution.
   (e) elute peptides with 1.2 mL of elution solution.
   (f) Dry eluted peptides in a SpeedVac concentrator (*see* **Note 5**).

3.2 SRM Method Development and Data Analysis

1. Select proteotypic peptides for FVIII (*see* Table 1 and **Note 6**).
2. Program Skyline software with the list of proteotypic peptides and perform rounds of iteration for selecting the most intense peptide ions and peptide fragment ions (transitions) for monitoring of human FVIII (*see* **Note 7**).
3. Synthesize or obtain commercially the proteotypic peptides selected for an absolute quantitative analysis (*see* **Note 8**).
4. Suspend samples already digested in reconstitution buffer.

Table 1
Peptides selected for monitoring human FVIII by SMR

Location	Peptide sequence	Retention time (min)	Precursor *m/z*	Fragment ion	Product *m/z*	Collision energy
FVIII—heavy chain (A2)	DFPILPGEIFK	24.7	638.35	y9	1013.60	16
FVIII—heavy chain (A2)	DFPILPGEIFK	24.7	638.35	y7	803.47	25
FVIII—heavy chain (A2)	DFPILPGEIFK	24.7	638.35	y6	690.38	25
FVIII—light chain (C2)	VTGVTTQGVK	4.8	495.28	y9	890.49	14
FVIII—light chain (C2)	VTGVTTQGVK	4.8	495.28	y8	789.45	14
FVIII—light chain (C2)	VTGVTTQGVK	4.8	495.28	y6	633.36	14

The data indicates the two best proteotypic peptide, fragment ions and collision energy defined in the described platform. These two peptides were selected to specifically represent heavy and light chain independently

5. Add 5 μL mix of standard synthetic isotopically labeled peptides in the sample (*see* **Note 9**).
6. Vortex samples thoroughly, centrifuge for 15 min at 12,000 × *g*, and transfer the supernatants (45 μL) to a mass spectrometry-compatible injection vial (*see* **Note 10**).
7. Inject 10 μL sample in triplicate to facilitate statistical analysis (*see* **Note 11**).
8. Analyze the data using the Skyline software (Fig. 1) (*see* **Note 12**).

4 Notes

1. Sk-Hep-1 cell line is cultivated in DMEM media without addition of FBS. All cells were incubated at 37 °C and 5% CO_2. Culture medium of cells producing rFVIII is collected after 24 h of culture, and centrifuged at 3000 × *g* for 10 min a 4 °C. Cell culture supernatant is concentrated by centrifugation at 1500 × *g* at 4 ° C using filters VivaSpin 10 kDa MWCO, according to the manufacturer's recommendations. After concentration of the supernatant, aliquots of 100 μL are separated and stored at –80 °C until processing.
2. Always prepare fresh reducing and alkylation solutions to guarantee efficient reactions with cysteine residues. Do not store these solutions.
3. Quantitation of total protein is important to allow efficient reactions of reduction, alkylation and enzymatic protein digestion.
4. Dilution of sample solution guarantees good trypsin activity, which is close to 100% in urea solutions at concentration less than 1 M, according the supplier.
5. After dried, samples can be stored at −80 °C until ready for SRM analysis.
6. A theoretical list of tryptic peptides produced from human FVIII based on UniProt protein sequence (P00451) (*see* Ref. 9). This initial list containing 115 peptides was reduced to 15 peptides based on specific selection criteria The initial list of all tryptic peptides for FVIII was reduced to 15 peptides based on the following criteria: (a) peptides with 10–15 amino acids, (b) structural constitution, (c) absence of post-translational modifications, (d) number of peptides observed in other experiments using PeptideAtlas database.
7. Selected proteotypic peptide targets for SRM can be analyzed by exploratory LC-MS/MS runs of standard material, such as a tryptic digestion of SARP. Vary collision energy parameters and chromatographic conditions to select for most intense transitions.

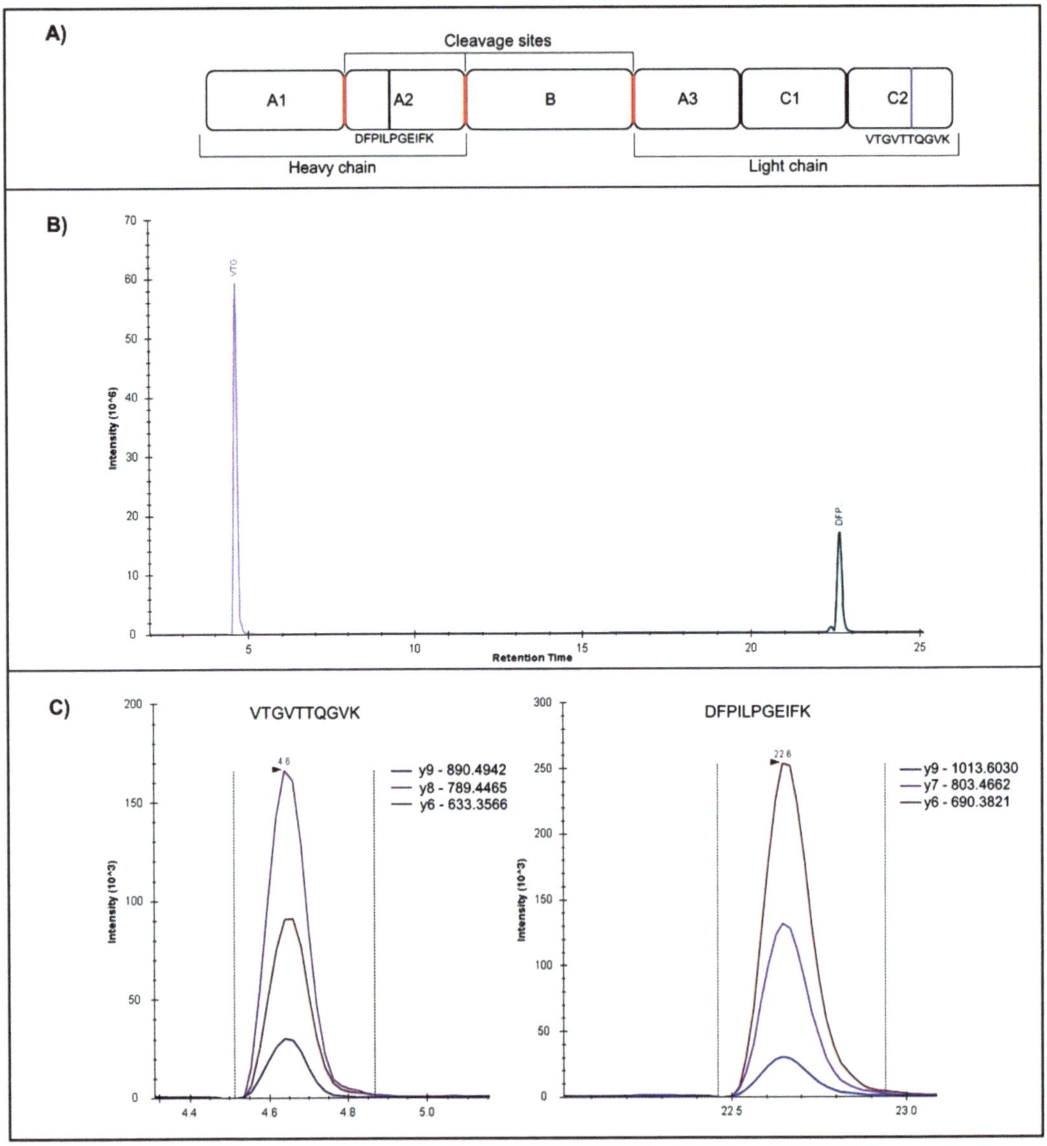

Fig. 1 Quantification of human FVIII by SRM. Panel **a** presents a schematic structure of human FVIII, indicating endogenous cleavage sites and position of two monitored peptides. The peptides selected allow the independent quantification of heavy and light chain. Panel **b** chromatogram showing eluted peptides monitored by MS in the described chromatographic conditions. Panel **c** shows the signal obtained for rFVIII produced by sk-HEP-1 engineered cells, indicating concentrations around 1 U (˜200 ng) of pdFVIII, based on correlation with SARP intensities. The method has sensitivity to detect less than 0.1 U of FVIII

8. Although the synthetic peptides representing FVIII are not obligatory for relative quantification, for instance when comparing with SARP, they are very useful for method development and refinement. Peptides can be quickly synthesized in house using standard solid-phase Fmoc chemistry. Preselection

of short peptides (<20 amino acids) facilitates the synthesis and provides good synthesis yield. If an absolute quantitation is of interest, purify peptides by reversed-phase chromatography and use an accurate method to quantify them, using amino acid analysis for example. Also, the use of isotopically labeled peptides (e.g., $Lys^{13}C_6$), allow more precise quantification relying on internal standards.

9. A mixture of standard peptides, preferably isotopically labeled, can be spiked in samples for an accurate quantitation. Carefully prepare spiking standard solution in order to obtain ion intensities in scales close to those expected for unknown samples. Excess of standard peptides can cause ion suppression and loss of sensitivity.
10. Chromatography in microscale or nanoscale is susceptible to microparticles originating from insoluble material or solid-phase extraction column leaking. This simple centrifugation step improves the column lifetime.
11. The injection of 10 μL corresponds approximately to 10 μg of protein digest. This is kept below the loading capacity of the UPLC T3 Column (1 mm I.D. × 150 mm long) in order to obtain the best chromatographic separations.
12. The multiplatform Skyline software can integrate, calibrate, and quantitate samples analyzed by SRM. This streamlines the development and application of the method.

Acknowledgments

This research was supported by FAPESP (Young Scientist Grants—Proc. No. 2011/09347-1), Center for Cell Based Thereapy—CTC-CEPID (Proc. FAPESP 2013/08135-2) and CISBi-NAP. G.P.L. received a fellowship from CNPq Proc. No. 130142/2015-8, and V.M.F. received a fellowship from CNPq, Proc. No. 308561/2014-7.

References

1. CA SB, Lim A (2007) Applications of mass spectrometry for the structural characterization of recombinant protein pharmaceuticals. Mass Spectrom 26:370–388
2. Picotti P, Aebersold R (2012) Selected reaction monitoring-based proteomics: workflows, potential, pitfalls and future directions. Nat Methods 9:555–566
3. Liebler D, Zimmerman LJ (2013) Targeted quantification of proteins by mass spectrometry. Biochemistry 52:3797–3806
4. Kuster B, Schirle M, Mallick P, Aebersold R (2005) Scoring proteomes with proteotypic peptide probes. Nat Rev Mol Cell Biol 6 (7):577–583
5. Deutsch EW, Lam H, Aebersold R (2008) PeptideAtlas: a resource for target selection for emerging targeted proteomics workflows. EMBO Rep 9:429–434
6. Maclean B, Tomazela DM, Shulman N, Chambers M, Finney GL, Frewen B, Kern R, Tabb DL, Liebler DC, Maccoss MJ (2010) Skyline:

an open source document editor for creating and analyzing targeted proteomics experiments. Bioinformatics 26:966–968

7. Russo-Carbolante EMS, Swiech K, Covas DT, Picanço-Castro V (2015) Large scale and stable production of human FVIII in the human cell line Sk-Hep-1. US Patent 8,969,041 B2

8. Rosa NG, Swiech K, Picanço-castro V, Russo-carbolante EMS, Neto MSA, Fernandes AC, Faça VM, Covas DT (2012) SK-HEP cells and lentiviral vector for production of human recombinant factor VIII. Biotechnol Lett 34 (8):1435–1443

9. Apweiler R, Bairoch A, Wu CH, Barker WC, Boeckmann B, Ferro S, Gasteiger E, Huang H, Lopez R, Magrane M, Martin MJ, Natale DA, O'donavan C, Redashi N, Yeh LSL (2004) UniProt: the universal protein knowledgebase. Nucleic Acids Res 32:115–119

INDEX

A

B

C

D

E

F

G

H

Virgínia Picanço-Castro and Kamilla Swiech (eds.), *Recombinant Glycoprotein Production: Methods and Protocols*, Methods in Molecular Biology, vol. 1674, DOI 10.1007/978-1-4939-7312-5, © Springer Science+Business Media LLC 2018

MIX
Papier aus verantwortungsvollen Quellen
Paper from responsible sources
FSC® C105338

If you have any concerns about our products,
you can contact us on
ProductSafety@springernature.com

In case Publisher is established outside the EU,
the EU authorized representative is:
Springer Nature Customer Service Center GmbH
Europaplatz 3, 69115 Heidelberg, Germany

Printed by Libri Plureos GmbH
in Hamburg, Germany